Multivariate Public Key Cryptosystems

Advances in Information Security

Sushil Jajodia
Consulting Editor
Center for Secure Information Systems
George Mason University
Fairfax, VA 22030-4444
email: jajodia@gmu.edu

The goals of the Springer International Series on ADVANCES IN INFORMATION SECURITY are, one, to establish the state of the art of, and set the course for future research in information security and, two, to serve as a central reference source for advanced and timely topics in information security research and development. The scope of this series includes all aspects of computer and network security and related areas such as fault tolerance and software assurance.

ADVANCES IN INFORMATION SECURITY aims to publish thorough and cohesive overviews of specific topics in information security, as well as works that are larger in scope or that contain more detailed background information than can be accommodated in shorter survey articles. The series also serves as a forum for topics that may not have reached a level of maturity to warrant a comprehensive textbook treatment.

Researchers, as well as developers, are encouraged to contact Professor Sushil Jajodia with ideas for books under this series.

Additional titles in the series:

UNDERSTANDING INTRUSION DETECTION THROUGH VISUALIZATION by Stefan Axelsson; ISBN-10: 0-387-27634-3
QUALITY OF PROTECTION: Security Measurements and Metrics by Dieter Gollmann, Fabio Massacci and Artsiom Yautsiukhin; ISBN-10: 0-387-29016-8
COMPUTER VIRUSES AND MALWARE by John Aycock; ISBN-10: 0-387-30236-0
HOP INTEGRITY IN THE INTERNET by Chin-Tser Huang and Mohamed G. Gouda; ISBN-10: 0-387-22426-3
CRYPTOGRAPHICS: Exploiting Graphics Cards For Security by Debra Cook and Angelos Keromytis; ISBN: 0-387-34189-7
PRIVACY PRESERVING DATA MINING by Jaideep Vaidya, Chris Clifton and Michael Zhu; ISBN-10: 0-387- 25886-8
BIOMETRIC USER AUTHENTICATION FOR IT SECURITY: From Fundamentals to Handwriting by Claus Vielhauer; ISBN-10: 0-387-26194-X
IMPACTS AND RISK ASSESSMENT OF TECHNOLOGY FOR INTERNET SECURITY:Enabled Information Small-Medium Enterprises (TEISMES) by Charles A. Shoniregun; ISBN-10: 0-387-24343-7
SECURITY IN E-LEARNING by Edgar R. Weippl; ISBN: 0-387-24341-0
IMAGE AND VIDEO ENCRYPTION: From Digital Rights Management to Secured Personal Communication by Andreas Uhl and Andreas Pommer; ISBN: 0-387-23402-0
INTRUSION DETECTION AND CORRELATION: Challenges and Solutions by Christopher Kruegel, Fredrik Valeur and Giovanni Vigna; ISBN: 0-387-23398-9
THE AUSTIN PROTOCOL COMPILER by Tommy M. McGuire and Mohamed G. Gouda; ISBN: 0-387-23227-3

Additional information about this series can be obtained from
http://www.springer.com

Multivariate Public Key Cryptosystems

by

Jintai Ding
Jason E. Gower
Dieter S. Schmidt
University of Cincinnati
USA

Jintai Ding
University of Cincinnati
Dept. Mathematical Sciences
P.O.Box 210025
Cincinnati OH 45221-0025
ding@math.uc.edu

Jason E. Gower
University of Cincinnati
Dept. Mathematical Sciences
P.O.Box 210025
Cincinnati OH 45221-0025
gowerj@math.uc.edu

Dieter S. Schmidt
Dept. of ECECS
P.O.Box 210030
Cincinnati, Ohio 45021-0030
dieter.schmidt@uc.edu

Multivariate Public Key Cryptosystems
by Jintai Ding, Jason E. Gower and Dieter S. Schmidt

ISBN 978-1-4419-4077-3

e-ISBN-10: 0-387-36946-5
e-ISBN-13: 978-0-387-36946-4

Printed on acid-free paper.

9 8 7 6 5 4 3 2 1

springer.com

This book is dedicated to our families,

in particular,

Romeliza Villegas-Ding

Contents

List of Figures

List of Tables

Introduction

In the last ten years, multivariate public key cryptosystems, or MPKCs for short, have increasingly been seen by some as a possible alternative to the public key cryptosystem RSA, which is widely in use today. The security of RSA depends on the difficulty of factoring large integers on a conventional computer. Shor's polynomial-time integer factorization algorithm for a quantum computer means that eventually such alternatives will be necessary, provided that we can build a quantum computer with enough quantum bits.

A result from complexity theory states that solving a set of randomly chosen nonlinear polynomial equations over a finite field is NP-hard. So far quantum computers have not yet been shown to be able to solve a set of multivariate polynomial equations efficiently, and the consensus is that quantum computers are unlikely to provide an advantage for this type of problem. Moreover, MPKC schemes are in general much more computationally efficient than number theoretic-based schemes. This has led to many new cryptographic schemes and constructions such as the Matsumoto-Imai cryptosystem (C^* or MI), the Hidden Field Equations cryptosystem (HFE), the Oil-Vinegar signature scheme, the Tamed Transformation Method cryptosystem (TTM), and cryptosystems derived from internal perturbation. Some of these schemes seem to be very suitable for use in the ubiquitous computing devices with limited computing capacity, such as smart cards, wireless sensor networks, and active RFID tags. Indeed, Flash, also known as Sflash^{v2}, a multivariate signature scheme, was recently accepted as a security standard for use in low-cost smart cards by the New European Schemes for Signatures, Integrity and Encryption (NESSIE): IST-1999-12324.

In general, multivariate public key cryptosystem is a public key cryptosystem in which the public key is a set of multivariate polynomials $f_1, \ldots, f_m$ in $k[x_1, \ldots, x_n]$, where k is a given finite field. If Alice wants to send the message $(x'_1, \ldots, x'_n) \in k^n$ to Bob, she looks up Bob's public key, computes $y'_i = f_i(x'_1, \ldots, x'_n)$ for $i = 1, \ldots, m$, and sends the encrypted message $(y'_1, \ldots, y'_m)$. Bob's secret key will be some information about the construction of the f_i without which it is computationally infeasible to solve the system $f_1(x_1, \ldots, x_n) = y'_1, \ldots, f_m(x_1, \ldots, x_n) = y'_m$ for $x_1, \ldots, x_n$.

Of course, Bob will need a secret key to recover Alice's message, and this indicates that the NP-hardness of the multivariate polynomial equation solving problem does not necessarily guarantee the security of practical schemes, though intuitively it does suggest that the more we can make the polynomial appear to be "random," the more secure the scheme is likely to be.

Research on MPKCs has undergone rapid development in the last decade, providing many interesting results in designing and attacking the MPKCs with examples as previously stated. In addition, the study of MPKCs has also resulted in new ideas in solving systems of multivariate polynomial equations over a finite field, a purely mathematical problem that lies in the area of algebraic geometry. This has also attracted a lot of attention. New work in this direction includes the linearization equations, the XL family of algorithms, the new Gröbner basis algorithms, and the Zhuang-Zi algorithm.

We believe that this area has developed to the point where a book is needed to systematically present the subject matter to a broad audience, including information security experts in industry, computer scientists and mathematicians. We hope that this book can be used in the following ways: by industry experts as a guide for understanding the basic mathematical structures needed to implement these cryptosystems for practical applications, as a starting point for researchers in both computer science and mathematics looking to explore this exciting new field, or as a textbook for a course in MPKC suitable for beginning graduate students in mathematics or computer science. Due to the above considerations, this book has been written more from the computational perspective, though we have tried to provide the necessary mathematical background.

It should be noted that there are usually several improvements on the schemes that we present, in particular in terms of the efficiency of the computation in both implementation and attacks. However, to keep the size of this book reasonable and to keep the book more focused, we have chosen not to cover some of these details. Instead, we have tried to present the essential ideas, methods, and examples so that a reader will not be distracted by technical details that can be found in the references provided. Nevertheless, for those readers interested in the practical side of the MPKCs, we highly recommend reading through the details in order to discover improvements. Improving the performance of a cryptosystem by even a small factor may not be significant from a mathematical perspective, but can be very important in practice.

This book is arranged not in historical order but rather in terms of the mathematical ideas behind each topic. We begin with an overview of the basic ideas and early development of both multivariate public key cryptography and signature schemes. We next present the main families of multivariate schemes: MI, Oil-Vinegar, HFE, and TTM. We also present the concept of perturbation, the means by which the security of various schemes can be improved without much cost in efficiency. Each family is introduced in terms of the origin of the mathematical idea behind its construction, followed by generalizations and related attacks specific to that family. Generic attacks that can be applied to any MPKC, in particular methods for solving systems of multivariate polynomial equations over a finite field, are then addressed, followed by a discussion of the future of MPKCs. The reader will find one supplementary appendix at the end of the book where we have collected results from finite field theory needed in the main text of the book.

This book grew out of a survey paper written by Jintai Ding and Dieter Schmidt, and from the lecture notes for a graduate course at the University of Cincinnati taught by Jintai Ding during the 2004–2005 and 2005–2006 academic years. Indeed, we have written this book to be used as a text for a year-long course in advanced topics in cryptography or applied algebra, or as a supplementary text for a first course in cryptography. Students with some previous exposure to abstract algebra (groups, rings, fields and ideals) will be more than well-prepared to read and understand the various topics. For those with a programming background, we plan to develop a website where we will make our related software available at

`http://math.uc.edu/~aac/MPKC/software.html`

for public use. This will provide interested readers a starting point to further develop their understanding and computational intuition by experimenting with the software. Those readers new to the field of MPKC will be best served by first reading the introductory chapter, after which the chapters are written so as to be essentially self-contained. Readers with previous exposure to MPKC may use the text to learn more about a given scheme and as a guide to related articles. Although it was our intention to include all related references, we apologize to those we have missed. Also, the amount of space devoted to a given topic is not necessarily related to how important we consider it. Rather it is likely due to space constraints or to maintain the consistency and convenience of the structure and flow of the book.

We plan to maintain a webpage at

`http://math.uc.edu/~aac/MPKC/errata.html`

where we will list corrections to the book. Readers are encouraged to submit their findings to that website or send them via e-mail to `aac@math.uc.edu`.

We would like to thank Robert Hess, Timothy Hodges, Gregory Hull, Crystal Updegrove, and John Wagner for attending the lectures and giving thoughtful feedback about the lectures and the early stages of the book. We would like to also thank Jiun-ming Chen, Lei Hu, Christopher Wolf, Bo-yin Yang for reading the book and providing us with their valuable comments. Many thanks go to the staff at Springer for their constant support and help, and to the Department of Mathematical Sciences at the University of Cincinnati for their support. Finally, we would like to thank our families for their constant support and encouragement.

Chapter 1

OVERVIEW

1.1 Public Key Cryptosystems

The revolutionary idea of a public key cryptosystem, which has fundamentally changed our modern communication systems, was first suggested by Diffie and Hellman [Diffie and Hellman, 1976]. The first practical realization of this idea was the famous RSA cryptosystems proposed by Rivest, Shamir and Adleman [Rivest et al., 1978; Rivest et al., 1982], whose security is based on the difficulty of factoring a large integer into a product of prime numbers. Diffie and Hellman also suggested the famous Diffie-Hellman key exchange protocol, whose security is based on the difficulty of the discrete-logarithm problem over a large prime field [Diffie and Hellman, 1976].

A public key cryptosystem, unlike a traditional symmetric cryptosystem where the two parties in the communication process have exactly the same key, is an asymmetric cryptosystem where the encryption key is different from the decryption key. The encryption key should be made public so that anyone can use it to send an encrypted messages. However, the decryption key should be kept private so that only the intended recipient can decrypt the secret message. Similarly, signature schemes based on public key cryptosystems come with two keys: one is public and is used to verify signatures, while the other is private and is used to produce an electronic signature.

Symmetric cryptosystems use the same key for both encryption and decryption and thus require a prior secret key exchange in order to communicate securely in an open communication channel. However, due to the asymmetric property, especially the disclosure of the public key, currently symmetric cryptosystems work far more efficiently than the asymmetric ones. Therefore, it is preferable to use symmetric schemes if a secure key exchange can be accomplished efficiently. Without any prior secret key exchange, public key schemes can be used to securely exchange a symmetric scheme key that can then be used to securely communicate over any open communication channel. Since electronic transmissions and the Internet are totally open communication systems where anyone can eavesdrop on essentially any communication, public key cryptosystems have become critical for providing a reasonable level of security and privacy.

Despite all its strengths, the RSA cryptosystem also has its weaknesses. The security of RSA relies on the fact that we do not have any fast algorithm for factoring large integers. Due to recent developments in the area of integer factorization (such as the number field sieve and algorithms based on elliptic curves), RSA must use increasingly larger parameters in order to maintain a necessary level of security. By today's standards, a secure RSA public key cryptosystem requires an integer N, a product of two prime numbers p and q, such that $N = pq$ has at least 1000 binary digits. Working with numbers of this size requires a huge amount of calculations, which makes the entire encryption and decryption process slow and therefore inefficient. This kind of performance is intolerable for "small" ubiquitous computing devices with limited computing power and memory, such as cell phones and smart cards, let alone wireless network sensors and RFID tags.

Moreover, quantum computers have recently emerged as a threat to the RSA cryptosystem. Peter Shor [Shor, 1999] discovered an algorithm that can be used to factor an integer in polynomial time on a quantum computer. This means that once we have quantum computers that can deal with a large number of quantum bits, the RSA cryptosystem can no longer be considered secure. Because of the tremendous effort directed towards developing quantum computers, this threat should be taken seriously, particularly in consideration of the appearance of the first small-scale but real quantum computer in 2001, which factored 15 into 3×5 using Shor's algorithm [Vandersypen et al., 2001]. Even though we do not presently have quantum computers of the desired capacity, we have a very strong motivation to search for efficient and secure alternative public key cryptosystems.

1.2 Multivariate Public Key Cryptosystems

There are several directions in which to search for alternative public key cryptosystems. Some examples include: elliptic curve cryptography, where the Abelian group structure of the set of points on an elliptic curve is used; lattice-based cryptography, which exploits the metric structure of a lattice; and error-correcting code-based cryptography. Another direction is multivariate public key cryptography where the building blocks are multivariate polynomials over a finite field. Typically these polynomials are of total degree two; that is, quadratic polynomials. This new direction was very much inspired by the knowledge that solving a set of multivariate polynomial equations over a finite field, in general, is proven to be an NP-hard problem [Garey and Johnson, 1979]. By itself this fact does not guarantee that any such cryptosystem is secure. Quantum computers do not appear to have an advantage when dealing with NP-hard problems and we do not expect them to find a solution to a set of polynomial equations efficiently even in the future. On the other hand no proof exists that the integer factorization problem is NP-hard, nor is it known to which class of problems it belongs. Nevertheless, a quantum computer is perfectly suited for the integer factorization problem.

Multivariate public key cryptosystems have increasingly been seen by some as a possible alternative to number theoretic-based cryptosystems such as RSA, where the security assumption is based on the difficulty of factoring large integers or finding a discrete logarithm on a standard computer. Moreover, MPKC schemes are in general much more computationally efficient than number theoretic-based schemes. By now, there are many new cryptographic schemes and constructions, and some of these schemes seem to be very suitable for use in "small" ubiquitous computing devices with limited computing capacity, such as smart cards, wireless sensor networks, and active RFID tags. Indeed, Flash, a multivariate signature scheme, was recently accepted as an European security standard for use in low-cost smart cards by the New European Schemes for Signatures, Integrity and Encryption [NESSIE, 1999].

Mathematically speaking, the security of RSA-type cryptosystems relies on the complexity of integer factorization and is based on results in number theory developed in the 17th and 18th centuries. Elliptic curve cryptosystems employ the use of mathematics from the 19th century. Multivariate cryptography goes one step further, using results in algebraic geometry developed in the 20th century. This perception was first clearly stated by Diffie (`http://www.minrank.org`).

Currently all existing multivariate cryptosystems can be divided into two categories with one exception, the Isomorphism of Polynomial (IP) authentication scheme. The first category is called *bipolar* and the second is called *mixed*.

Bipolar Systems

Let k be a finite field. In a bipolar multivariate public key cryptosystem, the cipher is given as a map $\bar{F}$ from k^n to k^m

$$\bar{F}(x_1, \ldots, x_n) = (\bar{f}_1, \ldots, \bar{f}_m)$$

where each $\bar{f}_i$ is a polynomial in $k[x_1, \ldots, x_n]$. A typical construction of this type of system begins with first building a map F from k^n to k^m such that:

1.) $F(x_1, \ldots, x_n) = (f_1, \ldots, f_m)$, where $f_i \in k[x_1, \ldots, x_n]$;

2.) Any equation

$$F(x_1, \ldots, x_n) = (y'_1, \ldots, y'_m),$$

can be easily solved. Equivalently, we can efficiently find a pre-image of $(y'_1, \ldots, y'_m)$, which will be unique for the case of encryption, and is denoted by

$$F^{-1}(y'_1, \ldots, y'_m).$$

Note that here F^{-1} means finding the pre-image and should not be taken to mean that the map F is invertible according to the strict mathematical definition of the invertibility of a map.

Once such a map is found, the cipher $\bar{F}$ is constructed as a composition of three maps:

$$\bar{F} = L_1 \circ F \circ L_2, \tag{1.1}$$

where L_1 is a randomly-chosen invertible affine transformations from k^m to k^m, and L_2 is a randomly-chosen invertible affine transformation from k^n to k^n. In this case, the public key consists of the m polynomial components of $\bar{F}$ and the field structure of k, while the secret key consists of L_1 and L_2. The map F may or may not be part of the secret key depending on its precise nature.

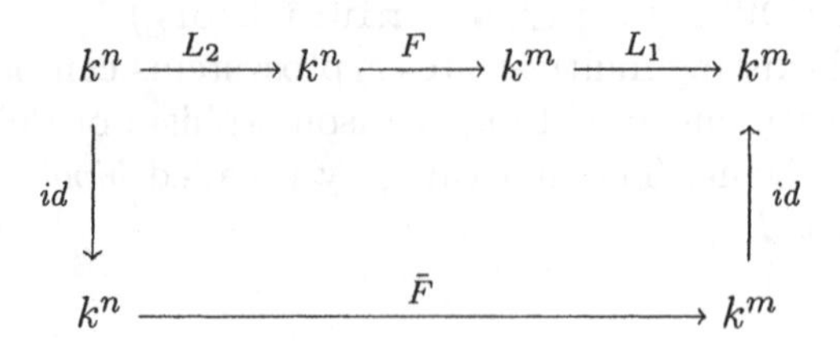

Figure 1.1. Composition of maps.

To encrypt the message $X' = (x'_1, \ldots, x'_n)$, we calculate $\bar{F}(X')$. To decrypt a ciphertext $Y' = (y'_1, \ldots, y'_m)$, we solve the system of equations defined by

$$\bar{F}(x_1, \ldots, x_n) = Y'. \tag{1.2}$$

This is accomplished by first finding $Y_1 = L_1^{-1}(Y')$, then $Y_2 = F^{-1}(Y_1)$, followed by $L_2^{-1}(Y_2)$.

To sign a message Y', one has to find any solution to (1.2), which we denote by $X' = (x'_1, \ldots, x'_n)$. Anyone can verify if it is indeed a legitimate signature by checking to see whether or not

$$\bar{F}(x'_1, \ldots, x'_n) = Y'.$$

We can see that one of the key ideas involved in the bipolar multivariate schemes is that L_1 and L_2 serve the purpose of "hiding" or "masking" the map F that could otherwise easily be inverted. In some cases, the choices for F are relatively limited and otherwise known to the attacker. This explains why there is sometimes no substantial security advantage in keeping F as part of the secret key. Currently, the majority of multivariate schemes are of bipolar type.

Mixed Systems

A mixed multivariate public key scheme uses the mapping $\bar{H}$ from k^{n+m} to k^l as its public key

$$\bar{H}(x_1, \ldots, x_n, y_1, \ldots, y_m) = (\bar{h}_1, \ldots, \bar{h}_l), \tag{1.3}$$

where each $\bar{h}_i$ is a polynomial in $k[x_1, \ldots, x_n, y_1, \ldots, y_m]$. To build such a scheme, we find a mapping $H : k^{n+m} \longrightarrow k^l$

$$H(x_1, \ldots, x_n, y_1, \ldots, y_m) = (h_1, \ldots, h_l),$$

where each h_i is a polynomial in $k[x_1, \ldots, x_n, y_1, \ldots, y_m]$ such that:

1.) For any given specific $(x'_1, \ldots, x'_n)$, the system of equations

$$H(x'_1, \ldots, x'_n, y_1, \ldots, y_m) = (0, \ldots, 0) \tag{1.4}$$

can be easily solved. In most cases, Equation (1.4) is a set of linear (affine) equations in the variables $y_1, \ldots, y_m$.

2.) For any given specific element $(y'_1, \ldots, y'_m)$, the system of equations

$$H(x_1, \ldots, x_n, y'_1, \ldots, y'_m) = (0, \ldots, 0) \tag{1.5}$$

can easily be solved. Equation (1.5) is a set of specially designed nonlinear equations.

Once such a mapping is found, $\bar{H}$ is constructed as

$$\bar{H} = L_3 \circ H \circ (L_1 \times L_2)$$

where $L_1 : k^n \longrightarrow k^n$ and $L_2 : k^m \longrightarrow k^m$ are defined as in the bipolar case, and $L_3 : k^l \longrightarrow k^l$ is an invertible linear map.

To encrypt the message $X' = (x'_1, \ldots, x'_n)$, we substitute into Equation (1.3) and solve the system of equations

$$\bar{H}(x'_1, \ldots, x'_n, y_1, \ldots, y_m) = (0, \ldots, 0),$$

denoting the solution by $Y' = (y'_1, \ldots, y'_m)$. This Y' is the encrypted message. To decrypt a ciphertext $Y' = (y'_1, \ldots, y'_m)$, we first calculate $\bar{Y} = L_3^{-1}(Y')$. Then, letting $\bar{Y} = (\bar{y}_1, \ldots, \bar{y}_m)$, we substitute $\bar{Y}$ into Equation (1.5), and solve the system of equations

$$H(x_1, \ldots, x_n, \bar{y}_1, \ldots, \bar{y}_m) = (0, \ldots, 0).$$

If the solution of this equation is denoted by $\bar{X}$, then the plaintext is given as $X' = L_1^{-1}(\bar{X})$.

To sign a message $Y' = (y'_1, \ldots, y'_m)$, we must go through the decryption process above to find an element $X' = (x'_1, \ldots, x'_n)$ in k^n. Anyone can verify that it is indeed a legitimate signature by checking:

$$\bar{H}(x'_1, \ldots, x'_n, y'_1, \ldots, y'_m) = (0, \ldots, 0).$$

The public key consists of the l polynomial components of $\bar{H}$ and the field structure of k. The secret key mainly consists of L_1, L_2 and L_3. The equation $H(X, Y) = (0, \ldots, 0)$, depending on different cases, can be either part of the secret key or the public key.

The key idea is that L_1, L_2, L_3 serve the purpose to "hide" the equation $H(X, Y) = 0$, which otherwise could be easily solved if given the value of Y. As with bipolar schemes, it is usually not necessary or always possible to hide the form of H. Mixed type schemes are relatively rare, one example being Patarin's Dragon cryptosystem [Patarin, 1996a].

IP schemes

The Isomorphism of Polynomials (IP) problem originated by trying to attack MPKCs by finding the secret keys. Let F_1, F_2 with

$$F_i(x_1, \ldots, x_n) = (\bar{f}_{i1}, \ldots, \bar{f}_{im}),$$

be two polynomial maps from k^n to k^m. The IP problem is to look for two invertible affine linear transformations L_1 on k^n and L_2 over k^m (if they exist) such that

$$F_1(x_1, \ldots, x_n) = L_2 \circ F_2 \circ L_1(x_1, \ldots, x_n).$$

This problem is closely related to the attack of finding private keys for a MPKC, for example the Matsumoto-Imai cryptosystems, and the IP cryptosystem was first proposed by Patarin [Patarin, 1996b], where the verification process is performed through showing the equivalence (or isomorphism) of two different maps. A simplified version is called the isomorphism of polynomials with one secret (IP1s) problem, where we only need to find the map L_1 (if it exists), while the map L_2 is known to be the identity map.

Later, more works were devoted to this direction [Patarin et al., 1998; Levy-dit-Vehel and Perret, 2003; Geiselmann et al., 2003; Perret, 2005; Faugère and Perret, 2006]. However, the emphasis of this book is on quadratic bipolar systems, so we will not discuss the IP schemes in any detail.

1.3 Basic Security and Efficiency Assumptions

As with any public key cryptosystem, multivariate schemes must be efficient and secure if they are to be of any practical value. We will now discuss the basic aspects of these features relevant to all schemes in the context of bipolar encryption cryptosystems. The case of signature schemes is very similar.

Essentially, any encryption process applies a polynomial map, often quadratic, to an element in k^n to produce an element in k^m. The decryption process is a process to find its "inverse," for example by solving (1.2). This means that (1.2) must be hard to solve for anyone without some additional information. The main reason this is generally believed to be such a hard problem is the well-known fact that essentially the only method to solve polynomial equations are general methods, such as the Gröbner basis method, which is normally expected to be of exponential complexity. If the encryption map has an inverse in terms of strict mathematical definition, then it can also be expressed as a polynomial map, hence we must ensure that this inverse map has a substantially high degree. Otherwise we can use the public key to generate enough plaintext-ciphertext pairs to find the inverse of the cipher easily, and defeat the cryptosystem.

We must also ensure that it is hard to factorize the encryption map in terms of the composition that defines it, as given in (1.1). Otherwise the secret key becomes easy to retrieve from the public key. In general, this is known to be very difficult. Perhaps more precisely, we know extremely little about how to factor maps and it is generally thought to be a very hard mathematical problem for multivariate maps. The hardness of this problem is closely related to the well-known Jacobian conjecture about invertible maps, which essentially asks if the Jacobian of a multivariate polynomial map over $\mathbb{R}^n$, for $n > 1$, is a nonzero constant also implies the map is invertible. Here, $\mathbb{R}$ stands for the set of real numbers.

Of course any public key cryptosystem is intended for practical application. This requires that both the encryption and decryption processes can be performed efficiently. The public key, a set of multivariate polynomials, has to be stored somewhere, must be publicly and easily accessible, and its values must be swiftly calculated. Therefore, the polynomial components must all be of relatively small degree, but of course not linear, since otherwise the system will be useless due to the security problem. As we will see, it seems best to use schemes where the component polynomials have total degree two.

1.4 Early Attempts

Before we start the main body of the book, we would like to present a brief history of the early pioneering work in MPKC.

The first attempt to construct a multivariate signature was given in [Ong et al., 1984; Ong et al., 1985]. This system is based on a quadratic equation

$$y = x_1^2 + \alpha x_2^2 \bmod n, \tag{1.6}$$

where n is a large composite integer that is difficult to factorize. To sign a message y, we need to find one of the many (about n) solutions (x_1, x_2) to (1.6), which is easy if we know the factorization of n. The public key is essentially the integer n and (1.6). Since the security relies on the factorization of n, this system is in some sense still in the shadow of the RSA cryptosystem, though it indeed initiated the idea of multivariate cryptosystems.

Shortly after its introduction, Pollard and Schnorr broke this cryptosystem [Pollard and Schnorr, 1987]. In particular, they found an algorithm to solve (1.6) for any given y without knowing the factors of n. With the assumption of the generalized Riemann hypothesis, a solution can be found by a probabilistic algorithm that has a complexity of $O((\log n)^2 \log\log |k|)$ in $O(\log n)$-bit integer operations.

An attempt to build a true multivariate (with four variables) public key cryptosystem was also made by Matsumoto, Imai, Harashima and Miyagawa[Matsumoto et al., 1985], where the public keys are given by quadratic polynomials. However it was soon defeated [Okamoto and Nakamura, 1986].

Another early attempt to build a multivariate cryptosystem was made by Fell and Diffie [Fell and Diffie, 1986]. Their idea was to build a cryptosystems using the composition of many invertible linear maps and simple triangular maps of the form

$$T(x_1, \ldots, x_n) = (x_1 + g(x_2, \ldots, x_n), x_2, \ldots, x_n), \tag{1.7}$$

where g is any polynomial. Clearly T is invertible, assuming g is known, and therefore the decryption process can be done easily. However, due to efficiency considerations such as the key size, the authors concluded in [Fell and Diffie, 1986] that it appeared very difficult to build such a cryptosystem with practical value that is both secure and has a public key of practical size. Here one should notice that the simple triangular maps described above belong to the family of de Jonquières maps from algebraic geometry, which are more generally defined by:

$$J(x_1, \ldots, x_n) = (x_1 + g_1(x_2, \ldots, x_n), \ldots, x_{n-1} + g_{n-1}(x_n), x_n), \tag{1.8}$$

where the g_i are arbitrary polynomial functions. We note that J can be easily inverted assuming that the g_i are known.

The invertible affine linear maps over k^n together with the de Jonquières maps belong to the family of so-called tame transformations from algebraic geometry, including all transformations that arise as a composition of elements of these two types of transformations. Tame transformations are elements of the group of automorphisms of the polynomial ring $k[x_1, \ldots, x_n]$. Elements in this automorphism group that are not tame are call wild. Given a polynomial map, it is in general very difficult to decide whether or not the map is tame, or even if there is indeed any wild map [Nagata, 1972], a question closely related to the Jacobian conjecture. This problem was solved in 2003 [Shestakov and Umirbaev, 2003], which claims to prove that the Nagata map is indeed wild.

Triangular constructions were also pursued unsuccessfully in Japan [Tsujii et al., 1986; Tsujii et al., 1987; Hasegawa and Kaneko, 1987]. Their construction is actually even more general in the sense that they use rational functions instead of just polynomials. However, these works are not so well-known, partially because these papers were written in Japanese. The cryptosystems are called sequential solution type systems. The birational construction by Shamir [Shamir, 1993] in some way also belongs to this family. A much more complicated attempt was pursued much later when Moh presented TTM, the tame transformation method [Moh, 1999a].

1.5 Quadratic Constructions

The majority of multivariate public key cryptosystems belong to the bipolar type, while only a very few are of mixed type. Also, it is usually the case that the public key is given as a set of polynomials of total degree two; i.e., they are all quadratic polynomials of the form:

$$\sum_{i \leq j} a_{ij} x_i x_j + \sum b_i x_i + c.$$

The reason for this comes mainly from efficiency and security considerations. For a fixed n, the number of possible terms in a polynomial of degree d in n variables is

$$\binom{n+d}{d} = \frac{(n+d)!}{n!\, d!}$$

which grows rapidly as d increases. If d is large, then the public key size will be too great for fast computation and efficient manipulation and storage. It is obvious that linear constructions are not an option, so quadratic constructions seem to be the best choice from the perspective of both security and efficiency. In fact, from a purely mathematical point of view, any system of large total degree polynomials can always be transformed into a larger system of total degree two polynomials by adding more variables. Therefore the increase in security we gain by increasing the total degree is not necessarily a good trade-off due to the loss in efficiency. Because of these considerations, we will concentrate our attention on quadratic multivariate public key cryptosystems of bipolar type, as they represent the majority of all known multivariate schemes used for either encryption or signature purposes.

Chapter 2

MATSUMOTO-IMAI CRYPTOSYSTEMS

In the previous chapter we discussed some early attempts to build MPKCs. However, these attempts were not very successful and it became very clear that new mathematical ideas were needed. The first such new idea was proposed by Matsumoto and Imai [Matsumoto and Imai, 1988]. Their key idea was to utilize both the vector space and the hidden field structure of k^n, where k is a finite field. More specifically, instead of searching for invertible maps over the vector space k^n directly, they looked for invertible maps on a field K, a degree n field extension of k, which can also be identified as an n dimensional vector space over k. This map could then be transformed into an invertible map over k^n.

One such cryptosystem, known as C^* or MI, attracted a lot of attention due to its high efficiency and potential use in practical applications. In fact, the MI cryptosystem was submitted as a candidate for security standards of the Japanese government. However, before the final selection, MI was broken by Jacques Patarin using an algebraic attack that utilizes linearization equations [Patarin, 1995]. This method takes advantage of certain specific hidden algebraic structures in MI.

Normally one would conclude that this is the end of MI, though in fact the subsequent story goes into the opposite direction. One reason is that the MI cryptosystem represents a fundamental breakthrough on the conceptual level in that it brought a totally new mathematical idea into the field and consequentially was widely explored and extended. Another reason is that there are many new variants of the MI cryptosystems that seem to have great potential, including the Sflash signature scheme [Akkar et al., 2003; Patarin et al., 2001], which was accepted in 2004 as one of the final selections for the New European Schemes for Signatures,

Integrity, and Encryption project [NESSIE, 1999] for use in low cost smart cards.

Indeed, the work of Matsumoto and Imai has played a critical role as a catalyst in this new area and has stimulated the subsequent development. In this chapter, we will present the MI cryptosystem in detail, Patarin's cryptanalysis of MI, the Plus-Minus variants, related attacks and security analysis.

2.1 Construction of a Matsumoto-Imai System

Let k be a finite field of characteristic two and cardinality q, and take $g(x) \in k[x]$ to be any irreducible polynomial of degree n. Define the field $K = k[x]/g(x)$, a degree n extension of k. In general the $\mathrm{char}(k) = 2$ condition is not necessary for the following construction, though we would need to modify the system slightly due to the loss of bijectivity in the final map used for the construction of the corresponding public key.

Let $\phi : K \longrightarrow k^n$ be the standard k-linear isomorphism between K and k^n given by

$$\phi(a_0 + a_1 x + \cdots + a_{n-1}x^{n-1}) = (a_0, a_1, \ldots, a_{n-1}).$$

The subfield k of K is embedded in k^n in the standard way:

$$\phi(a) = (a, 0, \ldots, 0), \qquad \forall a \in k.$$

Note that here ϕ is a k-linear map if we treat k as a subfield in K.

Choose θ so that $0 < \theta < n$ and

$$\gcd(q^\theta + 1, q^n - 1) = 1,$$

and define the map $\tilde{F}$ over K by

$$\tilde{F}(X) = X^{1+q^\theta}. \tag{2.1}$$

The conditions on θ insure that $\tilde{F}$ is an invertible map; indeed, if t is an integer such that

$$t(1 + q^\theta) \equiv 1 \bmod (q^n - 1),$$

then $\tilde{F}^{-1}$ is simply

$$\tilde{F}^{-1}(X) = X^t.$$

Now let F be the map over k^n defined by

$$F(x_1, \ldots, x_n) = \phi \circ \tilde{F} \circ \phi^{-1}(x_1, \ldots, x_n) = (f_1, \ldots, f_n),$$

where $f_1, \ldots, f_n \in k[x_1, \ldots, x_n]$. To finish the description of the construction of Matsumoto-Imai, let us now choose L_1 and L_2 to be two invertible affine transformations over k^n. Define the map over k^n by

$$\bar{F}(x_1, \ldots, x_n) = L_1 \circ F \circ L_2(x_1, \ldots, x_n) = (\bar{f}_1, \ldots, \bar{f}_n), \qquad (2.2)$$

where $\bar{f}_1, \ldots, \bar{f}_n \in k[x_1, \ldots, x_n]$. See Figure 2.1 for a commutative diagram that captures the essence of the MI construction.

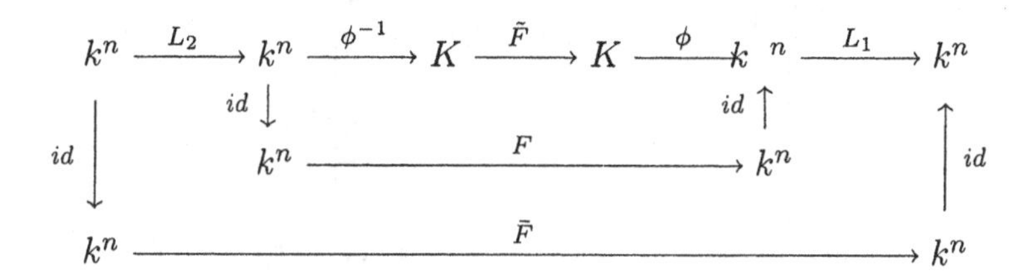

Figure 2.1. Composition of maps in the construction of MI.

We can now fully describe the Matsumoto-Imai public key cryptosystem.

The Public Key

The public key of MI includes the following:

1.) The field k including its additive and multiplicative structure;

2.) The n polynomials $\bar{f}_1, \ldots, \bar{f}_n \in k[x_1, \ldots, x_n]$.

The Private Key

The private key includes the two invertible affine transformations L_1 and L_2. The parameter θ can be kept private, though this is not critical. Since there are fewer than n choices for θ and n is typically not very large, hiding θ has no substantial effect on attack complexities (only a factor of n).

Encryption

Given a plaintext message $(x'_1, \ldots, x'_n)$, the associated ciphertext is $(y'_1, \ldots, y'_n)$, where

$$y'_i = \bar{f}_i(x'_1, \ldots, x'_n),$$

for $i = 1, \ldots, n$. This can be done by anyone, since the public key is available to anyone.

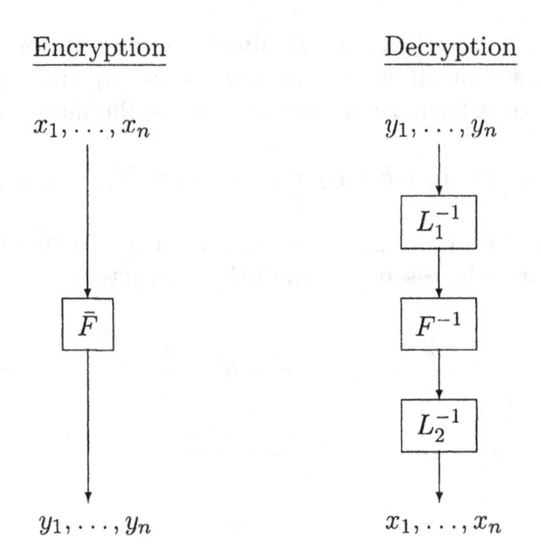

Figure 2.2. Single-branch MI encryption and decryption.

Decryption

We can decrypt the ciphertext $(y_1', \ldots, y_n')$ by computing

$$\begin{aligned}\bar{F}^{-1}(y_1', \ldots, y_n') &= L_2^{-1} \circ F^{-1} \circ L_1^{-1}(y_1', \ldots, y_n') \\ &= L_2^{-1} \circ \phi \circ \tilde{F}^{-1} \circ \phi^{-1} \circ L_1^{-1}(y_1', \ldots, y_n').\end{aligned}$$

In general the components of $\bar{F}^{-1}$ will be of very high degree, and therefore in practice we decrypt the ciphertext $(y_1', \ldots, y_n')$ by executing the following steps:

1.) First compute $(z_1', \ldots, z_n') = L_1^{-1}(y_1', \ldots, y_n')$;

2.) Then compute $(\bar{z}_1, \ldots, \bar{z}_n) = \phi \circ \tilde{F}^{-1} \circ \phi^{-1}(z_1', \ldots, z_n')$;

3.) Finally compute $(x_1', \ldots, x_n') = L_2^{-1}(\bar{z}_1, \ldots, \bar{z}_n)$.

If the corresponding cryptosystem is secure, then this decryption process can be performed only by those who have access to the private key. See Figure 2.2 for a graphical representation of the encryption and decryption process.

Degree of the Public Key Components

The components of the map F are polynomials in $k[x_1, \ldots, x_n]$. In fact, since we are thinking of the variables $x_1, \ldots, x_n$ as the plaintext message "bits" in the field k, we will identify $f_1, \ldots, f_n$ with the corresponding representative of minimal total degree in the ring of functions from k^n to k

$$\mathrm{Fun}(k^n, k) = k[x_1, \ldots, x_n]/(x_1^q - x_1, \ldots, x_n^q - x_n),$$

where total degree is defined as usual. For notational convenience, we will abuse notation and use $k[x_1, \ldots, x_n]$ instead of $\mathrm{Fun}(k^n, k)$. We shall never use the notation $k[x_1, \ldots, x_n]$ for the polynomial ring in the variables $x_1, \ldots, x_n$ with coefficients in k unless explicitly announced beforehand. Similarly, the notation $K[X]$ will be used for the ring of functions from K to K; that is, we identify $K[X]$ with $K[X]/(X^{q^n} - X)$, unless announced otherwise. As such, we shall use the terms "polynomial" and "function" interchangeably. Let us now explore the relationship between the degree of $\tilde{F}$ and the degrees of $f_1, \ldots, f_n$.

The maps $T_i(X) = X^{q^i}$ on K, for $i = 0, 1, \ldots, n-1$, are the well-known Frobenius maps. In fact, the set of these maps is exactly the Galois group $G = \mathrm{Gal}(K/k)$, and the group ring $KG = \{\sum_{i=0}^{n-1} \alpha_i T_i \mid \alpha_i \in K\}$ is the set of all k-linear maps on K (see Appendix A). But from this it is easy to see that for any $L(X) \in KG$ we have that $\phi \circ L \circ \phi^{-1}$ is a k-linear map over k^n, hence the components of $\phi \circ L \circ \phi^{-1}$ each have total degree one in $k[x_1, \ldots, x_n]$.

In order to better see the relationship between the degree of $H(X) \in K[X]$ and the degree of the components of $\phi \circ H \circ \phi^{-1}$, let us define the q-Hamming weight degree of the monomial $X^e \in K[X]$, where $0 \le e < q^n$, to be the sum of the coefficients in the base-q expansion of e, also known as the q-Hamming weight of e. The q-Hamming weight degree of a function $H(X) \in K[X]$ is then defined to be the largest q-Hamming weight degree over all monomials of $H(X)$.

Now suppose we have a function $H(X) \in K[X]$ of q-Hamming weight degree d. Then the components of $\phi \circ H \circ \phi^{-1}$ will be of total degree d. In particular, since the q-Hamming weight degree of $\tilde{F}$ is two, it follows that the total degree of each of the $f_1, \ldots, f_n$ is two. Since L_1 and L_2 are invertible affine transformations, the total degree of each of the $\bar{f}_1, \ldots, \bar{f}_n$ is two as well.

A Toy Example

We now illustrate the MI cryptosystem using a toy example with small parameters.

Let $k = GF(2^2)$ be the finite field with $q = 2^2 = 4$ elements. The multiplicative group for the nonzero elements of this field can be generated by the field element α which satisfies $\alpha^2 + \alpha + 1 = 0$. The field elements

+	0	1	α	α^2
0	0	1	α	α^2
1	1	0	α^2	α
α	α	α^2	0	1
α^2	α^2	α	1	0

*	0	1	α	α^2
0	0	0	0	0
1	0	1	α	α^2
α	0	α	α^2	1
α^2	0	α^2	1	α

Table 2.1. Addition and multiplication table of $GF(2^2)$.

of k can be presented as $\{0, 1, \alpha, \alpha^2\}$ and the addition and multiplication tables are given in Table 2.1.

Next choose $n = 3$ and $g(x) = x^3 + x + 1$, an irreducible polynomial in $k[x]$. Set $K = k[x]/(x^3 + x + 1)$. There are only two possible choices for θ; namely $\theta = 1$ or $\theta = 2$. We will use $\theta = 2$. The map $\tilde{F}$ and its inverse are given by

$$\tilde{F}(X) = X^{1+4^2} \qquad \tilde{F}^{-1}(X) = X^{26}.$$

Let L_1 and L_2 be given by

$$L_1(x_1, x_2, x_3) = \begin{pmatrix} \alpha^2 & \alpha & \alpha \\ \alpha & 1 & 0 \\ 1 & 0 & 1 \end{pmatrix} \begin{pmatrix} x_1 \\ x_2 \\ x_3 \end{pmatrix} + \begin{pmatrix} 0 \\ 1 \\ \alpha \end{pmatrix}$$

and

$$L_2(x, x_2, x_3) = \begin{pmatrix} 1 & 0 & \alpha \\ 0 & 1 & \alpha \\ 1 & \alpha & 0 \end{pmatrix} \begin{pmatrix} x_1 \\ x_2 \\ x_3 \end{pmatrix} + \begin{pmatrix} \alpha \\ \alpha^2 \\ \alpha^2 \end{pmatrix}$$

To derive the public key polynomials in terms of the plaintext message variables x_1, x_2, x_3 we begin by computing $\phi^{-1} \circ L_2(x_1, x_2, x_3)$, which we find to be

$$(\alpha + x_1 + \alpha x_3) + (\alpha^2 + x_2 + \alpha x_3)x + (\alpha^2 + x_1 + \alpha x_2)x^2.$$

If we denote this by X, then we next compute $\tilde{F}(X) = X^{1+4^2} = X \cdot X^{16}$. The exponentiation is easily done since we only have to apply it to each term of X. There are no degrees higher than two since we are working in the finite field k of characteristic two. Thus $\tilde{F}(X)$ is

$$\begin{aligned} &1 + \alpha^2 x_1 + \alpha x_2 + x_3 + x_1 x_2 + \alpha x_1 x_3 + \alpha^2 x_2 x_3 \\ &+ (\alpha + \alpha x_1 + x_2 + \alpha^2 x_3 + x_1^2 + \alpha^2 x_1 x_2 + x_2^2 + x_2 x_3)x + (\alpha^2 + \alpha^2 x_1 \\ &\quad + \alpha x_2 + \alpha x_3 + x_1^2 + x_1 x_2 + \alpha x_1 x_3 + \alpha^2 x_2^2 + \alpha x_2 x_3 + \alpha^2 x_3^2)x^2. \end{aligned}$$

Finally we compute $L_1 \circ \phi(X)$ to get the public key polynomials

$$\begin{aligned}
\bar{f}_1(x_1, x_2, x_3) &= 1 + x_3 + \alpha x_1 x_3 + \alpha^2 x_2^2 + \alpha^2 x_2 x_3 + x_3^2 \\
\bar{f}_2(x_1, x_2, x_3) &= 1 + \alpha^2 x_1 + \alpha x_2 + x_3 + x_1^2 + x_1 x_2 + \alpha^2 x_1 x_3 + x_2^2 \\
\bar{f}_3(x_1, x_2, x_3) &= \alpha^2 x_3 + x_1^2 + \alpha^2 x_2^2 + x_2 x_3 + \alpha^2 x_3^2,
\end{aligned}$$

which will be used to encrypt plaintext messages. If, for example, we wish to encrypt the plaintext $(x_1', x_2', x_3') = (1, \alpha, \alpha^2)$, then we compute

$$\begin{aligned}
y_1' &= \bar{f}_1(1, \alpha, \alpha^2) = 0 \\
y_2' &= \bar{f}_2(1, \alpha, \alpha^2) = 0 \\
y_3' &= \bar{f}_3(1, \alpha, \alpha^2) = 1
\end{aligned}$$

to get the ciphertext $(0, 0, 1)$.

The person in charge of decrypting this ciphertext knows L_1^{-1}, $\tilde{F}^{-1}$ and L_2^{-1}. With

$$L_1^{-1}(y_1, y_2, y_3) = \begin{pmatrix} \alpha^2 & 1 & 1 \\ 1 & \alpha^2 & \alpha \\ \alpha^2 & 1 & 0 \end{pmatrix} \begin{pmatrix} y_1 - 0 \\ y_2 - 1 \\ y_3 - \alpha \end{pmatrix}$$

and the given ciphertext we first find

$$L_1^{-1}(0, 0, 1) = \begin{pmatrix} \alpha \\ \alpha \\ 1 \end{pmatrix},$$

from which $X = \alpha + \alpha x + x^2$ follows. In this toy example

$$\tilde{F}^{-1}(X) = X^{26} = \alpha + x^2,$$

which can easily be computed by the binary method (also known as the square-and-multiply method). In real applications this approach would be too time consuming, since the exponent t for X is typically very large. Instead one selects a θ where the binary representation of t exhibits a pattern, which then can be exploited to speed up the process of evaluating X^t.

Continuing with the toy example, we now have $(\bar{z}_1, \bar{z}_2, \bar{z}_3) = (\alpha, 0, 1)$. From

$$L_2^{-1}(y_1, y_2, y_3) = \begin{pmatrix} \alpha^2 & \alpha^2 & \alpha \\ \alpha & \alpha & \alpha \\ 1 & \alpha & 1 \end{pmatrix} \begin{pmatrix} y_1 - \alpha \\ y_2 - \alpha^2 \\ y_3 - \alpha^2 \end{pmatrix}.$$

we obtain $L_2^{-1}(\alpha, 0, 1) = (1, \alpha, \alpha^2)^T$, the original plaintext.

Multiple-Branch MI

A multiple-branch cryptosystem is one essentially composed of several basic (single-branch) cryptosystems. The input is partitioned first, with each part sent to its own single branch cipher. The outputs of each branch are then combined into a single output. The input is first transformed, usually in the form of an invertible affine transformation, before being partitioned in order to hide the branches. Similarly, the combination of the outputs from the branches usually undergoes a transformation. See Figure 2.3 for a pictorial illustration of this general idea. Note that if the single-branch ciphers $C_1, C_2, \ldots, C_b$ and the input-output transformations are invertible, then the multi-branch cipher will be invertible as well.

In the case of multi-branch MI, each branch will be a basic single-branch MI as described in the previous section. Let b be the number of branches and pick positive integers $n_1, \ldots, n_b$ such that $n_1 + \cdots + n_b = n$. For each i, pick an irreducible polynomial $g_i(x) \in k[x]$ of degree n_i and define $K_i = k[x]/g_i(x)$. Then K_i is a degree n_i field extension of k, with k-linear isomorphism

$$\phi_i : K_i \longrightarrow k^{n_i}$$

such that

$$\phi_i(a_0 + a_1 x + \cdots + a_{n_i-1} x^{n_i-1}) = (a_0, a_1, \ldots, a_{n_i-1}).$$

As in the case of a single branch, if we choose (independently) the $\theta_1, \ldots, \theta_b$ such that $0 < \theta_i < n_i$ and $\gcd(q^{\theta_i} + 1, q^{n_i} - 1) = 1$ for each i, then we can construct the invertible maps

$$\tilde{F}_i(X) = X^{1+q^{\theta_i}}$$

and then

$$F_i = \phi_i \circ \tilde{F}_i \circ \phi_i^{-1} = (f_{i1}, \ldots, f_{in_i}),$$

where each f_{1j} is a polynomial in $k[x_1, \ldots, x_{n_1}]$, for $j = 1, \ldots, n_1$; each f_{2j} is a polynomial in $k[x_{n_1+1}, \ldots, x_{n_1+n_2}]$, for $j = 1, \ldots, n_2$; ...; and each f_{bj} is a polynomial in $k[x_{n-n_b+1}, \ldots, x_n]$ for $j = 1, \ldots, n_b$.

We then combine the branches together to define a new map F over k^n by

$$\begin{aligned} F(x_1, \ldots, x_n) &= (F_1, F_2, \ldots, F_b) \\ &= (f_{11}, \ldots, f_{1n_1}, f_{21}, \ldots, f_{2n_2}, \ldots, f_{b1}, \ldots, f_{bn_b}), \end{aligned} \tag{2.3}$$

and choose L_1 and L_2 to be invertible affine transformations on k^n. Finally define the map $\bar{F}$ over k^n as before:

$$\bar{F}(x_1, \ldots, x_n) = L_1 \circ F \circ L_2(x_1, \ldots, x_n) = (\bar{f}_1, \ldots, \bar{f}_n),$$

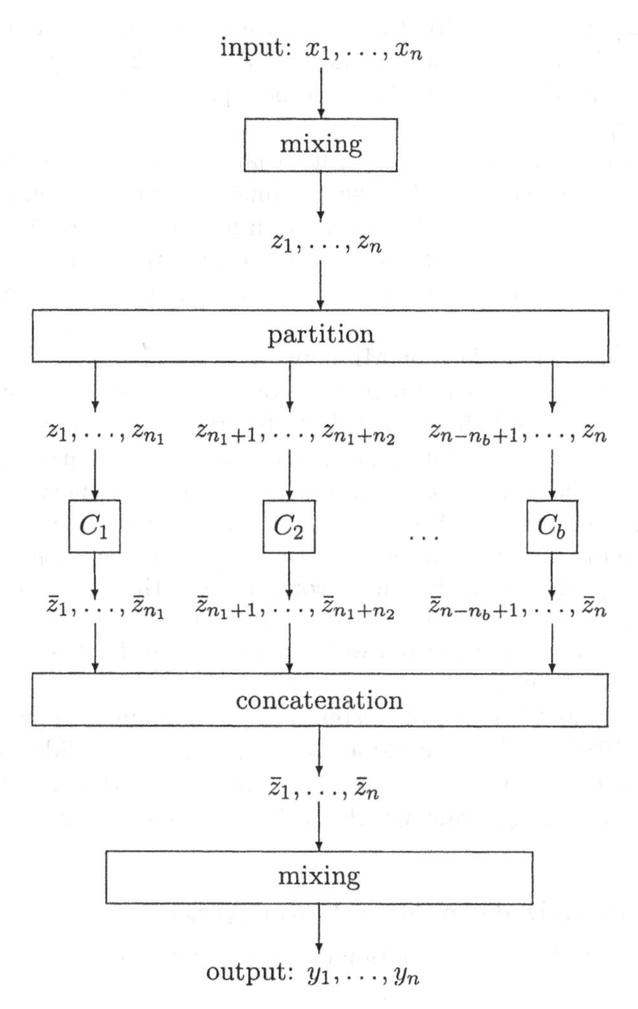

Figure 2.3. A multi-branch cipher composed of single-branch ciphers $C_1, C_2, \ldots, C_b$.

where each $\bar{f}_i$ is a degree two polynomial in $k[x_1, \ldots, x_n]$.

We can see that a multiple-branch implementation of MI is essentially the image of several single-branch MI implementations under an invertible affine transformation. Though it may seem that multiple branches provide more security, we shall see later that this is not the case.

2.2 Key Size and Efficiency of MI

The public key of the Matsumoto-Imai cryptosystem is a set of degree two polynomials $\bar{f}_1, \ldots, \bar{f}_n \in k[x_1, \ldots, x_n]$. Each polynomial has $1+n+$

$n(n+1)/2 = (n+1)(n+2)/2$ terms, hence the public key amounts to a set of $n(n+1)(n+2)/2$ coefficients in k when $q > 2$. For $q = 2$ the key size will be smaller because there are no square terms due to the fact that $x_i^2 = x_i$.

This is rather large compared with that of RSA, even if we choose k to be $GF(2^8)$ and $n = 32$, the parameters originally suggested by Matsumoto-Imai in 1988. However, with systems like RSA there are other considerations, in particular the implementation software, whereas with MPKCs the implementation requires minimum work beyond the public key.

Though the public key of MI may be large compared with other schemes such as RSA, the great advantage of MI lies in its computational efficiency. If we choose $q = |k|$ to be small, then we can store the multiplication table in memory using the fact that the nonzero elements of k form a cyclic multiplicative group. This makes the encryption much faster than schemes like RSA which must work with large integers. This technical detail can also be used in the decryption process, including the most expensive calculation in computing with $\tilde{F}^{-1}$. In fact, MI originally generated a lot of excitement precisely because the practical implementations first suggested were far faster than RSA and promised the same level of security.

The Matsumoto-Imai cryptosystem was proposed in 1988 [Matsumoto and Imai, 1988], and was considered as one of the candidates for the Japanese government security standard. However, MI was defeated in 1995 by Patarin's algebraic attack via linearization equations [Patarin, 1995].

2.3 Linearization Equations Attack

We begin by defining the notion of a linearization equation (LE) in a general way.

Definition 2.3.1. *Let $\mathcal{G} = \{g_1, \ldots, g_m\}$ be any set of m polynomials in $k[x_1, \ldots, x_n]$. A linearization equation for $\mathcal{G}$ is any polynomial in $k[x_1, \ldots, x_n, y_1, \ldots, y_m]$ of the form*

$$\sum_{i=1}^{n}\sum_{j=1}^{m} a_{ij}x_iy_j + \sum_{i=1}^{n} b_ix_i + \sum_{j=1}^{m} c_jy_j + d, \tag{2.4}$$

such that we obtain the the zero function in $k[x_1, \ldots, x_n]$ upon substituting in g_j for y_j, for $j = 1, \ldots, n$. Equivalently, a linearization equation *is any equation in* $k[x_1, \ldots, x_n]$ *of the form*

$$\sum_{i=1}^{n}\sum_{j=1}^{m} a_{ij}x_i g_j(x_1, \ldots, x_n) + \sum_{i=1}^{n} b_i x_i + \sum_{j=1}^{m} c_j g_j(x_1, \ldots, x_n) + d = 0$$

which holds for all $(x'_1, \ldots, x'_n) \in k^n$.

It is clear that for a given $\mathcal{G}$, the set of all linearization equations of $\mathcal{G}$ forms a k-vector space. This space will be referred to as the linearization equation space of $\mathcal{G}$.

Patarin keenly observed that the linearization equation space for the components of $\bar{F}$ can be used to attack the Matsumoto-Imai cryptosystems. To see this, let $\{\bar{f}_1, \ldots, \bar{f}_n\}$ be the set of components of $\bar{F}$, and suppose we have a linearization equation of this set of the form of (2.4). For a given ciphertext $(y'_1, \ldots, y'_n)$, substituting in y'_i for $\bar{f}_i$ produces a linear (hopefully nontrivial) equation in the variables $x_1, \ldots, x_n$ whose solution set contains the plaintext.

With enough linearization equations, we can hope to produce enough linear equations such that the resulting system has the desired plaintext as its unique solution. Even if we cannot find directly the plaintext from these linear equations for a given ciphertext, as long as the LEs can produce enough linearly independent linear equations for the corresponding plaintext, these linear equations can then be plugged into the quadratic public equations derived from the public key and the ciphertext to reduce the number of variables and make it much easier to solve it. To decide the feasibility of this attack, we must first find the number of linearly independent linear equations we can hope to derive from the space of linearization equations of the components of $\bar{F}$. We begin the analysis by considering the single-branch case of MI.

Linearization Equations of Single-Branch MI

The following theorem gives a lower bound on the number of linearly independent linear equations that we can generate from the components of $\bar{F}$.

Theorem 2.3.1. *Let* $\{\bar{f}_1, \ldots, \bar{f}_n\}$ *be the public key for a single-branch implementation of MI. Fix a ciphertext* $Y' = (y'_1, \ldots, y'_n) \in k^n$ *and let* $\bar{\mathcal{L}}$ *be the space of linearization equations of* $\{\bar{f}_1, \ldots, \bar{f}_n\}$. *If* $\bar{\mathcal{L}}_{Y'}$ *is the space of equations that are derived by substituting in* y'_i *for* y_i *(for* $i = 1, \ldots, n$*) in each equation of* $\bar{\mathcal{L}}$, *then the number of linearly independent linear equations in* $\bar{\mathcal{L}}_{Y'}$ *is at least*

$$n - \gcd(n, \theta) \geq \frac{2n}{3}.$$

The exceptional case is $L^{-1}(Y') = (0, \dots, 0)$ when there are only trivial equations.

To prove this theorem we will need the following two lemmas.

Lemma 2.3.1. *Let $\bar{F} = L_1 \circ F \circ L_2$ be as in the construction of single-branch MI. Let $\mathcal{L}$ be the space of linearization equations of $\{f_1, \dots, f_n\}$ and let $\bar{\mathcal{L}}$ be the space of linearization equations of $\{\bar{f}_1, \dots, \bar{f}_n\}$. Then these two k-vector spaces have the same dimension; i.e.,*

$$\dim_k \mathcal{L} = \dim_k \bar{\mathcal{L}}.$$

Proof. First suppose that L_2 is the identity, so that

$$\bar{f}_j(x_1, \dots, x_n) = \sum_{i=1}^{n} \alpha_{ij} f_i(x_1, \dots, x_n) + \beta_j.$$

Then

$$\begin{aligned}
0 &= \sum_{i=1}^{n}\sum_{j=1}^{n} a_{ij}x_i\bar{f}_j + \sum_{i=1}^{n} b_i x_i + \sum_{j=1}^{n} c_j \bar{f}_j + d \\
&= \sum_{i=1}^{n}\sum_{j=1}^{n} a_{ij}x_i\left(\sum_{l=1}^{n}\alpha_{jl}f_l + \beta_j\right) + \sum_{i=1}^{n} b_i x_i + \sum_{j=1}^{n} c_j\left(\sum_{l=1}^{n}\alpha_{jl}f_l + \beta_j\right) \\
&\quad + d \\
&= \sum_{i=1}^{n}\sum_{j=1}^{n} a'_{ij}x_i f_j + \sum_{i=1}^{n} b'_i x_i + \sum_{j=1}^{n} c'_j f_j + d',
\end{aligned}$$

a linearization equation for $f_1, \dots, f_n$.

Similarly, by looking at $F = L_1^{-1} \circ \bar{F}$ and starting with a linearization equation for $f_1, \dots, f_n$, we can derive a linearization equation for $\bar{f}_1, \dots, \bar{f}_n$. From this bijection we see that the dimension of the linearization equations for F and $L_1 \circ F$ are the same.

Now suppose that L_1 is the identity, and let

$$\bar{x}_j = \sum_{i=1}^{n} \alpha_{ij}x_i + \beta_j,$$

so that

$$\bar{f}_i(x_1, \dots, x_n) = f_i(\bar{x}_1, \dots, \bar{x}_n).$$

Then

$$0 = \sum_{i=1}^{n}\sum_{j=1}^{n} a_{ij}x_i f_j(x_1, \dots, x_n) + \sum_{i=1}^{n} b_i x_i + \sum_{j=1}^{n} c_j f_j(x_1, \dots, x_n) + d$$

which gives

$$0 = \sum_{i=1}^{n}\sum_{j=1}^{n} a_{ij}\bar{x}_i f_j(\bar{x}_1, \ldots, \bar{x}_n) + \sum_{i=1}^{n} b_i\bar{x}_i + \sum_{j=1}^{n} c_j f_j(\bar{x}_1, \ldots, \bar{x}_n) + d,$$

since the invertible change of variables amounts to a permutation on k^n. But then we have

$$0 = \sum_{i=1}^{n}\sum_{j=1}^{n} a_{ij}\bar{x}_i \bar{f}_j(x_1, \ldots, x_n) + \sum_{i=1}^{n} b_i\bar{x}_i + \sum_{j=1}^{n} c_j \bar{f}_j(x_1, \ldots, x_n) + d,$$

which, as above, can be rewritten as

$$0 = \sum_{i=1}^{n}\sum_{j=1}^{n} a'_{ij} x_i \bar{f}_j + \sum_{i=1}^{n} b'_i x_i + \sum_{j=1}^{n} c'_j \bar{f}_j + d',$$

a linearization equation for $\bar{f}_1, \ldots, \bar{f}_n$.

Similarly, by looking at $F = \bar{F} \circ L_2^{-1}$ and starting with a linearization equation for $\bar{f}_1, \ldots, \bar{f}_n$, we can derive a linearization equation for $f_1, \ldots, f_n$. From this bijection we see that the dimension of the linearization equations for F and $F \circ L_2$ are the same.

Finally, we conclude that $\dim_k \mathcal{L} = \dim_k \bar{\mathcal{L}}$. □

Lemma 2.3.2. *Let $\mathcal{L}$ and $\bar{\mathcal{L}}$ be as in the previous lemma, fix a ciphertext $Y' = (y'_1, \ldots, y'_n) \in k^n$, and let $Z = L_1^{-1}(Y') = (z_1, \ldots, z_n)$. Let $\mathcal{L}_Z$ be the space of linear equations that arise from substituting in z_i for y_i (for $i = 1, \ldots, n$) in each linearization equation in $\mathcal{L}$, and let $\bar{\mathcal{L}}_{Y'}$ be the space of linear equations that arise from substituting in y'_i for y_i (for $i = 1, \ldots, n$) in each linearization equation in $\bar{\mathcal{L}}$. Then these two k-vector spaces have the same dimension; i.e.,*

$$\dim_k \mathcal{L}_Z = \dim_k \bar{\mathcal{L}}_{Y'}.$$

Proof. In the proof of the previous lemma we constructed a bijection between $\mathcal{L}$ and $\bar{\mathcal{L}}$. This induces a bijection between $\mathcal{L}_Z$ and $\bar{\mathcal{L}}_{Y'}$ from which the result follows. □

To see how Patarin first constructed linearization equations, we let $X, Y \in K$ be such that

$$Y = \tilde{F}(X) = X^{q^\theta+1}.$$

We then have

$$\begin{aligned} Y^{q^\theta-1} &= (X^{q^\theta+1})^{q^\theta-1} \\ &= X^{(q^\theta+1)(q^\theta-1)} \\ &= X^{q^{2\theta}-1}. \end{aligned}$$

If we multiply both sides by XY, we see that

$$XY^{q^\theta} = X^{q^{2\theta}}Y,$$

or equivalently,

$$XY^{q^\theta} - X^{q^{2\theta}}Y = 0.$$

Finally define $\tilde{R}(X,Y) \in K[X,Y]$ by

$$\tilde{R}(X,Y) = XY^{q^\theta} - X^{q^{2\theta}}Y,$$

and

$$R = \phi \circ \tilde{R} \circ (\phi^{-1} \times \phi^{-1}) \tag{2.5}$$

From this R we can derive n linearization equations for the components of F. Specifically, the n components of $R(x_1,\ldots,x_n,y_1,\ldots,y_n)$ are of the form (2.4), and, by construction, substituting in f_i for y_i (for $i = 1,\ldots,n$) yields the zero polynomial in $k[x_1,\ldots,x_n]$.

It is natural to ask how many linearly independent linear equations arise from R for a specific $(y'_1,\ldots,y'_n) \in k^n$. Let $(x'_1,\ldots,x'_n) \in k^n$ be $F^{-1}(y'_1,\ldots,y'_n)$, and let $Y' = \phi^{-1}(y'_1,\ldots,y'_n)$ and $X' = \phi^{-1}(x'_1,\ldots,x'_n)$. Then X' must be a solution of

$$X^{q^{2\theta}}Y' = X(Y')^{q^\theta}, \tag{2.6}$$

or

$$X^{q^{2\theta}-1} = (Y')^{q^\theta-1}, \tag{2.7}$$

if $Y' \neq 0$. But the second equation has at most $\gcd(q^{2\theta}-1, q^n-1)$ solutions in K. Furthermore, because of the condition $\gcd(q^\theta+1, q^n-1) = 1$, we have that

$$\gcd(q^{2\theta}-1, q^n-1) = \gcd(q^\theta-1, q^n-1),$$

hence (2.6) has at most $\gcd(q^\theta-1, q^n-1)+1$ solutions, including the trivial solution. To find this number explicitly we will need the following lemma, which is easily proved.

Lemma 2.3.3. *For any two positive integers a, b we have*

$$\gcd(q^a-1, q^b-1) = q^{\gcd(a,b)} - 1.$$

In particular, the lemma tells us that the total number of solutions for (2.6) is at most $q^{\gcd(\theta,n)}$. If λ is the number of linearly independent linear equations that arise from (2.6), then there will be $q^{n-\lambda}$ solutions to the corresponding system of linear equations. Therefore $q^{n-\lambda} \leq q^{\gcd(\theta,n)}$, and so $\lambda \geq n - \gcd(\theta, n)$.

The three largest possible values of $\gcd(\theta, n)$ are n, $n/2$ if n is even, and $n/3$ if 3 divides n, and the rest are of all smaller values. Therefore, if we show that the first two cases are impossible, then we can conclude that

$$n - \gcd(\theta, n) \geq \frac{2n}{3}.$$

First we know that it is impossible that $\gcd(\theta, n)$ is n, because of the choice of θ is larger than 0 and less than n. Second, if $\gcd(\theta, n) = n/2$, this means that θ must be $n/2$ itself. Then we know that

$$\gcd(q^{n/2} + 1, q^n - 1) = q^{n/2} + 1 > 1,$$

which contradicts the invertibility condition which requires that

$$\gcd(q^\theta + 1, q^n - 1) = 1.$$

Therefore $\gcd(\theta, n)$ cannot be $n/2$ either and the largest possible value for $\gcd(\theta, n)$ is $n/3$.

This proves the following theorem, which combined with Lemma 2.3.2 gives us a proof of Theorem 2.3.1. The exceptional case in Theorem 2.3.1 is $L_1^{-1}(Y') = (0, \ldots, 0)$ and all linear equations derived from the linearization equation are again trivial ones, $0 = 0$.

Theorem 2.3.2. *Let $\mathcal{L}$ be the space of linearization equations for the components of F and fix $Y' = (y_1', \ldots, y_n') \in k^n$. If $\mathcal{L}_{Y'}$ is the space of linear equations resulting from substituting in y_i' for y_i (for $i = 1, \ldots, n$) in each element of $\mathcal{L}$, then* $\dim_k \mathcal{L}_{Y'}$ *is at least*

$$n - \gcd(\theta, n) \geq \frac{2n}{3},$$

except when $Y' = (0, \ldots, 0)$.

If $\gcd(\theta, n) = 1$ then it is clear that we can defeat the system easily using linearization equations alone. More generally, we see that the single branch Matsumoto-Imai cryptosystem is not very secure since for a given ciphertext we can always find at least $2n/3$ linear equations satisfied by the plaintext, which is analogous to leaking $2/3$ of the information. More importantly, these equations can be used to eliminate $2/3$ of the variables of the quadratic public equations derived from the public key and the ciphertext, which should then be much easier to solve than before.

The next question we consider is how to actually generate linearization equations. We explain two different approaches: one based on plaintext-ciphertext pairs, and the other based on the structure of polynomial functions.

Plaintext-Ciphertext Pairs

Using the public key we can generate several plaintext-ciphertext pairs. For each pair given by $\bar{F}(x'_1, \ldots, x'_n) = (y'_1, \ldots, y'_n)$, we can substitute in x'_i for x_i and y'_j for y_j into the generic linearization equation

$$\sum a_{ij}x_iy_j + \sum b_ix_i + \sum c_jy_j + d = 0,$$

to get a linear equation in the $(n+1)^2$ unknowns $a_{ij}, b_i, c_j, d \in k$. Therefore, if we choose roughly $(n+1)^2$ plaintext-ciphertext pairs, then it is very likely that we can solve the resulting system for the unknown coefficients. The total cost of this process includes:

1.) Computation of $(n+1)^2$ plaintext-ciphertext pairs, which has complexity $O(n^4)$;

2.) Solving a set $(n+1)^2$ linear equations in $(n+1)^2$ variables, which has complexity $O(n^6)$.

This can be done relatively easily.

Structure of Polynomial Functions

We begin with a generic linearization equation for the components of $\bar{F}$:

$$\sum a_{ij}x_i\bar{f}_j + \sum b_ix_i + \sum c_j\bar{f}_j + d = 0.$$

As before, we treat the coefficients a_{ij}, b_i, c_j, d as variables taking values in k. After rewriting the left-hand side of this equation as a sum of monomials in the variables $x_1, \ldots, x_n$, we have an equation of the form:

$$\sum \alpha_{ijl}x_ix_jx_l + \sum \beta_{ij}x_ix_j + \sum \gamma_ix_i + \delta = 0, \qquad (2.8)$$

where the coefficients $\alpha_{ijl}, \beta_{ij}, \gamma_i, \delta$ are linear functions in the unknown coefficients a_{ij}, b_i, c_j, d.

Remark 2.3.1. *If $q = 2$, then we should make use of the fact that $x^3 = x^2 = x$ for any $x \in k$. In particular, any power of x_i occurring in (2.8) will be replaced by x_i, for $i = 1, \ldots, n$.*

From the theory of polynomials over a finite field, we know that each of the $\alpha_{ijk}, \beta_{ij}, \gamma_i, \delta$ must be equal to zero, which produces $\frac{(n+1)(n+2)(n+3)}{6}$

linear equations in the unknown coefficients a_{ij}, b_i, c_j, d, when $q > 2$. The solution set for this system of equations is then used to construct linearization equations.

It is very likely that we will not need to use all $(n+1)(n+2)(n+3)/6$ linear equations, and that we probably only need roughly $(n+1)^2$ of them. We can also confirm easily if indeed we have the right solution space, if we know the dimension of the space of linearization equations (we will say more in the next subsection about how to calculate this dimension). If the dimension of the space is too large, we can always add more equations until the right solution space is found.

Here the main cost is to solve a set of $(n+1)^2$ linear equations in $(n+1)^2$ variables. As before, the complexity of this is $O(n^6)$.

Dimension of the Space of Linearization Equations for Basic MI

Now we will present the results related to calculation of the dimension of the space of linearization equations as presented in [Diene et al., 2006].

Theorem 2.3.3. *Let $\mathcal{L}$ be the space of linearization equations associated with the components of a given invertible Matsumoto-Imai map $\bar{F}$ (hence we may assume that $\theta \neq n/2$). If $q > 2$, then*

$$\dim_k \mathcal{L} = \begin{cases} 2n/3, & \text{if } \theta = n/3,\ 2n/3; \\ n, & \text{otherwise.} \end{cases}$$

If $q = 2$ and $\theta = n/3, 2n/3$, then

$$\dim_k \mathcal{L} = \begin{cases} 7, & \text{if } n = 6,\ \theta = 2,\ 4; \\ 8, & \text{if } n = 3,\ \theta = 1,\ 2; \\ 2n/3, & \text{otherwise.} \end{cases}$$

If $q = 2$ and $\theta \neq n/3, 2n/3$, then

$$\dim_k \mathcal{L} = \begin{cases} 10, & \text{if } n = 4,\ \theta = 1,\ 3; \\ 2n, & \text{if } \theta = 1,\ n-1,\ (n \pm 1)/2; \\ 3n/2, & \text{if } \theta = (n \pm 2)/2; \\ n, & \text{otherwise.} \end{cases}$$

The key idea used in the calculation of $\dim_k \mathcal{L}$ is to lift the problem to an extension field. The approach is very similar to that used by Kipnis and Shamir in [Kipnis and Shamir, 1999]. We present only a sketch of the proof of the case where $q > 2$; see [Diene et al., 2006] for the complete proof of Theorem 2.3.3.

The proof in [Diene et al., 2006] uses some very abstract mathematical concepts and theorems, which look simple but may be difficult for people who are not very familiar with the related mathematical theory. Our proof here is more direct and more from the point of computation.

Recall $\tilde{R} : K \times K \longrightarrow K$ is defined by

$$\tilde{R}(X,Y) = XY^{q^\theta} - X^{q^{2\theta}}Y,$$

and $R : k^{2n} \longrightarrow k^n$ is defined by

$$R = \phi \circ \tilde{R} \circ (\phi^{-1} \times \phi^{-1}) = (r_1, \ldots, r_n),$$

where $r_1, \ldots, r_n \in k[x_1, \ldots, x_n, y_1, \ldots, y_n]$. The first step is to show that the n linearization equations derived from R are linearly independent if $q > 2$ and $\theta \neq n/3, 2n/3$. We will show this by way of contradiction, so let us assume that these n linearization equations are not linearly independent. In this case there must exist a nonzero vector $(\alpha_1, \ldots, \alpha_n) \in k^n$ such that $\alpha_1 r_1 + \cdots + \alpha_n r_n = 0$ in the polynomial ring $k[x_1, \ldots, x_n, y_1, \ldots, y_n]$.

Let $L : k^n \longrightarrow k^n$ be the linear map defined by

$$L(x_1, \ldots, x_n) = (\alpha_1 x_1 + \cdots + \alpha_n x_n, 0, \ldots, 0),$$

hence $L \circ R$ is the zero function from k^{2n} to k^n. From this it follows that $\phi^{-1} \circ L \circ \phi \circ \tilde{R}$ is the zero function from $K \times K$ to K since

$$\begin{aligned}\phi^{-1} \circ L \circ \phi \circ \tilde{R} &= (\phi^{-1} \circ L \circ \phi) \circ (\phi^{-1} \circ R \circ (\phi \times \phi)) \\ &= \phi^{-1} \circ (L \circ R) \circ (\phi \times \phi) \\ &= \phi^{-1} \circ 0 \circ (\phi \times \phi).\end{aligned}$$

Now from Lemma A.0.1 and its corollary, there exists a nonzero vector in K^n, say $(A_0, \ldots, A_{n-1})$, such that

$$\phi^{-1} \circ L \circ \phi(X) = \sum_{i=0}^{n-1} A_i X^{q_i},$$

hence

$$\sum_{i=0}^{n-1} A_i (XY^{q^\theta} - X^{q^{2\theta}}Y)^{q^i} = 0.$$

It is not hard to see that if $q > 2$ and $i \neq 0$ then

$$XY^{q^\theta} \neq (YX^{q^{2\theta}})^{q^i} = Y^{q^i} X^{q^{2\theta+i}},$$

unless $3\theta = n, 2n$. Since we have assumed otherwise, the monomials in this polynomial are linearly independent, and hence all A_i are zero. This contradicts our assumption, and thus the n linearization equations are linearly independent.

To prove that there are no other linearization equations is very similar. Pick any linearization equation, say

$$\sum_{i=1}^{n} a_{ij}x_i y_j + \sum_{i=1}^{n} b_i x_i + \sum_{i=1}^{n} c_j y_j + d = 0$$

so that

$$\sum_{i=1}^{n} a_{ij}x_i \bar{f}_j + \sum_{i=1}^{n} b_i x_i + \sum_{i=1}^{n} c_j \bar{f}_j + d = 0$$

in $k[x_1, \ldots, x_n]$, and not all the $a_{ij}, b_i, c_j, d \in k$ are zero.

The map Q taking $(x_1, \ldots, x_n, y_1, \ldots, y_n)$ to

$$\left(\sum a_{ij}x_i y_j + \sum b_i x_i + \sum c_i y_i + d, 0, \ldots, 0\right) \tag{2.9}$$

is a nonzero map from k^{2n} to k^n. Hence by Lemma A.0.3 in Appendix A, there exists a corresponding unique map $\bar{Q}$ from $K \times K$ to K:

$$\bar{Q} = \phi \circ Q \circ (\phi^{-1} \times \phi^{-1})$$

such that

$$\bar{Q}(X, Y) = \sum_{i=0}^{n-1}\sum_{j=0}^{n-1} A_{ij}X^{q^i}Y^{q^j} + \sum_{i=0}^{n-1} B_i X^{q^i} + \sum_{j=0}^{n-1} C_j Y^{q^j} + D,$$

where not all the $A_{ij}, B_i, C_j, D \in K$ are zero, and $X = \phi^{-1}(x_1, \ldots, x_n)$ and $Y = \phi^{-1}(y_1, \ldots, y_n)$.

Because $\bar{Q}$ is derived from a linearization equation, when we substitute in Y for $X^{q^\theta+1}$ in this expression, then we will have the zero function from $K \times K$ to K. Via a direct computation we can show that it will be in the form

$$\sum_{i=0}^{n-1} A_i (XY^{q^\theta} - X^{q^{2\theta}}Y)^{q^i} = 0,$$

if $q > 2$ and $\theta \neq n/3, 2n/3$. From this we conclude that all linearization equations for $\bar{F}$ are linear combinations of the n components of R, and that the dimension of the space of linearization equations is n in the case of $q > 2$ and $\theta \neq n/3, 2n/3$.

Linearization Equations Toy Example

We will illustrate the use of the linearization equations with a small example. We will again use the field $GF(2^2)$ for k, whose field operations are given in Table 2.1. The plaintext is given by $n = 5$ variables $(x_1, x_2, x_3, x_4, x_5) \in k^5$. In order to represent the public key in a more compact form we introduce the additional value $x_0 = 1$, so that the public key can be written as a sum of quadratic terms. With the row vector $\mathsf{x} = (x_0, x_1, x_2, x_3, x_4, x_5)$ the public key is given by

$$y_1 = \mathsf{x} \begin{pmatrix} 0 & 0 & \alpha & 1 & 1 & 1 \\ & \alpha & \alpha & \alpha^2 & \alpha & 0 \\ & & 1 & \alpha^2 & 0 & \alpha \\ & & & \alpha^2 & \alpha & \alpha \\ & & & & \alpha & \alpha^2 \\ & & & & & 1 \end{pmatrix} \mathsf{x}^T, \tag{2.10}$$

$$y_2 = \mathsf{x} \begin{pmatrix} \alpha & 0 & 0 & 0 & \alpha^2 & 1 \\ & \alpha & 0 & \alpha^2 & \alpha^2 & 1 \\ & & 1 & \alpha^2 & 0 & 0 \\ & & & 0 & \alpha & 0 \\ & & & & 1 & \alpha^2 \\ & & & & & 1 \end{pmatrix} \mathsf{x}^T, \tag{2.11}$$

$$y_3 = \mathsf{x} \begin{pmatrix} 1 & \alpha^2 & \alpha & \alpha^2 & 1 & \alpha^2 \\ & \alpha^2 & 0 & 0 & 1 & \alpha^2 \\ & & \alpha^2 & 0 & 1 & 1 \\ & & & \alpha^2 & \alpha & \alpha \\ & & & & 0 & \alpha^2 \\ & & & & & \alpha^2 \end{pmatrix} \mathsf{x}^T, \tag{2.12}$$

$$y_4 = \mathsf{x} \begin{pmatrix} 1 & \alpha^2 & 1 & \alpha^2 & 0 & 0 \\ & \alpha^2 & \alpha & \alpha^2 & \alpha^2 & \alpha^2 \\ & & 1 & 0 & \alpha^2 & \alpha \\ & & & 1 & \alpha^2 & \alpha \\ & & & & 0 & \alpha^2 \\ & & & & & 1 \end{pmatrix} \mathsf{x}^T, \tag{2.13}$$

$$y_5 = \mathsf{x} \begin{pmatrix} \alpha^2 & \alpha^2 & \alpha & 1 & 1 & \alpha^2 \\ & 0 & 0 & \alpha^2 & \alpha^2 & 0 \\ & & \alpha^2 & \alpha & 1 & 1 \\ & & & 1 & 0 & \alpha^2 \\ & & & & \alpha & \alpha \\ & & & & & \alpha^2 \end{pmatrix} \mathsf{x}^T. \tag{2.14}$$

The entries left blank in the matrices are zero, and they will not be stored in a real life application. Assume that a plain text produced the cipher text $(1, 0, 0, 0, 1)$. We will show how to recover the plain text with the help of the linearization equations.

Introduce the value $y_0 = 1$ so that the public key can be represented by the row vector $\mathsf{y} = (y_0, y_1, y_2, y_3, y_4, y_5)$. The linearization equations (2.4), which in our case use $m = 5$ and $n = 5$, can now be written in matrix form

$$\mathsf{x}\mathsf{A}\mathsf{y}^T = 0 \tag{2.15}$$

where A is a 6×6 matrix with unknown coefficients $\mathsf{A}_{i,j}$, $i, j = 0, \ldots, 5$. For setting up the system of linear equations it is easier if the $(m + 1)(n + 1)$ unknowns are represented by a one dimensional array. With a notation commonly used in programming we introduce the correspondence

$$\mathsf{A}_{i,j} \Longleftrightarrow A[(m+1)i + j] = 0$$

so that we have the following correspondence for the unknowns appearing in (2.4)

$$\begin{array}{rcll} a_{ij} & \Longleftrightarrow & A[(m+1)i+j] & \text{for} \quad i = 1, \ldots, n;\ j = 1, \ldots, m; \\ b_i & \Longleftrightarrow & A[(m+1)i] & \text{for} \quad i = 1, \ldots, n; \\ c_j & \Longleftrightarrow & A[j] & \text{for} \quad j = 1, \ldots, m; \\ d & \Longleftrightarrow & A[0]. & \end{array}$$

Substituting the public key into (2.15) produces a homogeneous polynomial, which is cubic in x_i for $i = 0, \ldots, 5$. Collecting the coefficients of the 56 different terms, we obtain a homogeneous system of linear equations in the 36 unknowns $A[0]$ to $A[35]$. The rank of the corresponding matrix is 31, so that the dimension of the linearization equations is $36 - 31 = 5$, which is the common case as predicted by Theorem 2.3.3.

Reducing the matrix to row echelon form we obtain the following

$$\begin{array}{rcl} A[0] & = & \alpha A[29] + \alpha^2 A[32] + A[34] + A[35] \\ A[1] & = & A[29] + \alpha^2 A[32] + A[33] + A[34] + A[35] \\ A[2] & = & \alpha A[29] + A[32] + A[35] \\ A[3] & = & A[29] + \alpha A[32] + \alpha^2 A[33] + \alpha^2 A[34] + \alpha A[35] \\ A[4] & = & \alpha A[32] + \alpha^2 A[33] + \alpha A[34] + A[35] \\ A[5] & = & \alpha A[32] \\ A[6] & = & A[29] + A[32] + \alpha^2 A[33] + A[34] + A[35] \\ A[7] & = & A[32] + A[33] + \alpha^2 A[34] \\ A[8] & = & \alpha^2 A[32] + \alpha A[35] \end{array}$$

$$
\begin{aligned}
A[9] &= A[29] + A[32] + \alpha^2 A[33] + A[34] + A[35] \\
A[10] &= A[29] + \alpha^2 A[34] \\
A[11] &= \alpha^2 A[29] + \alpha A[35] \\
A[12] &= \alpha^2 A[34] + \alpha^2 A[35] \\
A[13] &= A[29] + A[32] + A[33] + \alpha^2 A[34] \\
A[14] &= \alpha A[32] + \alpha^2 A[35] \\
A[15] &= \alpha^2 A[29] + \alpha^2 A[32] + \alpha^2 A[33] + \alpha^2 A[34] + \alpha^2 A[35] \\
A[16] &= \alpha^2 A[29] + \alpha A[32] + \alpha^2 A[33] + A[34] + A[35] \\
A[17] &= \alpha A[32] \\
A[18] &= \alpha^2 A[29] + \alpha^2 A[32] + \alpha^2 A[33] \\
A[19] &= A[32] + A[35] \\
A[20] &= A[29] + A[32] + \alpha^2 A[33] + \alpha^2 A[34] \\
A[21] &= A[29] + A[32] + \alpha A[33] \\
A[22] &= \alpha^2 A[32] + A[33] + \alpha A[34] + \alpha^2 A[35] \\
A[23] &= \alpha^2 A[29] + \alpha^2 A[32] + \alpha A[34] \\
A[24] &= A[32] + A[34] + A[35] \\
A[25] &= \alpha^2 A[32] + \alpha^2 A[35] \\
A[26] &= \alpha A[29] + A[32] + A[35] \\
A[27] &= \alpha^2 A[32] + \alpha A[33] + \alpha A[34] + \alpha^2 A[35] \\
A[28] &= A[29] + A[34] \\
A[30] &= \alpha A[33] + \alpha A[34] + \alpha A[35] \\
A[31] &= A[32] + \alpha^2 A[33] + A[34] + \alpha^2 A[35]
\end{aligned}
$$

where $A[29]$, $A[32]$, $A[33]$, $A[34]$ and $A[35]$ are free parameters. These values and the given cipher text

$$\mathsf{y} = (1, y_1', y_2', y_3', y_4', y_5') = (1, 1, 0, 0, 0, 1)$$

are now substituted back into (2.15), and the coefficients of the free parameters $A[29], A[32], A[33], A[34], A[35]$ are set to zero to give the following set of equations for the plaintext:

$$
\begin{aligned}
\alpha x_1 + x_2 + x_4 + \alpha^2 &= 0, \\
\alpha^2 x_2 + x_3 + \alpha x_4 + x_5 + \alpha &= 0, \\
\alpha x_1 + x_2 + \alpha^2 x_3 + x_5 + 1 &= 0, \\
\alpha x_1 + \alpha x_3 + x_4 + \alpha^2 x_5 &= 0, \\
\alpha^2 x_1 + \alpha^2 x_2 + x_3 + \alpha x_4 &= 0.
\end{aligned}
$$

The system of equations has the following solution

$$x_1 = \alpha x_5 + \alpha^2, \quad (2.16)$$
$$x_2 = x_5 + \alpha, \quad (2.17)$$
$$x_3 = x_5 + \alpha^2, \quad (2.18)$$
$$x_4 = \alpha x_5. \quad (2.19)$$

Finally we can find the value of the plaintext in one of two ways.

In the first method we try all possible values of $x_5 \in k$ in order to find out which of the possible plaintexts produced the given ciphertext. With the different values for x_5 in the solutions (2.16) to (2.19) and the public key in (2.10) to (2.14) we find the following possibilities:

plaintext		ciphertext
$(\alpha^2, \alpha, \alpha^2, 0, 0)$	$\Longrightarrow$	$(1, 0, 0, 0, 1)$
$(1, \alpha^2, \alpha, \alpha, 1)$	$\Longrightarrow$	$(0, \alpha, 0, \alpha^2, \alpha)$
$(0, 0, 1, \alpha^2, \alpha)$	$\Longrightarrow$	$(\alpha, 1, 0, \alpha, \alpha^2)$
$(\alpha, 1, 0, 1, \alpha^2)$	$\Longrightarrow$	$(\alpha^2, \alpha^2, 0, 1, 0)$

Only the first case produces the given ciphertext and thus we know that the original plaintext was $(\alpha^2, \alpha, \alpha^2, 0, 0)$.

In the other method we substitute the linear equations (2.16) to (2.19) into the public key (2.10) to (2.14) and set it equal to the given ciphertext, that is

$$y_1 = 1,$$
$$y_2 = 0,$$
$$y_3 = 0,$$
$$y_4 = 0,$$
$$y_5 = 1.$$

This results in quadratic equations, which the free parameter has to satisfy. In our case the free parameter is x_5. Some of the resulting equations are trivial, but others are $x_5^2 = 0$. From this we conclude that $x_5 = 0$ and find the remaining plaintext from (2.16) to (2.19).

Linearization Equations for Multiple-Branch MI

Using the notation of the multiple-branch case discussed above, it is evident we have the following theorem.

Theorem 2.3.4. *Let $\mathcal{L}$ be the space of linearization equations for a given implementation of MI and fix a ciphertext $(y'_1, \ldots, y'_n) \in k^n$. Let*

$Y' = \phi^{-1}(y'_1, \ldots, y'_n)$ and define $\mathcal{L}_{Y'}$ to be the space of linear equations (in the plaintext variables $x_1, \ldots, x_n$) obtained by substituting in y'_j in for y_j (for $j = 1, \ldots, n$) in every element of $\mathcal{L}$. Then with probability

$$\frac{(q^{n_1} - 1)(q^{n_2} - 1) \cdots (q^{n_b} - 1)}{q^n}$$

$\dim_k \mathcal{L}_{Y'}$ *is at least*

$$n - \sum_{i=1}^{b} \gcd(n_i, \theta_i) \geq \frac{2n}{3}$$

Therefore the linearization attack for the single-branch case can also be applied to the multiple-branch case. Additionally, there are refined methods suggested by Patarin [Patarin, 2000] that improve the efficiency of the algorithm where one separates the branches before attacking the system.

From a mathematical point view one can see that it is possible to separate the different branches using the idea of finding a common invariant subspace. This idea was pursued in [Felke, 2005] for the more general case of multi-branch HFE.

Remark 2.3.2. *It is not difficult to see that the attack of Kipnis-Shamir [Kipnis and Shamir, 1999] on the HFE cryptosystem can also be used to attack the Matsumoto-Imai cryptosystem. In this case one can actually recover the private key, and it applies to both single- and multiple-branch cases. One can also see that the linearization attack can be viewed as the prototype and the origin of the XL algorithm for solving polynomial equations.*

2.4 Another Attack on Matsumoto-Imai

In this section, we will present an attack that is an extension of the Kipnis-Shamir attack on HFE for use against the Matsumoto-Imai cryptosystem. Unlike the linearization attack, this attack will allow us to recover the private key. This attack has not been published before, though it is probably known to the experts in this area. The importance of this new approach is that it may lead to a new attack on MI-Minus, which then can be used to attack Sflash^{v2}.

The key idea of the Kipnis-Shamir attack on HFE is to attack the problem from its origin. The constructions of MI and HFE are based on the idea that we can construct a map on a k-vector space from a map on an extension field. Their idea was to use the structure of the map on the extension field to design the attack on the k-vector space mapping.

With this point of view, if $\bar{F} : k^n \longrightarrow k^n$ is a given Matsumoto-Imai public key mapping, then the first step of the attack is to lift $\bar{F}$ back to a map over K; i.e., we must study $\phi^{-1} \circ \bar{F} \circ \phi$, in order to use the underlying algebraic structures from the extension field, not the vector space over the small field.

To simplify the exposition we assume that $q > 2$ and that L_1, L_2 are linear instead of affine, in effect ignoring the constant terms. In other words, we assume that the $\bar{f}_1, \ldots, \bar{f}_n$ are degree two homogeneous polynomials in $k[x_1, \ldots, x_n]$. Also, we assume that we know the field K and hence the map $\phi : K \longrightarrow k^n$. If we do not have this information, then we will produce L_1', L_2' and F' such that $\bar{F} = L_1' \circ F' \circ L_2'$. We now justify this claim.

As before, the legitimate user picks an degree n irreducible polynomial $g(x) \in k[x]$ in order to construct $K = k[x]/g(x)$ and $\phi : K \longrightarrow k^n$. Suppose the attacker has chosen another degree n irreducible polynomial $h(y) \in k[y]$ and constructs $K' = k[y]/h(y)$ and $\psi : K' \longrightarrow k^n$. Of course, K and K' are isomorphic, and in fact, k-linear field isomorphisms exist between K and K'. Let $\alpha(y) \in K'$ be such that

$$g(\alpha(y)) = 0 \bmod h(y),$$

and let $\iota : K \longrightarrow K'$ be defined by

$$\iota(p(x)) = p(\alpha(y)) \bmod h(y),$$

for $p(x) \in K$. It is easy to check that ι is a k-linear field isomorphism between K and K'.

Observe that

$$\begin{aligned}
\bar{F} &= L_1 \circ F \circ L_2 \\
&= L_1 \circ (\phi \circ \tilde{F} \circ \phi^{-1}) \circ L_2 \\
&= L_1 \circ \phi \circ (\iota^{-1} \circ \iota) \circ \tilde{F} \circ (\iota^{-1} \circ \iota) \circ \phi^{-1} \circ L_2 \\
&= (L_1 \circ \phi \circ \iota^{-1}) \circ (\iota \circ \tilde{F} \circ \iota^{-1}) \circ (\iota \circ \phi^{-1} \circ L_2).
\end{aligned}$$

Define $M_1 : K' \longrightarrow k^n$, $\tilde{F}' : K' \longrightarrow K'$, and $M_2 : k^n \longrightarrow K'$ by

$$\begin{aligned}
M_1 &= L_1 \circ \phi \circ \iota^{-1} \\
\tilde{F}' &= \iota \circ \tilde{F} \circ \iota^{-1} \\
M_2 &= \iota \circ \phi^{-1} \circ L_2.
\end{aligned}$$

Observe that M_1 and M_2 are k-linear vector space isomorphisms, and that

$$\begin{aligned}\tilde{F}'(X) &= \iota\left(\left(\iota^{-1}(X)\right)^{q^\theta+1}\right)\\ &= \iota\left(\iota^{-1}\left(X^{q^\theta+1}\right)\right)\\ &= X^{q^\theta+1}\\ &= \tilde{F}(X).\end{aligned}$$

We consider whether or not there exists L_1' and L_2' such that

$$\bar{F} = L_1' \circ \psi \circ \tilde{F}' \circ \psi^{-1} \circ L_2',$$

or equivalently, for a given M_1 and M_2, whether or not there exists L_1' and L_2' such that

$$\begin{aligned}L_1' \circ \psi &= M_1\\ \psi^{-1} \circ L_2' &= M_2.\end{aligned}$$

Solving for L_1' and L_2' we see that

$$\begin{aligned}L_1' &= M_1 \circ \psi^{-1}\\ L_2' &= \psi \circ M_2.\end{aligned}$$

The attacker cannot know ι, and thus cannot know M_1 and M_2. Therefore the attacker cannot know which L_1' and L_2' will be obtained. However, if some L_1' and L_2' can be found, then these are just as useful as the original L_1 and L_2 for the attack since we can then easily invert the map F anyway. Therefore it does not really matter if the attacker knows the extension field K or not, and thus, there is no advantage in hiding the field structure of K. From now on, we assume that we know the field K.

Now, from Lemma A.0.2 in Appendix A, we know that

$$\phi^{-1} \circ \bar{F} \circ \phi(X) = \sum_{i=0}^{n-1}\sum_{j=0}^{i-1} A_{ij} X^{q^i+q^j}, \tag{2.20}$$

for some $A_{ij} \in K$. We also know that

$$\begin{aligned}\phi^{-1} \circ \bar{F} \circ \phi &= \phi^{-1} \circ (L_1 \circ F \circ L_2) \circ \phi\\ &= \phi^{-1} \circ (L_1 \circ (\phi \circ \tilde{F} \circ \phi^{-1}) \circ L_2) \circ \phi\\ &= (\phi^{-1} \circ L_1 \circ \phi) \circ \tilde{F} \circ (\phi^{-1} \circ L_2 \circ \phi),\end{aligned} \tag{2.21}$$

where the parentheses are added to show the composition is of three maps defined on K. In particular, we know $\tilde{F}$ and we can construct $\phi^{-1} \circ \bar{F} \circ \phi$ from the known $\bar{F}$. Our attack will focus on using the properties of these known maps to find both of the unknown linear maps $\phi^{-1} \circ L_1 \circ \phi$ and $\phi^{-1} \circ L_2 \circ \phi$. In particular, we will study

$$(\phi^{-1} \circ L_1^{-1} \circ \phi) \circ (\phi^{-1} \circ \bar{F} \circ \phi) = \tilde{F} \circ (\phi^{-1} \circ L_2 \circ \phi),$$

and the properties of the functions in this formula.

From Lemma A.0.1 in Appendix A we have the following equations:

$$\phi^{-1} \circ L_1 \circ \phi(X) = \sum_{j=0}^{n-1} L_{1j} X^{q^j} \tag{2.22}$$

$$\phi^{-1} \circ L_1^{-1} \circ \phi(X) = \sum_{j=0}^{n-1} L_{1j}^{-1} X^{q^j} \tag{2.23}$$

$$\phi^{-1} \circ L_2 \circ \phi(X) = \sum_{j=0}^{n-1} L_{2j} X^{q^j}. \tag{2.24}$$

where $L_{ij} \in K$. Our attack comes down to finding the L_{ij}, from which we can then construct L_1 and L_2.

Remark 2.4.1. *We make special note that the notation L_{1j}^{-1} represents the coefficient of X^{q^j} in the polynomial representation of L_1^{-1}. This is to be distinguished from $(L_{1j})^{-1}$, the multiplicative inverse of the coefficient of X^{q^j} in the polynomial representation of L_1. In general these two notations will not refer to the same value in K. All other exponent notations will be written as usual without parentheses.*

Now for any polynomial $G(X) \in K[X]$ of the form

$$G(X) = \sum_{i=0}^{n-1} \sum_{j=0}^{i} G_{ij} X^{q^i + q^j},$$

we can associate a unique $n \times n$ symmetric matrix G defined by

$$[\mathsf{G}]_{ij} = \begin{cases} 2G_{ii} & \text{if } i = j; \\ G_{ij} & \text{if } i > j; \\ G_{ji} & \text{if } i < j. \end{cases}$$

Note that this matrix is such that

$$G(X + Y) - G(X) - G(Y) = \mathsf{x}\, \mathsf{G}\, \mathsf{y}^T,$$

where $\mathsf{x} = (X, X^q, \ldots, X^{q^{n-1}})$ and $\mathsf{y} = (Y, Y^q, \ldots, Y^{q^{n-1}})$. We make a special note here that the index of the rows and column range in $0, \ldots, n-1$, and not $1, \ldots, n$. We also note that because the characteristic of K is two, the entries on the diagonal of G are all zero.

Remark 2.4.2. *The trouble with the case $q = 2$ is that in $\bar{F}$ the square terms and the linear terms are now the same and therefore mixed. But because of the symmetrization process, we realize that these linear terms are only related to the diagonal elements in the matrix, which are annihilated here anyway. Therefore there is no problem with this attack for the case $q = 2$.*

With this correspondence between homogeneous quadratic functions on K and $n \times n$ matrices with entries in K, we will shift from the function point of view to that of matrices. In particular, let $\tilde{\mathsf{F}}$ be the matrix associated with $\tilde{F}$. Then clearly $\tilde{\mathsf{F}}$ has only two nonzero entries: $[\tilde{\mathsf{F}}]_{0\theta} = 1$ and $[\tilde{\mathsf{F}}]_{\theta 0} = 1$. To see the basic idea of the attack, we must first understand how the bilinear form behaves if we compose the function by a k-linear function from the left or right. The results are presented in the following two lemmas that deal with how these matrices behave under function composition.

Lemma 2.4.1. *Let $G(X)$ be as defined above, let $S(X) = \sum_{i=0}^{n-1} S_i X^{q^i}$ and let G' be the symmetric matrix associated with $G(S(X))$. Then*

$$\mathsf{G}' = \mathsf{W}^T \mathsf{G} \mathsf{W},$$

where W is an $n \times n$ matrix defined by

$$[\mathsf{W}]_{ij} = S_{j-i}^{q^i},$$

and $j - i$ is calculated modulo n.

Proof. We begin by expanding $G(S(X))$:

$$\begin{aligned} G(S(X)) &= \sum_{u=0}^{n-1} \sum_{v=0}^{u} G_{uv} \left(\sum_{l=0}^{n-1} S_l X^{q^l} \right)^{q^u + q^v} \\ &= \sum_{u=0}^{n-1} \sum_{v=0}^{u} G_{uv} \left(\sum_{l=0}^{n-1} S_l X^{q^l} \right)^{q^u} \left(\sum_{l=0}^{n-1} S_l X^{q^l} \right)^{q^v} \\ &= \sum_{u=0}^{n-1} \sum_{v=0}^{u} G_{uv} \left(\sum_{l=0}^{n-1} S_l^{q^u} X^{q^{l+u}} \right) \left(\sum_{l=0}^{n-1} S_l^{q^v} X^{q^{l+v}} \right) \end{aligned}$$

$$= \sum_{u=0}^{n-1}\sum_{v=0}^{u} G_{uv}\left(\sum_{i=0}^{n-1} S_{i-u}^{q^u} X^{q^i}\right)\left(\sum_{j=0}^{n-1} S_{j-v}^{q^v} X^{q^j}\right)$$

$$= \sum_{u=0}^{n-1}\sum_{v=0}^{u} G_{uv}\left(\sum_{i=0}^{n-1}\sum_{j=0}^{n-1} S_{i-u}^{q^u} S_{j-v}^{q^v} X^{q^i+q^j}\right)$$

$$= \sum_{i=0}^{n-1}\sum_{j=0}^{n-1}\left(\sum_{u=0}^{n-1}\sum_{v=0}^{u} G_{uv} S_{i-u}^{q^u} S_{j-v}^{q^v}\right) X^{q^i+q^j}$$

$$= \sum_{i=0}^{n-1}\sum_{j=0}^{i}\left(\sum_{u=0}^{n-1}\sum_{v=0}^{u} G_{uv}\left(S_{i-u}^{q^u} S_{j-v}^{q^v} + S_{j-u}^{q^u} S_{i-v}^{q^v}\right)\right) X^{q^i+q^j}$$

$$- \sum_{w=0}^{n-1}\left(\sum_{u=0}^{n-1}\sum_{v=0}^{u} G_{uv} S_{w-u}^{q^u} S_{w-v}^{q^v}\right) X^{2q^w}.$$

Thus the coefficient of $X^{q^i+q^j}$ for $i > j$ is

$$\sum_{u=0}^{n-1}\sum_{v=0}^{u} G_{uv}\left(S_{i-u}^{q^u} S_{j-v}^{q^v} + S_{j-u}^{q^u} S_{i-v}^{q^v}\right).$$

This is the same as $[\mathsf{G}']_{ij}$ for $i > j$, since:

$$[\mathsf{G}']_{ij} = [\mathsf{W}^T\,\mathsf{G}\,\mathsf{W}]_{ij} = \sum_{u=0}^{n-1} [\mathsf{W}^T]_{iu}\,[\mathsf{G}\,\mathsf{W}]_{uj}$$

$$= \sum_{u=0}^{n-1} [\mathsf{W}]_{ui}\left(\sum_{v=0}^{n-1} [\mathsf{G}]_{uv}\,[\mathsf{W}]_{vj}\right)$$

$$= \sum_{u=0}^{n-1} S_{i-u}^{q^u}\left(\sum_{v=0}^{n-1} [\mathsf{G}]_{uv} S_{j-v}^{q^v}\right)$$

$$= \sum_{u=0}^{n-1}\sum_{v=0}^{n-1} [\mathsf{G}]_{uv} S_{i-u}^{q^u} S_{j-v}^{q^v}$$

$$= \sum_{u=0}^{n-1}\sum_{v=0}^{u}\left([\mathsf{G}]_{uv} S_{i-u}^{q^u} S_{j-v}^{q^v} + [\mathsf{G}]_{vu} S_{i-v}^{q^v} S_{j-u}^{q^u}\right) - \sum_{l=0}^{n-1} [\mathsf{G}]_{ll} S_{i-l}^{q^l} S_{j-l}^{q^l}$$

$$= \sum_{u=0}^{n-1}\sum_{v=0}^{u} G_{uv}\left(S_{i-u}^{q^u} S_{j-v}^{q^v} + S_{i-v}^{q^v} S_{j-u}^{q^u}\right).$$

□

Lemma 2.4.2. *Let $G(X)$ and $S(X)$ be defined as in Lemma 2.4.1. Define G'' to be the symmetric matrix associated with $S(G(X))$. Then*

$$\mathsf{G}'' = \sum_{l=0}^{n-1} S_l \mathsf{G}_l,$$

where G_l is the $n \times n$ matrix defined by

$$[\mathsf{G}_l]_{ij} = G_{i-l\, j-l}^{q^l},$$

with both $i - l$ and $j - l$ calculated modulo n.

Proof. As with Lemma 2.4.1, we expand $G(S(X))$:

$$\begin{aligned} G(S(X)) &= \sum_{l=0}^{n-1} S_l \left(\sum_{u=0}^{n-1} \sum_{v=0}^{u} G_{uv} X^{q^u + q^v} \right)^{q_l} \\ &= \sum_{l=0}^{n-1} S_l \left(\sum_{u=0}^{n-1} \sum_{v=0}^{u} G_{uv}^{q^l} X^{q^{u+l} + q^{v+l}} \right) \\ &= \sum_{l=0}^{n-1} S_l \left(\sum_{i=0}^{n-1} \sum_{j=0}^{i} G_{i-l\, j-l}^{q^l} X^{q^i + q^j} \right) \\ &= \sum_{i=0}^{n-1} \sum_{j=0}^{i} \left(\sum_{l=0}^{n-1} S_l G_{i-l\, j-l}^{q^l} \right) X^{q^i + q^j}. \end{aligned}$$

Thus the coefficient of $X^{q^i + q^j}$ for $i > j$ is

$$\sum_{l=0}^{n-1} S_l G_{i-l\, j-l}^{q^l}.$$

This is the same as $[\mathsf{G}'']_{ij}$ for $i > j$, since:

$$[\mathsf{G}'']_{ij} = \sum_{l=0}^{n-1} S_l [\mathsf{G}_l]_{ij} = \sum_{l=0}^{n-1} S_l G_{i-1\, j-l}^{q^l}.$$

□

Suppose that $\tilde{\mathsf{F}}'$ is the matrix associated with $\tilde{F} \circ (\phi^{-1} \circ L_2 \circ \phi)$. Then from Lemma 2.4.1 we see that

$$\tilde{\mathsf{F}}' = \mathsf{L}_2^T \tilde{\mathsf{F}} \mathsf{L}_2, \tag{2.25}$$

where the $n \times n$ matrix L_2 is defined by

$$[\mathsf{L}_2]_{ij} = L_{2\,j-i}^{q^i}. \tag{2.26}$$

Now suppose that $\bar{\mathsf{F}}$ is the matrix associated with $\phi^{-1} \circ \bar{F} \circ \phi$, and that $\bar{\mathsf{F}}''$ is the matrix associated with $(\phi^{-1} \circ L_1^{-1} \circ \phi) \circ (\phi^{-1} \circ \bar{F} \circ \phi)$. Then from Lemma 2.4.2 we see that

$$\bar{\mathsf{F}}'' = \sum_{l=0}^{n-1} L_{1l}^{-1} \bar{\mathsf{F}}_l, \tag{2.27}$$

where

$$[\bar{\mathsf{F}}_l]_{ij} = [\bar{\mathsf{F}}]_{i-l\,j-l}^{q^l}. \tag{2.28}$$

However, we have seen that

$$(\phi^{-1} \circ L_1^{-1} \circ \phi) \circ (\phi^{-1} \circ \bar{F} \circ \phi) = \tilde{F} \circ (\phi^{-1} \circ L_2 \circ \phi), \tag{2.29}$$

and hence

$$\tilde{\mathsf{F}}' = \mathsf{M} = \bar{\mathsf{F}}'', \tag{2.30}$$

where M denotes the common value of $\tilde{\mathsf{F}}'$ and $\bar{\mathsf{F}}''$.

Clearly the matrix $\tilde{\mathsf{F}}$ has rank equal to two. Since L_2 is invertible, we see that $\mathsf{M} = \mathsf{L}_2^T \tilde{\mathsf{F}} \mathsf{L}_2$ has rank equal to two as well. But this means that the K-linear combination

$$\mathsf{M} = \sum_{l=0}^{n-1} L_{1l}^{-1} \bar{\mathsf{F}}_l$$

of the n known matrices $\bar{\mathsf{F}}_0, \ldots, \bar{\mathsf{F}}_{n-1}$ has rank two, a condition we can use to find the values of L_{1l}^{-1}. In fact, this is a so-called "MinRank" problem.

Definition 2.4.1. *(MinRank Problem) Given $n \times n$ matrices $\mathsf{A}_1, \ldots, \mathsf{A}_m$ over a finite field K and $r < n$, find a non-trivial linear combination of*

$$\mathsf{A} = \alpha_1 \mathsf{A}_1 + \cdots + \alpha_m \mathsf{A}_m$$

such that the rank of A is less than or equal to r.

The general MinRank problem has been studied by Shallit, Frandsen and Buss [Shallit et al., 1996], among others. It generalizes the so-called "Rank Distance Coding" problem posed by Gabidulin [Gabidulin, 1985], which has been studied in [Stern and Chabaud, 1996; Chen, 1996]. This problem is a generalization of the "Minimal Weight" problem of error correcting codes [Berlekamp et al., 1978]. The general MinRank problem was proven to be NP-complete in [Shallit et al., 1996] for the case where $r = n - 1$, which in this case corresponds to the problem of finding a linear combination of $\mathsf{A}_1, \ldots, \mathsf{A}_m$ which is singular.

Their proof uses the technique of writing a given set of multivariate equations as an instance of MinRank. This result can be extended to other cases like $r = n - 2, n - 3, \ldots$, however MinRank is not too hard when r is very small, as is our case.

The approach of Kipnis and Shamir is to use a new relinearization method to solve this problem. Later, Courtois [Courtois, 2001] proposed a more standard and straightforward method to solve this problem that originated from an idea of Coppersmith, Stern and Vaudenay [Coppersmith et al., 1997].

In the most general case, we treat the $\mathsf{A}_1, \ldots, \mathsf{A}_m$ as known, and the $\alpha_1, \ldots, \alpha_m$ as variables. If $\mathsf{A} = \alpha_1\mathsf{A}_1 + \cdots + \alpha_m\mathsf{A}_m$ is to have rank r, then each $(r+1) \times (r+1)$ submatrix minor must be equal to zero. This means that each $(r+1) \times (r+1)$ submatrix yields a total degree $r+1$ polynomial equation in the m variables $\alpha_1, \ldots, \alpha_m$.

In the case under consideration we have $r = 2$. We also know that the $\mathsf{A}_l = \bar{\mathsf{F}}_l$ are symmetric with diagonal entries equal to zero. This means that the number of nonzero degree three polynomials in the variables $L_{10}^{-1}, \ldots, L_{1n-1}^{-1}$ is $\binom{n}{3}(\binom{n}{3} - 1)/2$, where the equation obtained by choosing indices i_1, i_2, i_3 for the rows and j_1, j_2, j_3 for the columns is the same as the equation gotten by choosing indices j_1, j_2, j_3 for the rows and i_1, i_2, i_3 for the columns, and we discard the trivial equations gotten by taking $i_1 = j_1$, $i_2 = j_2$, and $i_3 = j_3$.

Since the equations are homogeneous, solutions should be thought of in the projective space of K^n. This means that if we find a solution vector $(L_{10}^{-1}, \ldots, L_{1n-1}^{-1})$, then $(\alpha L_{10}^{-1}, \ldots, \alpha L_{1n-1}^{-1})$ will also be a solution vector for any nonzero $\alpha \in K$. We may as well then take $L_{10}^{-1} = 1$ and substitute this into all the equations to arrive at a system of $\binom{n}{3}(\binom{n}{3} - 1)/2$ degree three equations in the $n - 1$ variables $L_{11}^{-1}, \ldots, L_{1n-1}^{-1}$, which we expect will be easy to solve [Courtois, 2001].

At this point we have $\phi^{-1} \circ L_1^{-1} \circ \phi$, and thus L_1, so we still need to to find L_2. Along the way we have found $\mathsf{M} = \tilde{\mathsf{F}}' = \mathsf{L}_2^T \tilde{\mathsf{F}} \mathsf{L}_2$, which we will now use to find L_2. We have two ways to proceed. First, if $\tilde{F}$ is easily inverted (i.e., if the q-Hamming weight degree of $\tilde{F}(X) = X^t$ is relatively small), then we can directly compute $\phi^{-1} \circ L_2 \circ \phi$, and hence, L_2, from (2.29). Otherwise, we proceed as did Kipnis and Shamir.

Let $\mathsf{u}_1, \ldots, \mathsf{u}_{n-D}$ be a basis of the left kernel of M, where D is the rank of M which we expect to be two. This means that for $i = 1, \ldots, n - D$

we have

$$0 = \mathsf{u}_i\,\mathsf{M} = \mathsf{u}_i\,\mathsf{L}_2^T\,\tilde{\mathsf{F}}\,\mathsf{L}_2.$$

The invertibility of L_2 implies that

$$0 = \mathsf{u}_i\,\mathsf{L}_2^T\,\tilde{\mathsf{F}},$$

and so, because of the special form of $\tilde{\mathsf{F}}$, we know that

$$(0, a_1, a_2, \ldots, a_{\theta-1}, 0, a_{\theta+1}, \ldots, a_{n-1}) = \mathsf{u}_i\,\mathsf{L}_2^T,$$

for some $a_1, \ldots, a_{\theta-1}, a_{\theta+1}, \ldots, a_{n-1} \in K$. Since the u_i are known, we evidently have $2(n-D)$ linear equations in the n^2 entries of L_2^T (or equivalently L_2) by taking the dot product of u_i with the 1st and θ^{th} columns of L_2^T, for $i = 1, \ldots, n-D$. In fact, the equations are of the form

$$\sum_{j=0}^{n-1} u_{ij} L_{2j} = 0$$

$$\sum_{j=0}^{n-1} u_{ij} L_{2\,j-\theta}^{q^\theta} = 0.$$

The first equation is linear in the variables L_{2j}. The second equation can be transformed into a linear equation by raising both sides to the $q^{n-\theta}$ power, yielding

$$\begin{aligned} 0 &= \left(\sum_{j=0}^{n-1} u_{ij} L_{2\,j-\theta}^{q^\theta}\right)^{q^{n-\theta}} \\ &= \sum_{j=0}^{n-1} u_{ij}^{q^{n-\theta}} \left(L_{2\,j-\theta}^{q^\theta}\right)^{q^{n-\theta}} \\ &= \sum_{j=0}^{n-1} u_{ij}^{q^{n-\theta}} L_{2\,j-\theta}^{q^n} \\ &= \sum_{j=0}^{n-1} u_{ij}^{q^{n-\theta}} L_{2\,j-\theta}. \end{aligned}$$

Thus we have $2(n-D)$ equations

$$\sum_{j=0}^{n-1} u_{ij} L_{2j} = 0$$

$$\sum_{j=0}^{n-1} u_{i\,j+\theta}^{q^{n-\theta}} L_{2j} = 0,$$

in the n unknowns $L_{20}, L_{21}, \ldots, L_{2n-1}$. Assuming these equations are linearly independent, and that $2(n-D) \geq n$, or equivalently $D \leq n/2$, we will be able to solve this system and finally obtain $\phi^{-1} \circ L_2 \circ \phi$, and thus L_2.

For more details of this attack, including time and memory complexities, the interested reader should check the related HFE case in [Courtois, 2001].

2.5 Matsumoto-Imai Variants

Two methods have been proposed to improve the security of the Matsumoto-Imai cryptosystem. One is called the "Minus" method, and is designed to resist the linearization attacks proposed by Patarin. The other is called the "Plus" method, and is used to make a cipher injective, thus enabling us to decrypt the ciphertext. Among all the Matsumoto-Imai variants proposed for practical use, the most successful is the Minus variant Sflash^{v2}.

The Minus Method

The Minus method was first suggested in [Shamir, 1993] and discovered independently by Patarin and Matsumoto. This method was utilized by Patarin and his collaborators in [Patarin et al., 1998] and elsewhere. As we will see in the case of Matsumoto-Imai, the application of this method clearly eliminates the possibility of the linearization equation attack, if the Minus number r is not too small.

The Minus method consists of deleting a few, say r, polynomial components from a given multivariate public key. For example, suppose $\bar{F} : k^n \longrightarrow k^l$ is a public key cryptosystem with polynomial components $\bar{f}_1, \ldots, \bar{f}_l \in k[x_1, \ldots, x_n]$. In most cases we have $l = n$, but the Minus method can also be used in other cases. Once we apply the Minus method to $\bar{F}$, for example by deleting the last r components, we will have a new map $\bar{F}^- : k^n \longrightarrow k^{l-r}$ defined by

$$\bar{F}^-(x_1, \ldots, x_n) = (\bar{f}_1, \ldots, \bar{f}_{l-r}). \tag{2.31}$$

The cryptosystem for signatures is, in general, set as follows.

The Public Key

The public key includes:

1.) The field structure of k;

2.) The set of polynomials: $(\bar{f}_1, \ldots, \bar{f}_{l-r}) \in k[x_1, \ldots, x_n]$.

The Private Key

The private key is the same as in the original cryptosystem.

The Signing Process

The document (or its hash value) is $Y'^{-} = (y'_1, \ldots, y'_{n-r})$, a vector in k^{n-r}. A legitimate user first chooses (or produces in some way) $n-r$ random elements $y'_{n-r+1}, \ldots, y'_n$ in k, which are appended to Y'^{-} to produce $Y' = (y'_1, \ldots, y'_n)$ in k^n. Then

$$X' = (x'_1, \ldots, x'_n) = \bar{F}^{-1}(Y'),$$

is calculated using the same decryption process as in the original cryptosystem. Finally, X' is the signature of the document Y'^{-}.

The Verifying Process

Anyone who receives the document Y'^{-} and its signature X' first obtains the public key and checks if indeed

$$(\bar{f}_1(X'), \ldots, \bar{f}_{l-r}(X')) = Y'.$$

If equality holds, then the signature is accepted as legitimate, otherwise it is rejected.

In the signing process it is very important that the appended values $y'_{n-r+1}, \ldots, y'_n$ are kept secret, otherwise they could be used to recover the missing polynomials to attack the systems as was shown in [Okeya et al., 2005].

The Minus method is particularly useful for converting an encryption scheme (which must be one-to-one) into a signature scheme since we no longer need injectivity. The security of this family of signature schemes is based on the assumption that to solve such a set of $l-r$ nonlinear equations in n variables is very difficult.

In order to illustrate a signature scheme we continue with the toy example, which we used to show how the linearization equation attack works. This time only the polynomials (2.10) to (2.13) are made public, that is (2.14) is hidden and not part of the public key.

The person signing a document has the secret key and with it the linear transformations or their inverses:

$$L_1^{-1}(y_1, y_2, y_3, y_4, y_5) = \begin{pmatrix} \alpha^2 & \alpha & \alpha^2 & 1 & 1 \\ 0 & 0 & \alpha^2 & \alpha^2 & 1 \\ \alpha & \alpha & 1 & \alpha^2 & 1 \\ \alpha & \alpha^2 & 0 & \alpha & 1 \\ 0 & 0 & \alpha & \alpha & 1 \end{pmatrix} \begin{pmatrix} y_1 - \alpha^2 \\ y_2 - \alpha^2 \\ y_3 - 0 \\ y_4 - 1 \\ y_5 - 0 \end{pmatrix}, \quad (2.32)$$

$$L_2^{-1}(y_1, y_2, y_3, y_4, y_5) = \begin{pmatrix} \alpha^2 & 1 & \alpha^2 & 0 & 1 \\ \alpha & 1 & \alpha & 1 & \alpha \\ 0 & \alpha^2 & \alpha & \alpha & 0 \\ \alpha & 1 & 1 & \alpha & \alpha \\ 0 & 1 & 0 & \alpha & \alpha^2 \end{pmatrix} \begin{pmatrix} y_1 - 1 \\ y_2 - 0 \\ y_3 - \alpha^2 \\ y_4 - \alpha^2 \\ y_5 - \alpha^2 \end{pmatrix}. \quad (2.33)$$

Also available for the signing process is $\theta = 3$ of the Matsumoto–Imai map, which gives $\tilde{F}^{-1}(X) = X^{362}$, and the irreducible polynomial $g(x) = x^5 + x^3 + x + \alpha^2$.

Assume that the document (plaintext) to be signed is

$$(\alpha^2, \alpha, \alpha^2, 0).$$

As mentioned above, the additional value should be chosen at random. In our toy example there are only four possibilities for y_5', and we will display them all

$$\begin{array}{rcl} Y' \text{ (Document)} & & X' \text{ (Signature)} \\ (\alpha^2, \alpha, \alpha^2, 0, 0) & \Longrightarrow & (0, \alpha, \alpha, 0, \alpha^2), \\ (\alpha^2, \alpha, \alpha^2, 0, 1) & \Longrightarrow & (1, 1, \alpha, \alpha, \alpha), \\ (\alpha^2, \alpha, \alpha^2, 0, \alpha) & \Longrightarrow & (1, 0, 1, 1, \alpha) \\ (\alpha^2, \alpha, \alpha^2, 0, \alpha^2) & \Longrightarrow & (\alpha^2, 1, 1, 0, \alpha^2) \end{array}$$

Any of these signatures, say the first one with $x_1 = 0$, $x_2 = \alpha$, $x_3 = \alpha$, $x_4 = 0$, and $x_5 = \alpha^2$, together with the public key (2.10) to (2.13) will verify that the signature is valid, since we find

$$(y_1, y_2, y_3, y_4) = (\alpha^2, \alpha, \alpha^2, 0).$$

If the four polynomials of the public key are used for an attack via the linearization equation, the attacker would see that $\dim_k \mathcal{L}_{Y'} = 1$ and would only find the equation

$$x_1 = \alpha^2 x_2 + \alpha x_3 + \alpha^2 x_4 + \alpha x_5 + \alpha^2,$$

a relationship satisfied by any of the four signatures. This is not enough to forge a signature. In general, when r becomes larger the linearization equations for the Minus cryptosystem disappear completely.

Flash and Sflash

The New European Schemes for Signatures, Integrity, and Encryption project (NESSIE) within the Information Society Technologies Programme of the European Commission made its final selections for cryptographic primitives at the beginning of 2004 after an evaluation process of more than two years [NESSIE, 1999]. Sflash^{v2}, a fast multivariate signature scheme, was selected by NESSIE as a security standard for use in low-cost smart cards. Sflash^{v2} is called Flash by NESSIE. The initial submission Sflash^{v1} was flawed, as a way was found to break it [Gilbert and Minier, 2002]. The flaw was due to the choice of $GF(2)$ for the field elements. It had been deliberately chosen to minimize the size of the public key. In any case it was not a fatal flaw and it could be corrected easily by choosing $GF(2^7)$ as the field elements in Sflash^{v2} [Patarin et al., 2001; Akkar et al., 2003]. The new version has a signature length of 259 bits and a public key of 15 KBytes.

The authors of the submission claimed that Sflash^{v2} is the fastest signature scheme in the world, and is the only digital signature scheme that can be used in practice for smart cards. Later, due to additional security concerns, the designers of Sflash recommended a new version called Sflash^{v3} [Courtois et al., 2003b], which is essentially Sflash^{v2} with a longer signature. Sflash^{v3} has a signature length of 469 bits and a public key of 112 KBytes. Later, the designers discovered that their security concerns are unfounded and so Sfalsh^{v2} is again recommended [Courtois, 2004]. At this point it seems that Sflash^{v2}, and with it Flash, should be considered secure.

For ease of exposition we give the basic implementation of Sflash^{v2}. The reader is referred to [Akkar et al., 2003] for technical details. Sflash is a Matsumoto–Imai Minus variant and it uses the single-branch map $\bar{F}$ as given in (2.1) with $\theta = 11$.

Furthermore, Sflash uses $n = 37$ and $r = 11$ so that $\bar{F}^- : k^{37} \longrightarrow k^{26}$ is defined by

$$\bar{F}^-(x_1, \ldots, x_n) = (\bar{f}_1, \ldots, \bar{f}_{n-r}),$$

where $\bar{f}_1, \ldots, \bar{f}_{26} \in k[x_1, \ldots, x_{37}]$. The Sflash scheme has the following structure.

Public Key

The following information can be made public, and is needed in order to verify a given Sflash signature:

1.) The field $k = GF(2^7)$, including its additive and multiplicative structure. In particular, $k = GF(2)[x]/(x^7 + x + 1)$.

2.) The 26 quadratic polynomials $\bar{f}_1, \ldots, \bar{f}_{26} \in k[x_1, \ldots, x_{37}]$.

Private Key

The following information should be kept private, and is needed in order to generate Sflash signatures:

1.) Δ, a randomly chosen 80-bit long secret key;

2.) The two invertible affine transformations L_1 and L_2 associated with the Matsumoto-Imai map $\bar{F}$.

Signature Generation

Let $\psi : k \longrightarrow GF(2)^7$ be the usual vector space isomorphism. The subscripts below refer to the position in the bit string, and "||" denotes the concatenation of bit strings. In order to sign a message M, we execute the following steps:

1.) Compute $M1 = \text{SHA-1}(M)$ and $M2 = \text{SHA-1}(M1)$, two 160-bit strings, using the SHA-1 hash function.

2.) Let

$$\begin{aligned} V &= M1||(M2_1, \ldots, M2_{22}) = (V_1, \ldots, V_{182}) \\ W &= \text{SHA-1}(V||\Delta) = (W_1, \ldots, W_{77}). \end{aligned}$$

3.) Let

$$\begin{aligned} M_1' &= \psi^{-1}(V_1, \ldots, V_7) \\ M_2' &= \psi^{-1}(V_8, \ldots, V_{14}) \\ &\vdots \\ M_{26}' &= \psi^{-1}(V_{176}, \ldots, V_{182}) \\ \\ M_{27}' &= \psi^{-1}(W_1, \ldots, W_7) \\ M_{28}' &= \psi^{-1}(W_8, \ldots, W_{14}) \\ &\vdots \\ M_{37}' &= \psi^{-1}(W_{71}, \ldots, W_{77}). \end{aligned}$$

Finally let $M' = (M_1', \ldots, M_{37}')$.

4.) Calculate the signature S of M by:

$$\begin{aligned} S &= \bar{F}^{-1}(M') \\ &= L_2^{-1} \circ F^{-1} \circ L_1^{-1}(M') \\ &= L_2^{-1} \circ \phi \circ \tilde{F}^{-1} \circ \phi^{-1} \circ L_1^{-1}(M'). \end{aligned} \tag{2.34}$$

The pair (M,S) represents the message M with signature S.

Signature Verification

Given the message-signature pair (M, S), we can verify the signature by executing the following steps:

1.) Signature verification begins in the same way as the generation. Compute

$$\begin{aligned} M1 &= \text{SHA-1}(M), \\ M2 &= \text{SHA-1}(M1) \\ V &= M1||(M2_1, \ldots, M2_{22}) = (V_1, \ldots, V_{182}). \end{aligned}$$

2.) Let

$$\begin{aligned} N_1' &= \psi^{-1}(V_1, \ldots, V_7) \\ N_2' &= \psi^{-1}(V_8, \ldots, V_{14}) \\ &\vdots \\ N_{26}' &= \psi^{-1}(V_{176}, \ldots, V_{182}) \end{aligned}$$

and $N' = (N_1', \ldots, N_{26}')$.

3.) If $N' = \bar{F}^-(S)$, then accept the signature S as valid; otherwise reject S.

It is clear that in order to forge a signature for the message M, we need to be able to find a single pre-image of N' under $\bar{F}^-$; i.e., find one solution (not necessarily all solutions) to a system of 26 equations in 37 variables. Here the secret key Δ is also very important in terms of security [Okeya et al., 2005]. Even if only this secret key Δ is leaked, one can defeat the system easily by using it to find the missing (Minus) polynomials. Finally, it is not hard to see that in the case of Matsumoto-Imai, the Minus method eliminates the possibility of the linearization equations attack.

As was previously mentioned, the Minus method is only suitable for signature schemes, where we need to find only a single element in the pre-image (as opposed to a unique pre-image required for encryption). The "Plus" method is one way in which we can modify a Minus scheme for use in encryption.

The Plus Method

The Plus method amounts to adding a few, say s, randomly chosen polynomial components to a given multivariate scheme, and then mixing them into the public key through an invertible affine transformation. Clearly the degree of the Plus polynomials should be chosen to be the same as the underlying scheme. For example, let us suppose that $\bar{F}$: $k^n \longrightarrow k^l$ is a mapping associated with some multivariate scheme. We append the s randomly chosen polynomials $p_1, \ldots, p_s \in k[x_1, \ldots, x_n]$ to create a new map $\bar{F}^+ : k^n \longrightarrow k^{l+s}$ defined by

$$\bar{F}^+ = L_3 \circ (\bar{f}_1, \ldots, \bar{f}_l, p_1, \ldots, p_s), \tag{2.35}$$

where $L_3 : k^{l+s} \longrightarrow k^{l+s}$ in an invertible affine transformation that mixes the Plus polynomials into the system.

We would like to point out that originally the main purpose of the Plus method was not to improve the security of the original scheme associated with $\bar{F}$, but rather to make the map $\bar{F}$, which is not injective, into an injective map, so that it can be used for encryption. In other words, if $\bar{F}^{-1}(y'_1, \ldots, y'_l)$ has multiple elements (q^r, in the case of Matsumoto-Imai-Minus), then the Plus polynomials can be used to reduce the number of pre-images to a single element if s is big enough. Equivalently, the Plus polynomials can help to differentiate which is the real plaintext from a set of possible candidates. From a mathematical point view, the Plus is a simple method to make a map M, which is not injective, into an injective map M^+ by adding more components (an embedding map). Roughly speaking, each additional Plus polynomial will reduce the probability of having multiple pre-images by a factor of q.

The Plus method does not improve the security of the Matsumoto-Imai public key cryptosystems when it is applied directly. It does nothing substantial to help in resisting the linearization equation attacks. The linearization equations are still there unlike in the case of the Minus method when there are not enough of them.

As an example of combining both the Plus and Minus methods, we now present the Matsumoto-Imai-Plus-Minus public key cryptosystem. Let $\bar{F} : k^n \longrightarrow k^n$ be a polynomial mapping whose components

$$\bar{f}_1, \ldots, \bar{f}_n \in k[x_1, \ldots, x_n]$$

form the public key of a Matsumoto-Imai public key cryptosystem. Delete the last r polynomials, add s randomly chosen degree two polynomials $p_1, \ldots, p_s \in k[x_1, \ldots, x_n]$, and define the map $\bar{F}^{\pm} : k^n \longrightarrow k^m$ by

$$\bar{F}^{\pm} = L_3 \circ (\bar{f}_1, \ldots, \bar{f}_{n-r}, p_1, \ldots, p_s) = (\bar{f}_1^{\pm}, \ldots, \bar{f}_m^{\pm}), \tag{2.36}$$

where $r \leq s$, $m = n - r + s$ and $L_3 : k^m \longrightarrow k^m$ is an invertible affine transformation. The Matsumoto-Imai-Plus-Minus scheme has the following structure.

Public Key

1.) The field k including its additive and multiplicative structure;

2.) The $m = n-r+s$ degree two polynomials $\bar{f}_1^{\pm}, \ldots, \bar{f}_m^{\pm} \in k[x_1, \ldots, x_n]$.

Private Key

1.) The degree two polynomials $p_1, \ldots, p_s \in k[x_1, \ldots, x_n]$;

2.) The three invertible affine transformations L_1, L_2, and L_3.

Encryption

Given a plaintext $(x'_1, \ldots, x'_n) \in k^n$, calculate $(y'_1, \ldots, y'_m) \in k^m$ with the public polynomials:

$$(y'_1, \ldots, y'_m) = \bar{F}^{\pm}(x'_1, \ldots, x'_n).$$

Decryption

To decrypt a message we execute the following steps:

1.) Calculate $(z_1, \ldots, z_{n-r+s}) = L_3^{-1}(y'_1, \ldots, y'_{n-r+s})$.

2.) For each $w = (w_1, \ldots, w_r) \in k^r$, compute

$$t_w = (t_1, \ldots, t_n) = \bar{F}^{-1}(z_1, \ldots, z_{n-r}, w_1, \ldots, w_r),$$

and define $T = \{(w, t_w) \mid w \in k^r\}$.

3.) For each $(w, t_w) \in T$, check if

$$p_i(t_w) = z_{n-r+i}$$

holds for all $i = 1, \ldots, s$. Keep each t_w that satisfy this criteria and discard the rest. If s is large enough, we should have only one element left, the plaintext $(x'_1, \ldots, x'_n)$.

Here the Plus method also serves the purpose of improving the security once the map L_3 is applied, since after the random polynomials are mixed into the system we cannot tell which are the original polynomials from the Matsumoto-Imai cryptosystem. This at least will make it too difficult to use any method that can be applied to the Matsumoto-Imai-Minus cryptosystems directly.

2.6 The Security of the Matsumoto-Imai Variants

Before using either the Plus or Minus method, we must decide how large (or small) the Plus and Minus should be. For security reasons we should not delete too few polynomials (r should not be too small), and for efficiency reasons we should not add too many polynomials (s should not be too big). The resulting problem of how to choose r and s optimally is not completely settled, though there are some results [Patarin et al., 1998], etc. In this section we will concentrate on the security analysis of the Minus variant of Matsumoto-Imai.

Cryptanalysis of Sflash^{v1}

Recall that for Sflash^{v1} the field k is chosen to be $GF(2)$, and in particular $k = GF(2)[x]/(x^7 + x + 1)$. The extension field K is chosen to be $k[x]/r(x)$, where $r(x) = x^{37} + x^{12} + x^{10} + x^2 + 1$ is irreducible in $k[x]$, and we know that $n = 37$, $\theta = 11$ and $r = 11$. The two secret maps $L_1, L_2 : k^n \longrightarrow k^n$ are specially chosen in that they are taken from a small subset of invertible affine transformations on k^n whose matrix representations have entries only from the subfield $GF(2)$.

Although we can use Sflash to sign documents from k^{26}, it is not hard to see that due to the special choice of $r(x)$, L_1 and L_2, the public signature verification polynomials all lie in the polynomial ring $GF(2)[x_1, \ldots, x_{37}]$. This reduces the required memory by a factor of seven from what it otherwise would be. On the other hand, it is straightforward to check that the public polynomial components obtained by taking $q' = 2$, $n' = n = 37$ and $\theta' = 3$ (so that the fields are $k' = GF(2)$ and $K' = GF(2^{37})$) will yield exactly those of $\bar{F}$. This is because

$$3 \equiv 7 \times 11 \bmod 37.$$

Furthermore, if we delete $r' = r = 11$ polynomials, we have a version of Sflash that is much easier to attack. The strategy of Gilbert and Minier [Gilbert and Minier, 2002] is to find the $GF(2)$-linear span of the deleted polynomials of this "smaller" version of Sflash. Any subset of eleven linearly independent polynomials from this span can be used with the original public polynomials to calculate signatures in the original Sflash signature scheme.

We may now think of $\bar{F}^-$ as a Matsumoto-Imai map from $GF(2^{37})$ to $GF(2^{26})$. Since $GF(2^{37})$ is a relatively small finite field, we can use brute force to the invert the map $\bar{F}^-$ over $GF(2^{37})$. In other words, for every $Y^- \in GF(2^{26})$ we can efficiently compute the set

$$U_{Y^-} = \{X \in GF(2^{37}) \mid \bar{F}^-(X) = Y^-\},$$

which can be stored for later use during the attack.

The strategy of the attack is to find r additional quadratic polynomials $q_1, \dots, q_r$ of the form

$$q_l(x_1, \dots, x_n) = \sum_{i=1}^{n} \sum_{j=1}^{i-1} \alpha_{ijl} x_i x_j + \sum_{i=1}^{n} \beta_{il} x_i, \tag{2.37}$$

where $\alpha_{ijl}, \beta_{il} \in GF(2)$, which together with the $n - r$ public quadratic polynomials from $\bar{F}^-$ will span the same linear space as all of the components of $\bar{F}$ except for some constant shift. This gives us an equivalent Matsumoto-Imai polynomial mapping $\bar{F}'$ that can then be subjected to the linearization attack by Patarin. For a given message we cannot use $\bar{F}'$ to produce the exact same signature as we would obtain by using $\bar{F}$. However, since the span of the components of $\bar{F}'$ is the same as the span of the components of $\bar{F}$, we can nevertheless produce valid signatures. In other words, if the legitimate user computes S as the signature of M, then at the end of this attack we will be able to compute S' such that $\bar{F}^-(S) = \bar{F}^-(S')$, and therefore can make a successful forgery of the legitimate signature.

The key step in the attack is the characterization of the coefficients of the $q_l(x_1, \dots, x_n)$ by using the fact that $\bar{F}$ is an invertible map and therefore one-to-one. This allows us to reduce the possible candidates for $q_l(x_1, \dots, x_n)$ from the space of all quadratic functions with coefficients in $GF(2)$ (a space with dimension $n(n-1)/2 + n = 703$) to a much smaller space of dimension $4 \times 37 = 148$. Though this space is still much too large, once we get to this point we will be able to reduce the dimension further to solve our problem.

The First Step of the Attack

We begin by noting that $\bar{F}$ is one-to-one, and therefore for each $Y^- = (y_1, \dots, y_{26}) \in GF(2^{26})$, the set U_{Y^-} will have exactly 2^{11} elements. Moreover, for each q_l of the form in (2.37) we must have

$$\sum_{X \in U_{Y^-}} q_l(X) = 0, \tag{2.38}$$

for $l = 1, \dots, 11$. This also follows from the injectivity of $\bar{F}$, which implies that exactly half of the elements $X \in U_{Y^-}$ are such that $q_l(X) = 0$, while the other half are such that $q_l(X) = 1$. Therefore, each U_{Y^-} provides one linear equation in the 703 coefficients of the quadratic function q_l. Generating U_{Y^-} for each Y^- can be done by simply calculating $\bar{F}^-(X)$ for each of the 2^{37} elements $X \in K'$.

According to Gilbert and Minier, it is often only necessary to compute U_{Y^-} for $N = 1000$ (a little more than 703) different Y^-. In any case, the N sets U_{Y^-} can be used to obtain an $N \times 703$ matrix with coefficients in $GF(2)$, whose kernel can be computed. This kernel, which we denote $\mathcal{Q}$, has dimension $37 \times 4 = 148$, and contains the $GF(2)$-vector space spanned by the 26 public polynomials and the 11 deleted polynomials (without constant terms). We now explain the appearance of spurious polynomials, polynomials not in the span of the components of $\bar{F}$. Before we do this, we first need to say a few words about discrete derivatives.

Discrete Derivatives

We consider only the case of a finite field of characteristic two. Let V be a vector space and let g be any function from V to V. The derivative of g with respect to the vector $v \in V$ is then defined to be:

$$\partial_v(g(x)) = g(x) + g(x+v).$$

More generally, if $W = \{v_1, \ldots, v_m\}$ is a subset of vectors in V, then the derivative of g with respect to the set of vectors W is defined to be:

$$\begin{aligned}\partial_W(g(x)) &= \partial_{v_1}(\partial_{v_2}(\cdots(\partial_{v_m}(g(x)))\cdots)) \\ &= \sum_{w \in W'} g(x+w),\end{aligned}$$

where W' is the set of all linear combinations $\alpha_1 v_1 + \cdots + \alpha_m v_m$ with $\alpha_1, \ldots, \alpha_m \in \{0,1\}$.

Now suppose W is an m-dimensional subspace of V, and that W has basis $B = \{v_1, \ldots, v_m\}$. Then we define the derivative of g with respect to the vector space W as just $\partial_B(g(x))$, though we will abuse notation and write $\partial_W(g(x))$. We note that if V is a $GF(2)$-vector space, then

$$\partial_W(g(x)) = \sum_{w \in W} g(x+w).$$

Finally, let A be an affine set of dimension m, so that $A = v + W$ for some vector $v \in V$ and m-dimensional subspace W. Then the derivative of g with respect to the affine set A is defined to be $\partial_B(g(x+v))$, where B is any basis of the subspace W. As before, we will abuse notation and write $\partial_A(g(x))$. If V is a $GF(2)$-vector space, then

$$\partial_A(g(x)) = \sum_{w \in W} g(x+v+w).$$

The following two results about the discrete derivative will be particularly useful when the vector space has an additional ring structure.

Lemma 2.6.1. *Suppose K is a degree n field extension of $GF(2)$, let $g(x)$ be a nonzero polynomial in $K[x]$, and pick any $a \in K$. Then the Hamming weight degree of $\partial_a(g(x))$ is strictly less than the Hamming weight degree of $g(x)$.*

Proof. Since the discrete derivative is clearly additive, it suffices to consider the case of $g(x) = x^l$ for $l \geq 0$. Suppose that there are m nonzero terms in the binary expansion of l:

$$l = 2^{i_1} + 2^{i_2} + \cdots + 2^{i_m}.$$

Then

$$\begin{aligned}
\partial_a(g(x)) &= g(x) + g(x+a) \\
&= x^l + (x+a)^l \\
&= x^l + (x+a)^{2^{i_1}+2^{i_2}+\cdots+2^{i_m}} \\
&= x^l + (x+a)^{2^{i_1}}(x+a)^{2^{i_2}} \cdots (x+a)^{2^{i_m}} \\
&= x^l + (x^{2^{i_1}} + a^{2^{i_1}})(x^{2^{i_2}} + a^{2^{i_2}}) \cdots (x^{2^{i_m}} + a^{2^{i_m}}) \\
&= x^l + x^{2^{i_1}+2^{i_2}+\cdots+2^{i_m}} + \text{lower weight terms} \\
&= 2x^l + \text{lower weight terms} \\
&= 0 + \text{lower weight terms},
\end{aligned}$$

where the last equality holds since the characteristic of K is two. □

Corollary 2.6.1. *Suppose K is a degree n field extension of $GF(2)$, and let $\phi : K \longrightarrow GF(2)^n$ be the usual identification. Pick $g(x) \in K[x]$ of Hamming weight degree d. If A is any m-dimensional affine set in $GF(2)^n$ with $d \leq m$, then*

$$\partial_{\phi^{-1}(A)}(g) = 0.$$

Proof. The proof follows directly from the previous lemma. □

Spurious Polynomials

Fix $Y^- = (y'_1, \ldots, y'_{26})$ and let

$$V_{Y^-} = \{(y_1, \ldots, y_{37}) \in GF(2)^{37} \mid (y_1, \ldots, y_{26}) = Y^-\},$$

an affine subset of $GF(2)^{37}$. Let Y be any element in V_{Y^-} and suppose $X = (x_1, \ldots, x_n)$ satisfies $\bar{F}(X) = Y = (y_1, \ldots, y_n)$. If q_l is in the span of the components of $\bar{F}$ (i.e., $q_l = \sum_{i=1}^n a_i \bar{f}_i$), then we must have that

$$q_l(x_1, \ldots, x_n) = q_l(\bar{F}^{-1}(y_1, \ldots, y_n)) = \sum_{i=1}^n a_i y_i, \qquad (2.39)$$

where the second equality comes from the fact that

$$y_i = \bar{f}_i(\bar{F}^{-1}(y_1, \ldots, y_n)),$$

for $i = 1, \ldots, n$. In this way we can associate with $q_l(x_1, \ldots, x_n)$ a new function

$$\tilde{q}_l(y_1, \ldots, y_n) = q_l \circ \bar{F}^{-1}(y_1, \ldots, y_n).$$

With this shift in perspective we have

$$\sum_{X \in U_{Y^-}} q_l(x_1, \ldots, x_n) = \sum_{Y \in V_{Y^-}} \tilde{q}_l(y_1, \ldots, y_n). \qquad (2.40)$$

Since V_{Y^-} is an affine subset in $GF(2)^{37}$, the sum $\sum_{X \in U_{Y^-}} q_l(x_1, \ldots, x_n)$ is now realized as a (discrete) derivative of the function $\tilde{q}_l(y_1, \ldots, y_n)$, which is itself a linear function in the $y_1, \ldots, y_n$, provided that $q_l = \sum_{i=1}^n a_i \bar{f}_i$.

Therefore, an equation of the form of (2.38) will be satisfied by any total degree two polynomial $q(x_1, \ldots, x_n)$ such that $\tilde{q}(y_1, \ldots, y_n) = q \circ \bar{F}^{-1}(y_1, \ldots, y_n)$ can be expressed as a polynomial of total degree at most 10 in the $y_1, \ldots, y_n$. Let us now explore how such functions occur.

Let $\tilde{F}_i : K' \longrightarrow K'$ be defined by

$$\tilde{F}_i(X) = X^{2^i+1},$$

for $i = 0, \ldots, 36$, and let $F_i : k^n \longrightarrow k^n$ be defined by

$$F_i = \phi \circ \tilde{F}_i \circ \phi^{-1} \circ L_2 = (f_{i1}, \ldots, f_{in}),$$

deviating slightly from the usual notation. Clearly $\tilde{F}$ is $\tilde{F}_3$.

Take $Y = \bar{F}(X) = L_1 \circ F_3(X)$. Then $F_3(X) = L_1^{-1}(Y)$. Also, $\tilde{F}_3^{-1}(X) = X^t$, where $t \equiv (2^3+1)^{-1} \bmod (2^{37}-1)$. Therefore, if any quadratic polynomial $q(X)$ (with total degree two in the components $x_1, \ldots, x_n$ of X) is equal to a linear combination of the components of some $F_i(X) = (f_{i1}, \ldots, f_{in})$, then $\tilde{q}$ can be expressed as a linear combination of the quadratic terms of the 37 $GF(2)$-components of $F_i \circ F^{-1}$. To see why this is true, consider the following. Assume

$$q(x_1, \ldots, x_n) = \sum_{j=1}^n a_j f_{ij},$$

and take

$$L(x_1, \ldots, x_n) = \sum_{j=1}^{n} a_j x_j.$$

We then clearly have

$$q(x_1, \ldots, x_n) = L \circ F_i,$$

and thus,

$$\begin{aligned}
\sum_{X \in U_{Y^-}} q(X) &= \sum_{Y \in V_{Y^-}} \tilde{q}(Y) \\
&= \sum_{Y \in V_{Y^-}} q \circ \bar{F}^{-1}(Y) \\
&= \sum_{Y \in V_{Y^-}} L \circ F_i \circ \bar{F}^{-1}(Y) \\
&= \sum_{Y \in V_{Y^-}} L \circ F_i \circ L_2^{-1} \circ \phi \circ \tilde{F}^{-1} \circ \phi^{-1} \circ L_1^{-1}(Y) \\
&= \sum_{Y \in V_{Y^-}} L \circ \phi \circ \tilde{F}_i \circ \tilde{F}^{-1} \circ \phi^{-1} \circ L_1^{-1}(Y),
\end{aligned}$$

the degree of the last expression in the components of $Y = (y_1, \ldots, y_n)$ being bounded above by the Hamming weight of the degree of $\tilde{F}_i \circ \tilde{F}^{-1}$, which is $t(2^i + 1) \bmod (2^{37} - 1)$.

One can easily compute $d_i = t(2^i + 1) \bmod (2^{37} - 1)$ for $i = 0, \ldots, 36$ and find that there are exactly four values of i such that the Hamming weight w_i of d_i is at most 10. In particular, we find that:

$$\begin{aligned}
d_3 = 1 = (1)_2 &\implies w_3 = 1 \\
d_9 = 57 = (111001)_2 &\implies w_9 = 4 \\
d_{15} = 3641 = (111000111001)_2 &\implies w_{15} = 7 \\
d_{21} = 233017 = (111000111000111001)_2 &\implies w_{21} = 10
\end{aligned}$$

and thus the components of $\tilde{F}_3, \tilde{F}_9, \tilde{F}_{15}$, and $\tilde{F}_{21}$ can all be expressed as functions of degree at most 10 in the components of Y. Therefore any linear combination of these $4 \times 37 = 148$ polynomials will satisfy an equations of the form in (2.38).

The Second Step of the Attack

We must now further characterize the coefficients of the desired $q_l(x)$. We will use the public knowledge we know about $\bar{F}$ to express additional conditions we can use to determine the $q_i(x)$ completely. Computer experiments confirm that these additional conditions do indeed determine the $q_i(x)$.

Choose a basis for $\mathcal{Q}$ using Gaussian elimination, say $\{q_1, \ldots, q_{148}\}$. We need a condition on the γ_i such that $q = \sum \gamma_i q_i(x)$ must belong to the space spanned by $\bar{f}_1, \ldots, \bar{f}_n$.

Let $q(x_1, \ldots, x_n) \in \mathcal{Q}$. From the condition imposed by (2.38) on q, we see that the total degree of $\tilde{q}(y_1, \ldots, y_n)$ cannot be more than 10, and that if $q(x_1, \ldots, x_n)$ belongs to the space spanned by $\bar{f}_1, \ldots, \bar{f}_n$ then the total degree of $\tilde{q}(y_1, \ldots, y_n)$ is 1, as we have seen from (2.39). Thus, if $q(x_1, \ldots, x_n)$ is indeed in the space spanned by $\bar{f}_1, \ldots, \bar{f}_n$, then for $i = 1, \ldots, 148$, the derivative with respect to any 12-dimensional affine set $\mathcal{A}$ of $\tilde{q}_i\tilde{q}$ (whose degree is at most $10 + 1 = 11$) will be zero. On the other hand, if $q(x_1, \ldots, x_n)$ does not belong to the space spanned by $\bar{f}_1, \ldots, \bar{f}_n$, then the degree of $\tilde{q}_i\tilde{q}$ is expected to be at least $10 + 4 = 14$, due to the fact that the Hamming weight of $t(2^i + 1) \bmod (2^{37} - 1)$ for $i = 9, 15, 21$ are of weight $4, 7, 10$, respectively. Therefore we do not expect that the derivative of $\tilde{q}_i\tilde{q}$ will be zero. We are now ready to formulate the desired conditions on the γ_i.

Let $Y^{--} = (y_1, \ldots, y_{25}) \in GF(2)^{25}$, and let us denote by $V_{Y^{--}}$ the affine subset of $GF(2)^{37}$

$$V_{Y^{--}} = \{(y_1, \ldots, y_{37}) \in GF(2)^{37} \mid (y_1, \ldots, y_{25}) = Y^{--}\}.$$

With this notation we have

$$\sum_{Y \in V_{Y^{--}}} \tilde{q}_i(Y)\tilde{q}(Y) = 0.$$

For each $Y^{--} = (y_1, \ldots, y_{25})$, define $Y_0^{-} = (y_1, \ldots, y_{25}, 0)$ and $Y_1^{-} = (y_1, \ldots, y_{25}, 1)$, and let $U(Y^{--}) = U_{Y_0^{-}} \cup U_{Y_1^{-}}$. The above equation gives rise to a linear equation in the 148 unknown $GF(2)$-coefficients γ_i of q in the form:

$$\sum_{X \in U(Y^{--})} \sum_{i=1}^{148} \gamma_i q_i(X) q(X) = 0. \tag{2.41}$$

In their computer experiments, Gilbert and Minier actually needed to use only two arbitrary quadratic polynomials, q_1 and q_2, which allowed them to collect $N' = 200$ (a little more than 148 equations) to obtain a solution space of dimension exactly 37. This completes step two of the attack.

Once this is done we have the space spanned by $\bar{f}_i$. After picking a basis for this much smaller space, we use the linearization attack to invert the Sflash public polynomials for any given image. This allows us to forge signatures.

Complexity

The most complex calculation required by the attack above is the exhaustive computation of the 2^{37} values of the public function $\bar{F}^{-}$, which is needed to obtain the (at most) $N + 2N'$ sets of 2^{11} pre-images required for the computations of the attack. The computations of Step 1 are the derivation of the $N = 1000$ linear equations in 703 variables and the Gaussian elimination of the resulting $N \times 703$ system, so the complexity of Step 1 is bounded above by $N \times 703 \times 2^{11} + N^3/3 < 2^{32}$. Similarly, the complexity of the derivation of the N' linear equations in 148 variables and the Gaussian elimination of the resulting $N' \times 148$ system in Step 2 is bounded above by 2^{27}. These are far lower than 2^{37} computations of the Sflash^{v1} public functions. We also note that the complexity of the linearization attack is about 2^{27} computations. Therefore the complexity of the entire attack is bounded above by 2^{37}.

The attack presented above is based on the fact that the Sflash^{v1} public function over k^{37} induces a restricted function over the much smaller vector space $GF(2)^{37}$. This attack does not seem to be applicable to more conservative instances of the Matsumoto-Imai-Minus scheme, such as Sflash^{v2}, since a much more efficient method would then have to be found to determine each set of q^r preimages under $\bar{F}^{-}$. In this case $q^r = (2^7)^{11} = 2^{77}$, which makes the brute force search for the set of pre-images by Gilbert and Minier above impossible.

Other Attacks on MI-Minus

In [Patarin et al., 1998], a general attack on the Matsumoto-Imai-Minus family was presented. This attack is essentially a differential type of attack where one uses the fact that $\bar{F}$ is an invertible map. The starting point is to use the so-called polar form of $\bar{F}$ given by

$$Q(X,T) = \bar{F}(X+T) - \bar{F}(X) - \bar{F}(T),$$

which in this case is related to bilinear forms of the polynomials components of $\bar{F}$. If we fix X to be a constant, then the equation above becomes linear in T. This method utilizes the fact that the public key polynomials come from a set of permutation polynomials, which allows us to use the general theory about permutation polynomials and the idea of orthogonal systems of equations [Lidl and Niederreiter, 1997]. Then we may look for a a value X such that solution space is of maximum dimension. The basic idea is to use this solution space to find a way to recover the lost (Minus) polynomials and then use again the linearization equations to break the system. From this we can see that this attack in essence is closely related to the attack by Gilbert and Minier above. We will omit the details of the attack here and refer the readers to the original paper [Patarin et al., 1998].

It is shown that such an attack should have complexity of $O(q^r)$, and therefore it is suggested that q^r should be at least 2^{64} in order to guarantee security against this attack. This attack is also very closely related to the differential attack [Fouque et al., 2005] on PMI [Ding, 2004a], which will be discussed later.

We believe that the new attack on MI in Section 2.4 can also be directly extended to attack the MI-Minus cryptosystem, especially when the Minus number r is small.

Security of MI-Plus-Minus

We believe that the security of MI-Plus-Minus is also still open, since it should be a much harder problem to attack MI-Plus-Minus than MI-Minus in general. Moreover, there is also a problem of how big the Plus can be before additional security concerns arise. In [Patarin et al., 1998], some attacks were suggested for MI-Plus-Minus that are actually prototypes of the XL-family of algorithms. We will leave the details of this discussion for the chapter on general methods for solving systems of polynomial equations.

Related work

First we like to point out that the Matsumoto-Imai cryptosystems we talk about in this chapter should not be confused with some of their other cryptosystems from 1983 [Matsumoto and Imai, 1983]. These were broken in 1984 [Delsarte et al., 1985] and are very different systems from what we study here.

The original ideas of the Matsumoto-Imai cryptosystems were first presented in [Imai and Matsumoto, 1985]. In the 1988 paper, two families of systems are discussed. The other one is the so-called Hidden Matrix (HM) scheme, where the key map uses matrix multiplications, and in particular the square of a matrix. These schemes were defeated by using the same method of linearization equations [Patarin et al., 1998]. In the 1985 paper [Imai and Matsumoto, 1985], there is also another scheme called the "B" scheme, and it was broken in 2001 [Youssef and Gong, 2001] using statistical methods.

In the process of developing a new differential method to attack PMI [Ding, 2004a], Fouque, Granboulan, and Stern also found a new differential attack to break the MI [Fouque et al., 2005].

From [Felke, 2005], we also see that the linearization attack was independently discovered by Dobbertin at the German Information Security Agency in 1993.

Chapter 3

OIL-VINEGAR SIGNATURE SCHEMES

One can see from the previous chapter that the generalization and extension of the Matsumoto-Imai cryptosystem has played a critical role in the recent rapid development of multivariate public key cryptosystems. Though defeated, we have not yet come to the end of the story of the Matsumoto-Imai cryptosystem. Surprisingly, Patarin started quite a different approach to the constructions of public key signature schemes by converting the linearization equation attack on the MI into the Oil-Vinegar public key signature schemes. Using an attack method to inspire a new scheme is unprecedented. It is indeed a surprise, but the connection of Oil-Vinegar construction with the linearization attack is very natural if one just takes a quick look at the basic ideas.

The Oil-Vinegar schemes can be grouped into three families: balanced Oil-Vinegar [Patarin, 1997], unbalanced Oil-Vinegar [Kipnis et al., 1999] and Rainbow, a multilayer construction using unbalanced Oil-Vinegar at each layer [Ding and Schmidt, 2005b]. Signature schemes from the first two families have been shown to possess security risks. The Rainbow scheme is a very efficient public key signature scheme with a very high security level. The Rainbow schemes are also closely related to the TTS and TRMC signature schemes in Chapter 6, which are derived by a very different method.

This chapter is arranged as follows. In the first section, we present the concepts and construction of the basic Oil-Vinegar signature schemes, encompassing both the balanced and unbalanced families of Oil-Vinegar. We then present the known attacks on the balanced and unbalanced Oil-Vinegar schemes, which include the Kipnis-Shamir attack on balanced Oil-Vinegar using the method of invariant subspaces, and the Kipnis-Patarin-Goubin attack on unbalanced Oil-Vinegar schemes, a generalization of the Kipnis-Shamir attack. We follow the general cryptanalysis with a practical example of the unbalanced Oil-Vinegar scheme, including its security and efficiency analysis. Finally we present Rainbow, its security and efficiency analysis, and a practical example.

3.1 The Basic Oil-Vinegar Signature Scheme

The basic building block of the Oil-Vinegar scheme is the Oil-Vinegar polynomial. Oil-Vinegar polynomials are quadratic polynomials in which Oil variables can only appear linearly. After fixing values for all Vinegar variables, the quadratic Oil-Vinegar polynomial becomes linear in the Oil variables. With a set of (not too many) Oil-Vinegar polynomials we can then solve for the Oil variables and produce a signature.

Let k be a finite field with q elements. The variables $x_1, \ldots, x_o$ will be called the Oil variables, and the variables $\check{x}_1, \ldots, \check{x}_v$ will be called the Vinegar variables. Let $n = o + v$.

Definition 3.1.1. *An Oil-Vinegar polynomial is any total degree two polynomial* $f \in k[x_1, \ldots, x_o, \check{x}_1, \ldots, \check{x}_v]$ *of the form*

$$f = \sum_{i=1}^{o}\sum_{j=1}^{v} a_{ij}x_i\check{x}_j + \sum_{i=1}^{v}\sum_{j=1}^{v} b_{ij}\check{x}_i\check{x}_j + \sum_{i=1}^{o} c_i x_i + \sum_{j=1}^{v} d_j\check{x}_j + e,$$

where $a_{ij}, b_{ij}, c_i, d_j, e \in k$.

The name for Oil-Vinegar polynomials comes from the fact that Oil-Vinegar variables are not fully mixed in the quadratic terms; i.e., there are no terms of the form $x_i x_j$.

Definition 3.1.2. *Let* $F : k^n \longrightarrow k^o$ *be a polynomial map of the form*

$$F(x_1, \ldots, x_o, \check{x}_1, \ldots, \check{x}_v) = (f_1, \ldots, f_o),$$

where the $f_1, \ldots, f_o \in k[x_1, \ldots, x_o, \check{x}_1, \ldots, \check{x}_v]$ *are Oil-Vinegar polynomials. Then* F *is called an Oil-Vinegar map.*

Note the similarity of the above formula with the linearization equations. The linearization equations are in some sense Oil-Vinegar polynomials as well, where the ciphertext components can be viewed as the Vinegar variables and the plaintext components can be viewed as Oil variables, or vice versa, because there are no cross terms among either the plaintext components or the ciphertext components.

The key property of the Oil-Vinegar map F is the following. If the coefficients of F are chosen randomly, then given a fixed vector $(y_1', \ldots, y_o') \in k^o$ we can"invert" F by randomly choosing $(\check{x}_1', \ldots, \check{x}_v') \in$

k^v to give the values of the Vinegar variables and solving the resulting system of linear equations in the Oil variables given by

$$F(x_1, \ldots, x_o, \check{x}'_1, \ldots, \check{x}'_v) = (y'_1, \ldots, y'_o).$$

Though the system of equations may not have a solution, the probability that it will have a solution is roughly $1 - q^{-1}$, essentially the same as the probability that a randomly chosen $o \times o$ matrix with entries in k is invertible. If the system has no solution, then a different choice for the values of the Vinegar variables can be tried. This process may be repeated a few times as necessary, but for reasonably large q and relatively much smaller n, the probability of success on the first attempt should be nearly one.

Remark 3.1.1. *The algorithm to find a pre-image of* $(y'_1, \ldots, y'_o)$ *works just like the linearization equations attack on the Matsumoto-Imai cryptosystem; that is, if we guess (or if we are given) the values of certain variables, the quadratic equations become linear and therefore easy to solve.*

Fix $(y'_1, \ldots, y'_o) \in k^o$ and let $(\check{x}'_1, \ldots, \check{x}'_v) \in k^v$ represent a choice of values for the Vinegar variables such that there exists $(x'_1, \ldots, x'_o)$ satisfying

$$F(x'_1, \ldots, x'_o, \check{x}'_1, \ldots, \check{x}'_v) = (y'_1, \ldots, y'_o).$$

We define the inverse of $(y'_1, \ldots, y'_o)$ under F with respect to $(\check{x}'_1, \ldots, \check{x}'_v)$ by

$$F^{-1}(y'_1, \ldots, y'_o) = (x'_1, \ldots, x'_o).$$

Although the notation $F^{-1}(y'_1, \ldots, y'_o)$ does not reflect the fact that the value depends on the choice of $(\check{x}'_1, \ldots, \check{x}'_v) \in k^v$, we shall only be concerned with whether or not $F^{-1}(y'_1, \ldots, y'_o)$ exists for a given choice of $(\check{x}'_1, \ldots, \check{x}'_v)$.

To construct an Oil-Vinegar signature scheme, we must first choose an Oil-Vinegar map F. Like the other constructions we have seen before, we then "hide" the Oil-Vinegar map by composing F with an invertible affine map $L : k^n \longrightarrow k^n$ of the form

$$(x_1, \ldots, x_o, \check{x}_1, \ldots, \check{x}_v) = L(z_1, \ldots, z_n).$$

The composition generates the quadratic map $\bar{F} : k^n \longrightarrow k^o$ defined by

$$\bar{F} = F \circ L = (\bar{f}_1, \ldots, \bar{f}_o).$$

Note that since the coefficients of the Oil-Vinegar map F are chosen at random, there is no need to compose on the left by an invertible affine transformation as was done with the Matsumoto-Imai constructions.

We are now ready to describe the basic Oil-Vinegar signature scheme.

Public key

The public key consists of the following items.

1.) The field k, including the additive and multiplicative structure;

2.) The map $\bar{F} = F \circ L$, or equivalently, its components

$$\bar{f}_1, \ldots, \bar{f}_o \in k[z_1, \ldots, z_n].$$

Private Information

The private key consists of the following items.

1.) The invertible affine transformation $L : k^n \longrightarrow k^n$;

2.) The Oil-Vinegar map F, or equivalently, its components

$$f_1, \ldots, f_o \in k[x_1, \ldots, x_o, \check{x}_1, \ldots, \check{x}_v].$$

Signature Generation

Let $(y'_1, \ldots, y'_o) \in k^o$ be the document to be signed. First the signer computes

$$(x'_1, \ldots, x'_o) = F^{-1}(y'_1, \ldots, y'_n),$$

for some random choice of $(\check{x}'_1, \ldots, \check{x}'_v) \in k^v$. Recall that this amounts to solving the linear system

$$F(x_1, \ldots, x_o, \check{x}'_1, \ldots, \check{x}'_v) = (y'_1, \ldots, y'_o).$$

The signer then computes the signature of $(y'_1, \ldots, y'_o)$ as

$$(z'_1, \ldots, z'_n) = L^{-1}(x'_1, \ldots, x'_o, \check{x}'_1, \ldots, \check{x}'_v).$$

Signature Verification

To verify that $(z'_1, \ldots, z'_n)$ is indeed a valid signature for the message $(y'_1, \ldots, y'_o)$, the recipient determines whether or not

$$\bar{F}(z'_1, \ldots, z'_n) = (y'_1, \ldots, y'_o).$$

A Toy Example

We will again use the finite field $k = GF(2^2)$ for which addition and multiplication were given in Table 2.1. The example will be a balanced ($o = v$) Oil-Vinegar signature scheme with $o = v = 3$ so that $n = 6$.

Let $\mathsf{x} = (x_1, x_2, x_3, \check{x}_1, \check{x}_2, \check{x}_3)^T$ be the vector for the Oil-Vinegar variables. With these variables we choose the following Oil-Vinegar polynomials:

$$\begin{aligned} f_1 &= x_1\check{x}_1 + \alpha^2 x_1\check{x}_2 + \alpha^2 x_1\check{x}_3 + x_2\check{x}_1 + \alpha x_2\check{x}_2 + x_2\check{x}_3 + \alpha^2 x_3\check{x}_1 \\ &\quad + \alpha^2 x_3\check{x}_2 + \alpha^2 x_3\check{x}_3 + \alpha\check{x}_1\check{x}_3 + \alpha^2\check{x}_2^2 + \check{x}_2\check{x}_3 + \check{x}_3^3, \end{aligned}$$

$$f_2 = \alpha x_1\check{x}_2 + \alpha x_1\check{x}_3 + x_2\check{x}_1 + \alpha^2 x_2\check{x}_2 + \alpha x_2\check{x}_3 + \alpha x_3\check{x}_1 + x_3\check{x}_2 + \alpha^2 x_3\check{x}_3 + \check{x}_1^2 + \alpha\check{x}_1\check{x}_2 + \check{x}_1\check{x}_3 + \check{x}_3^2,$$

$$f_3 = \alpha x_1\check{x}_1 + \alpha x_1\check{x}_2 + x_2\check{x}_1 + x_2\check{x}_3 + \alpha^2 x_3\check{x}_1 + x_3\check{x}_2 + \alpha^2 x_3\check{x}_3 + \alpha^2\check{x}_1\check{x}_2 + \check{x}_1\check{x}_3 + \check{x}_2\check{x}_3 + \alpha\check{x}_3^2.$$

When these functions are written in bilinear form $f_i = \mathsf{x}^T \mathsf{Q}i\mathsf{x}$ for $i = 1, 2, 3$ a possible choice for the matrices $\mathsf{Q}i$ is an upper triangular form:

$$\mathsf{Q}_1 = \begin{pmatrix} 0 & 0 & 0 & 1 & \alpha^2 & \alpha^2 \\ 0 & 0 & 0 & 1 & \alpha & 1 \\ 0 & 0 & 0 & \alpha^2 & \alpha^2 & \alpha^2 \\ 0 & 0 & 0 & \alpha & 0 & 0 \\ 0 & 0 & 0 & 0 & \alpha^2 & 1 \\ 0 & 0 & 0 & 0 & 0 & 1 \end{pmatrix},$$

$$\mathsf{Q}_2 = \begin{pmatrix} 0 & 0 & 0 & 0 & \alpha & \alpha \\ 0 & 0 & 0 & 1 & \alpha^2 & \alpha \\ 0 & 0 & 0 & \alpha & 1 & \alpha^2 \\ 0 & 0 & 0 & 1 & \alpha & 1 \\ 0 & 0 & 0 & 0 & 0 & 0 \\ 0 & 0 & 0 & 0 & 0 & 1 \end{pmatrix},$$

and

$$\mathsf{Q}_3 = \begin{pmatrix} 0 & 0 & 0 & \alpha & \alpha & 0 \\ 0 & 0 & 0 & 1 & 0 & 1 \\ 0 & 0 & 0 & \alpha^2 & 1 & \alpha^2 \\ 0 & 0 & 0 & 0 & \alpha^2 & 1 \\ 0 & 0 & 0 & 0 & 0 & \alpha \end{pmatrix}.$$

For simplicity, we will choose L to be the invertible *linear* transformation given in matrix form by

$$\mathsf{x} = \mathsf{L}\mathsf{z}$$

where x was given above, $\mathsf{z} = (z_1, z_2, z_3, z_4, z_5, z_6)^T$, and

$$\mathsf{L} = \begin{pmatrix} 1 & \alpha^2 & \alpha & \alpha & 0 & \alpha^2 \\ \alpha^2 & \alpha^2 & 1 & 1 & 1 & \alpha \\ 1 & 0 & 1 & \alpha^2 & 1 & \alpha^2 \\ \alpha & \alpha & 1 & \alpha & 0 & 1 \\ \alpha & 1 & \alpha & \alpha^2 & 0 & \alpha^2 \\ 1 & 1 & 1 & \alpha & \alpha & 0 \end{pmatrix}.$$

The public polynomials can then be computed via

$$\bar{f}_i = \mathsf{z}^T \bar{\mathsf{Q}}_i \, \mathsf{z} = \mathsf{z}^T \left(\mathsf{L}^T \mathsf{Q}_1 \, \mathsf{L}\right) \mathsf{z}$$

for $i = 1, 2, 3$ and they are:

$$\begin{aligned}
\bar{f}_1 &= z_1^2 + \alpha^2 z_1 z_2 + \alpha z_1 z_3 + z_1 z_6 + \alpha z_2^2 + z_2 z_3 + \alpha z_2 z_4 + z_2 z_5 \\
&\quad + \alpha^2 z_2 z_6 + \alpha^2 z_3 z_5 + z_3 z_6, \\
\bar{f}_2 &= z_1^2 + z_1 z_2 + \alpha^2 z_1 z_3 + z_1 z_4 + \alpha z_1 z_6 + z_2^2 + \alpha^2 z_2 z_4 + z_2 z_5 \\
&\quad + \alpha z_2 z_6 + z_3^2 + \alpha^2 z_3 z_4 + z_3 z_6 + \alpha z_4^2 + z_5^2 + z_6^2, \\
\bar{f}_3 &= \alpha z_1^2 + \alpha z_1 z_2 + \alpha z_1 z_4 + \alpha^2 z_1 z_5 + z_1 z_6 + z_2^2 + \alpha^2 z_2 z_6 \\
&\quad + \alpha^2 z_3 z_4 + \alpha^2 z_3 z_6 + z_4^2 + z_4 z_6 + \alpha z_5^2 + \alpha z_5 z_6 + \alpha z_6^2.
\end{aligned}$$

Suppose we want to send the hash value of a document as the message $M = (m_1, m_2, m_3) = (\alpha, 1, \alpha^2)$ with a signature $S = (s_1, \ldots, s_6)$ such that the signature verification process confirms that $\bar{F}(S) = M$. To find a valid signature, we begin by randomly choosing values for the Vinegar variables, say

$$(\check{x}_1, \check{x}_2, \check{x}_3) = (\alpha^2, \alpha^2, 1).$$

Substituting these values into the Oil-Vinegar polynomials yields the linear polynomials:

$$\begin{aligned}
f_1(x_1, x_2, x_3, \alpha^2, \alpha^2, 1) &= \alpha x_1 + \alpha^2 x_2 + \alpha^2 x_3 + \alpha, \\
f_2(x_1, x_2, x_3, \alpha^2, \alpha^2, 1) &= \alpha^2 x_1 + \alpha^2 x_2 + x_3 + \alpha^2, \\
f_3(x_1, x_2, x_3, \alpha^2, \alpha^2, 1) &= \alpha x_2 + \alpha x_3 + \alpha^2.
\end{aligned}$$

Setting $f_i(x_1, x_2, x_3, \alpha^2, \alpha^2, 1) = m_i$ for $i = 1, 2, 3$, we have the (simplified) linear system:

$$\begin{aligned}
x_1 + \alpha x_2 + \alpha x_3 &= 0 \\
x_1 + x_2 + \alpha x_3 &= \alpha^2 \\
x_2 + x_3 &= 0
\end{aligned}$$

which has the solution:

$$(x_1, x_2, x_3) = (0, 1, 1).$$

To check our work, we simply verify that:

$$F(0, 1, 1, \alpha^2, \alpha^2, 1) = (\alpha, 1, \alpha^2).$$

Finally, the signature is computed as:

$$S = L^{-1}(0, 1, 1, \alpha^2, \alpha^2, 1) = (\alpha^2, 1, \alpha, \alpha^2, 0, \alpha^2).$$

The signature-message pair (S, M) can then be sent, and the legitimacy of the signature can be verified with the computation:

$$\bar{F}(\alpha^2, 1, \alpha, \alpha^2, 0, \alpha^2) = (\alpha, 1, \alpha^2).$$

The original applications of Oil-Vinegar signature schemes used a balanced $(o = v)$ construction. However, any such construction can be defeated (thus making forgeries possible), as shown by Kipnis and Shamir [Kipnis and Shamir, 1998] using a matrix method related to the bilinear forms defined by quadratic polynomials, as detailed below.

For the unbalanced $(o < v)$ Oil-Vinegar scheme with v not much larger than o, a specific attack shown in [Kipnis et al., 1999] has a complexity of roughly $o^4 q^{v-o-1}$. This means that if o is not too large (say less than 100), we must carefully balance security (by taking $v - o$ to be large enough to thwart the attack) with efficiency (taking $v - o$ too large results in an inefficient scheme).

3.2 Cryptanalysis of the Oil-Vinegar schemes

In this section, we will present a detailed cryptanalysis of the Oil-Vinegar scheme.

Definition 3.2.1. *We define the Oil subspace $\mathcal{O}$ in k^n to be*

$$\mathcal{O} = \{(x_1, \ldots, x_o, 0, \ldots, 0) \mid x_i \in k\},$$

and the Vinegar subspace $\mathcal{V}$ to be

$$\mathcal{V} = \{(0, \ldots, 0, (\check{x})_1, \ldots, (\check{x})_v) \mid (\check{x})_i \in k\}.$$

Balanced Oil-Vinegar

We will start with the method to break the balanced $(o = v)$ case [Kipnis and Shamir, 1998]. Here

$$n = o + v = 2v = 2o.$$

The key observation of the attack is that the associated symmetric matrix for the corresponding quadratic form defined by the polynomial components of the public key are of a special form. This allows us to use its structure to recover another key that is equivalent to the original private key for the purpose of producing forgeries of legitimate signatures.

We first assume that the given field k has odd characteristic. At the end of our discussion we will explain the subtle difference for the case of characteristic two.

Let $\bar{F} : k^n \longrightarrow k^v$ be a public Oil-Vinegar signature mapping with components $\bar{f}_1, \ldots, \bar{f}_v \in k[z_1, \ldots, z_n]$. The private keys are the Oil-Vinegar map $F : k^n \longrightarrow k^v$, and the invertible affine transformation $L : k^n \longrightarrow k^n$. As usual, $\bar{F} = F \circ L$.

Let z be the n-dimensional column vector

$$\mathsf{z} = (z_1, \ldots, z_n)^T,$$

and let x be the n-dimensional column vector denoted by

$$\mathsf{x} = (x_1, \ldots, x_v, \check{x}_1, \ldots, \check{x}_v)^T.$$

Let

$$\mathsf{x} = \mathsf{L}\,\mathsf{z},$$

where L is the $n \times n$ matrix for L. Note that we will use the matrix notation $\mathsf{L}\,\mathsf{z}$ for the equivalent functional notation $L(z_1, \ldots, z_n)$ and for simplicity, we assume that L is linear.

Let $\mathcal{O}$ and $\mathcal{V}$ be the Oil-Vinegar spaces of F, respectively. We shall find an invertible linear map $L' : k^n \longrightarrow k^n$ such that

$$L' \circ L^{-1}(\mathcal{O}) = \mathcal{O}$$

and then compute a new Oil-Vinegar map $F' : k^n \longrightarrow k^v$ defined by

$$F' = \bar{F} \circ (L')^{-1}.$$

Since we have

$$F \circ L = \bar{F} = F' \circ L',$$

the attacker can use the equivalent private keys F', L' to forge signatures.

For each $i = 1, \ldots, v$ denote the quadratic part of $\bar{f}_i(z_1, \ldots, z_n)$ by $\bar{q}_i(z_1, \ldots, z_n)$. Since k is not of characteristic two, there exists a unique $n \times n$ symmetric matrix $\bar{\mathsf{Q}}_i$ such that $\bar{q}_i(z_1, \ldots, z_n)$ is given by

$$\mathsf{z}^T\,\bar{\mathsf{Q}}_i\,\mathsf{z}.$$

Similarly for each $f_i(x_1, \ldots, x_v, \check{x}_1, \ldots \check{x}_v)$, we denote its quadratic part by $q_i(x_1, \ldots, x_v, \check{x}_1, \ldots, \check{x}_v)$. We also have a symmetric $n \times n$ matrix Q_i such that $q_i(x_1, \ldots, x_v, \check{x}_1, \ldots, \check{x}_v)$ is given by

$$\mathsf{x}^T\,\mathsf{Q}_i\,\mathsf{x}$$

where Q_i has the special form

$$\mathsf{Q}_i = \begin{pmatrix} 0 & \mathsf{B}_{i1} \\ \mathsf{B}_{i1}^T & \mathsf{B}_{i2} \end{pmatrix},$$

Here $\mathsf{0}$ is the $v \times v$ zero matrix and $\mathsf{B}_{i1}, \mathsf{B}_{i2}$ are $v \times v$ matrices.

Since $\bar{q}_i(z_1, \ldots, z_n) = q_i(x_1, \ldots, x_v, \check{x}_1, \ldots, \check{x}_v)$ we have

$$\begin{aligned} \mathsf{z}^T \bar{\mathsf{Q}}_i \mathsf{z} &= \mathsf{x}^T \mathsf{Q}_i \mathsf{x} \\ &= (\mathsf{L}\,\mathsf{z})^T \mathsf{Q}_i (\mathsf{L}\,\mathsf{z}) \\ &= \mathsf{z}^T (\mathsf{L}^T \mathsf{Q}_i \mathsf{L})\, \mathsf{z}, \end{aligned}$$

hence

$$\bar{\mathsf{Q}}_i = \mathsf{L}^T \mathsf{Q}_i \mathsf{L},$$

or equivalently

$$\mathsf{Q}_i = (\mathsf{L}^{-1})^T \bar{\mathsf{Q}}_i \mathsf{L}^{-1}.$$

Since the $\bar{\mathsf{Q}}_i$ are all known, we have the following strategy. Find any invertible matrix M such that for all $i = 1, \ldots, v$, we have

$$\mathsf{M}^T \bar{\mathsf{Q}}_i \mathsf{M} = \begin{pmatrix} 0 & * \\ * & * \end{pmatrix}$$

simultaneously. Though we may not have $\mathsf{M} = \mathsf{L}^{-1}$, the M we shall generate can be used to construct an equivalent Oil-Vinegar map that will produce valid signatures. There are many matrices M that will do the job. In fact, let M' be an invertible $n \times n$ matrix with entries in k of the form

$$\mathsf{M}' = \begin{pmatrix} * & * \\ 0 & * \end{pmatrix},$$

where $\mathsf{0}$ is the $v \times v$ zero matrix. Then $\mathsf{M} = \mathsf{L}^{-1}\mathsf{M}'$ is such that

$$\mathsf{M}^T \bar{\mathsf{Q}}_i \mathsf{M} = (\mathsf{L}^{-1}\mathsf{M}')^T \bar{\mathsf{Q}}_i (\mathsf{L}^{-1}\mathsf{M}') = \mathsf{M}'^T \mathsf{Q}_i \mathsf{M}' = \begin{pmatrix} 0 & * \\ * & * \end{pmatrix}.$$

It follows that there is a large class of such matrices, any of which will allow the forgery of signatures.

To simplify the exposition, we assume that the Oil-Vinegar polynomials are homogeneous of degree two.

Lemma 3.2.1. *For any* $\mathsf{u}_1, \mathsf{u}_2 \in \mathcal{O}$

$$\mathsf{u}_1^T \mathsf{Q}_i \mathsf{u}_2 = 0.$$

Equivalently, for any $\mathsf{w}_1, \mathsf{w}_2 \in L^{-1}(\mathcal{O})$ *we have*

$$\mathsf{w}_1^T \bar{\mathsf{Q}}_i \mathsf{w}_2 = 0.$$

Proof. This follows from the definition of the Oil-Vinegar polynomials, since there are no quadratic terms consisting only of Oil variables. □

Therefore, the key problem is now to find the hidden Oil space for the public key polynomials. To solve this problem we will make use of the key fact that $\mathcal{O}$ is an invariant subspace of any matrix M with the shape as described above. Therefore, we need to find a v-dimensional subspace such that any two vectors u_1 and u_2 satisfy the property $\mathsf{u}_1 \bar{\mathsf{M}}_i \mathsf{u}_2^t = 0$ in the $\mathcal{O}$ space in Lemma 3.2.1, which is the image subspace of $\mathcal{O}$ under the action of L^{-1}.

Let $\bar{Q}$ be the linear subspace of matrices spanned by the $\bar{\mathsf{Q}}_i$ and let Q be the linear subspace of matrices spanned by the Q_i. Because the coefficients of each q_i are randomly chosen, if we randomly choose an element in $\bar{Q}$, then the probability of picking a nonsingular matrix is roughly $1 - q^{-1}$. Let $\bar{\mathsf{W}}_1, \bar{\mathsf{W}}_2$ be two nonsingular elements in $\bar{Q}$, and let $\mathsf{W}_1, \mathsf{W}_2$ be the corresponding matrices in Q. From our previous considerations it is clear that each of these matrices is of the form

$$\bar{\mathsf{W}}_i = \mathsf{L}^T \mathsf{W}_i \mathsf{L} = \mathsf{L}^T \begin{pmatrix} 0 & \mathsf{W}_{i1} \\ \mathsf{W}_{i1}^T & \mathsf{W}_{i2} \end{pmatrix} \mathsf{L},$$

and has the inverse

$$\bar{\mathsf{W}}_i^{-1} = \mathsf{L}^{-1} \begin{pmatrix} -(\mathsf{W}_{i1}^T)^{-1} \mathsf{W}_{i2} \mathsf{W}_{i1}^{-1} & (\mathsf{W}_{i1}^T)^{-1} \\ \mathsf{W}_{i1}^{-1} & 0 \end{pmatrix} (\mathsf{L}^T)^{-1}.$$

We now define $\bar{\mathsf{W}}_{ij} = \bar{\mathsf{W}}_i^{-1} \bar{\mathsf{W}}_j$, which has the form

$$\bar{\mathsf{W}}_{ij} = \mathsf{L}^{-1} \begin{pmatrix} \mathsf{V}_{11} & \mathsf{V}_{12} \\ 0 & \mathsf{V}_{22} \end{pmatrix} \mathsf{L}, \tag{3.1}$$

where

$$\begin{aligned} \mathsf{V}_{11} &= (\mathsf{W}_{i1}^T)^{-1} \mathsf{W}_{j1}^T, \\ \mathsf{V}_{12} &= -(\mathsf{W}_{i1}^T)^{-1} \mathsf{W}_{i2} \mathsf{W}_{i1}^{-1} \mathsf{W}_{j1} + (\mathsf{W}_{i1}^T)^{-1} \mathsf{W}_{j2}, \\ \mathsf{V}_{22} &= \mathsf{W}_{i1}^{-1} \mathsf{W}_{j1}, \end{aligned}$$

and set

$$\mathsf{V} = \begin{pmatrix} \mathsf{V}_{11} & \mathsf{V}_{12} \\ 0 & \mathsf{V}_{22} \end{pmatrix}. \tag{3.2}$$

Again, because the coefficients of q_i are randomly chosen, we expect to be able to generate many matrices $\bar{\mathsf{W}}_{ij}$. Our task now has been transformed into finding a matrix M such that for all pairs i, j with $\bar{\mathsf{W}}_j$ invertible, we have

$$\mathsf{M}^{-1} \bar{\mathsf{W}}_{ij} \mathsf{M} = \begin{pmatrix} * & * \\ 0 & * \end{pmatrix}$$

simultaneously.

Before we move to the next step, we will define the key notion of an invariant subspace.

Definition 3.2.2. *Let V be a k-vector space and let $H : V \longrightarrow V$ be a linear transformation on V. A linear subspace $S \subset V$ is called an invariant subspace for H if $H(s) \in S$ for all $s \in S$.*

Let $H : V \longrightarrow V$ be a linear transformation on V and let $S \subset V$ be an invariant subspace for H with basis $\{s_1, \ldots, s_m\}$. This basis can be extended to a basis $\{s_1, \ldots, s_m, v_1, \ldots, v_l\}$ of V. It is then clear that the corresponding matrix for H will have the form

$$\mathsf{H} = \begin{pmatrix} * & * \\ 0 & * \end{pmatrix}.$$

We note here that finding an invariant subspace is also a standard problem in representation theory. An irreducible representation corresponds to the case of a representation space which has no non-trivial invariant subspaces.

The following lemma gives us some very useful facts about the matrices W_i.

Lemma 3.2.2. *Let $W_i, W_j : k^n \longrightarrow k^n$ be the mappings with associated matrices $\mathsf{W}_i, \mathsf{W}_j$ as above. Then*

1.) $W_i(\mathcal{O}) \subset \mathcal{V}$; i.e., W_i maps the Oil subspace into the Vinegar subspace;

2.) If W_j^{-1} exists, then $W_j^{-1}(\mathcal{V}) \subset \mathcal{O}$; i.e., W_j^{-1} maps the Vinegar subspace into the Oil subspace;

3.) If W_j^{-1} exists, then $W_j^{-1} \circ W_i(\mathcal{O}) \subset \mathcal{O}$; i.e., $\mathcal{O}$ is an invariant subspace of $W_j^{-1} \circ W_i$.

Assume now that we have a large number of $\bar{\mathsf{W}}_{ij}$ as defined above. Let Ω be the linear space spanned by the $\bar{\mathsf{W}}_{ij}$. It is clear that all elements in Ω share the invariant subspace $\mathcal{O}$, which we know from linear algebra will give us the desired M. If we have enough $\bar{\mathsf{W}}_{ij}$, then the invariant subspace shared by all elements of Ω should be exactly $L^{-1}(\mathcal{O})$.

In [Kipnis and Shamir, 1998] two different methods are suggested to find invariant subspaces in a general setting. In the first method they reduce the problem to solving a set of over-determined quadratic equations. This is rather complicated, and we will not discuss it here.

The second method uses basic linear algebra. Choose $\bar{\mathsf{W}}_{ij}$ and let $\bar{W}_{ij}$ be the associated mapping on k^n. We could analyze the Jordan canonical form to characterize all possible invariant subspaces of $\bar{\mathsf{W}}_{ij}$. However, this is also too complicated, and is in any case not very useful when there are a large number of invariant subspaces.

Instead we know from (3.1) that the characteristic polynomial of $\bar{\mathsf{W}}_{ij}$ is the product of the characteristic polynomials of V_{11} and V_{22} each of degree v. This follows from the fact that the characteristic polynomial is not changed by a similarity transformation.

Let $C(\lambda)$ be the characteristic polynomial of $\bar{\mathsf{W}}_{ij}$. In the approach [Kipnis and Shamir, 1998], it is assumed that V_{11} and V_{22} are general matrices with characteristic polynomials $C_1(\lambda)$ and $C_2(\lambda)$. They consider the simplest case, in which

$$C(\lambda) = C_1(\lambda)C_2(\lambda)$$

and $C_1(\lambda)$ and $C_2(\lambda)$ are two distinct irreducible polynomials. In this case k^n can be decomposed into the direct sum of two irreducible subspaces of the algebra generated by $\bar{\mathsf{W}}_{ij}$. (Note that W_{ij} and the algebra generated by W_{ij} share the same invariant subspace.) In this case, these two invariant subspaces could be found easily.

Let K_1 be the kernel of $C_1(\bar{\mathsf{W}}_{ij})$ and let K_2 be the kernel of $C_2(\bar{\mathsf{W}}_{ij})$, where we treat $C_l(\bar{\mathsf{W}}_{ij})$ as a linear mapping on k^n. From basic linear algebra we know the following facts:

1.) $\dim K_1 + \dim K_2 = n$;

2.) $K_1 \cap K_2 = \{0\}$, hence $k^n = K_1 \oplus K_2$;

3.) $W_1(k^n) = K_2$ and $W_2(k^n) = K_1$;

4.) The only possible invariant subspaces of $\bar{\mathsf{W}}_{ij}$ are $\{0\}, K_1, K_2$, and k^n.

If the above scenario is true and $\bar{\mathsf{W}}_{ij}$ has a characteristic polynomial that factors into two degree v irreducible polynomials as described above, then our problem could be solved in the following way. After calculating the characteristic polynomial $C(\lambda)$, we factor it into two distinct irreducible factors $C_1(\lambda)$ and $C_2(\lambda)$ of degree v. The invariant subspace we seek is either the kernel of $C_1(\bar{\mathsf{W}}_{ij})$ or the kernel of $C_2(\bar{\mathsf{W}}_{ij})$.

However, we observe that the characteristic polynomial of the matrix in (3.1) is always a square, that is

$$C_1(\lambda) = C_2(\lambda),$$

and we can not factor the characteristic polynomials into two distinctive irreducible polynomials. This is clear from (3.1) since

$$\mathsf{V}_{11} = (\mathsf{W}_{i1}^T)^{-1}\mathsf{W}_{j1}^T,$$
$$\mathsf{V}_{22} = \mathsf{W}_{i1}^{-1}\mathsf{W}_{j1},$$

$$\mathsf{W}_{i1}^{-1}(\mathsf{V}_{11})^T\mathsf{W}_{i1} = \mathsf{V}_{22},$$

and V_{11} and V_{22} must have the same characteristic polynomials. This implies from (3.2) that

$$\mathsf{L}\,C_1(\bar{\mathsf{W}}_{ij})\mathsf{L}^{-1} = C_1(\mathsf{V}) = \begin{pmatrix} 0 & \bar{\mathsf{V}}_{12} \\ 0 & 0 \end{pmatrix},$$

since $C_1(\mathsf{V}_{11}) = C_1(\mathsf{V}_{22}) = 0$ by the Cayley-Hamilton theorem, but $\bar{\mathsf{V}}_{12}$ does not have to be a zero matrix, since $C_1(\lambda)$ may not be the minimal polynomial.

Remark 3.2.1. *Note that we have to use the standard association of a quadratic polynomial with a unique symmetric matrix, but in [Kipnis and Shamir, 1998] this point is not emphasized. The symmetric condition is critical, otherwise the attack cannot work. An attacker would not know how to exactly associate a matrix to a polynomial in the public key without the symmetric condition. In the original attack by Kipnis-Shamir the symmetric condition is not used. They state that it is possible to have* $C_1(\lambda)$ *and* $C_2(\lambda)$ *different, but this is not true, when the matrices associated with the quadratic form are symmetric. The case of characteristic two is even more subtle and it will be explained below.*

When $\bar{\mathsf{V}}_{12}$ is of rank v, then the kernel of $C_1(\bar{\mathsf{W}}_{ij})$ is the Oil space. This means that the kernel of $\bar{\mathsf{W}}_{ij}$ will give us the desired subspace, the transformed Oil space.

Let us consider the probability that $C_1(\lambda)$ is an irreducible polynomial of degree v. It is known that randomly chosen degree v polynomials with coefficients in a finite field are irreducible with probability approximately equal to v^{-1} [Lidl and Niederreiter, 1997]. Since the W_i are randomly selected, it is reasonable to assume that the sampling of resulting characteristic polynomials of V_{11} is approximately random. If this is the case, then we expect to find an irreducible $C_1(\lambda)$ after v tries. Of course, the characteristic polynomials many not be uniformly distributed, in which case we may need to do more (or less) sampling or even fail, in particular when q and $o = v$ are small.

Once we find the common invariant subspace of all the elements in Ω, we establish a new basis $e_1, \ldots, e_n$ for k^n in which the first v vectors

$e_1, \ldots, e_v$ form a basis of the invariant subspace. Our choice of basis, though not unique, serves the purpose of transforming the public key (now including linear and constant terms) into an Oil-Vinegar map for which it is clear which variables are Oil variables and which variables are Vinegar variables. This map will have the same subspaces $\mathcal{O}$, but it will not in general be the original Oil-Vinegar map F. Consequently, attackers can use the same fast algorithm of the legitimate user to generate forged signatures for arbitrary messages, even though they do not have the identical secret key.

Breaking a balanced Oil-Vinegar scheme requires the following steps.

1.) For $i = 1, \ldots, o$ find the symmetric matrix $\bar{\mathsf{Q}}_i$ associated with $\bar{f}_i$.

2.) Pick invertible $\bar{\mathsf{W}}_1, \bar{\mathsf{W}}_2$ in Ω and compute $\bar{\mathsf{W}}_{12} = \bar{\mathsf{W}}_1^{-1}\bar{\mathsf{W}}_2$.

3.) Compute the characteristic polynomial $C(\lambda)$ of $\bar{\mathsf{W}}_{12}$. If it has only quadratic factors write $C(\lambda) = C_1(\lambda)^2$ with $C_1(\lambda)$ square free, then go to the next step; otherwise go back to Step 2.

4.) Calculate $C_1(\bar{\mathsf{W}}_{12})$. Find a basis for the null space of this matrix and extend it to a basis for k^n.

5.) Use this basis to transform the public polynomials into the Oil-Vinegar form. This equivalent information can be used to forge signatures.

One must notice that this algorithm works under the condition that $\bar{\mathsf{V}}_{12}$ is of rank v and under the condition that $C_1(\lambda)$ is square free. This we can not prove to be true in all cases. Nevertheless, in our own computer experiments it seems to work most of the time. When a matrix has two repeated eigenvalues then it also very likely that the Jordan canonical has non-zero terms off the diagonal. Since $\bar{\mathsf{V}}_{12}$ is related to these off the diagonal it gives an intuitive explanation for our findings. An irreducible $C_1(\lambda)$ gives the best chance that $\bar{\mathsf{V}}_{12}$ has rank v. Even when $C_1(\lambda)$ has several distinct factors, there is a good chance that $\bar{\mathsf{V}}_{12}$ has rank v, except that now each repeated factor in $C(\lambda)$ must give rise to off diagonal terms in the Jordan canonical form.

The method above will definitely not work, if the rank of $\bar{\mathsf{V}}_{12}$ is lower than v. In this case the kernel of $C_1(\mathsf{V})$ will include the entire Oil space but also part of the Vinegar space. However as long as the rank of $\bar{\mathsf{V}}_{12}$ is not zero, the problem is easy to deal with since the image space of $C_1(\mathsf{V})$ is inside the Oil space. This means that the image space of $C_1(\bar{\mathsf{W}}_{12})$ is inside the desired invariant subspace. Thus we can use additional randomly chosen $\bar{\mathsf{W}}_{12}$ to act on this image space to generate more vectors in the desired invariant subspace. Since V is in a blockwise triangular form, it preserves the Oil space under this mapping. After we have chosen enough $\bar{W}_{12}$, all these vectors should span the desired invariant subspace. The algorithm is therefore modified as follows:

1.) For $i = 1, \ldots, o$ find the symmetric matrix $\bar{Q}_i$ associated with $\bar{f}_i$.

2.) Pick invertible $\bar{W}_1, \bar{W}_2$ in Ω and compute $\bar{W}_{12} = \bar{W}_1^{-1}\bar{W}_2$.

3.) Compute the characteristic polynomial $C(\lambda)$ of $\bar{W}_{12}$ and factor it into $C_1(\lambda)^2$. Calculate $C_1(\bar{W}_{12})$, if it is not a zero matrix, go to the next step; otherwise go back to Step 2.

4.) Find a basis for the image space of the matrix $C_1(\bar{W}_{12})$. Call this space T. If T is of dimension v, go to Step 6, otherwise go to the next step.

5.) Repeat this step until the dimension of T is v: Pick another pair of invertible $\bar{W}_1, \bar{W}_2$, compute $\bar{W}_{12} = \bar{W}_1^{-1}\bar{W}_2$ and calculate the image space of T under the action of this $\bar{W}_{12}$. Find a basis of the space spanned by T and the image space of T under the action of this $\bar{W}_{12}$. Again call this space T.

6.) Extend the basis of T to a basis for k^n and use it to transform the public polynomials into the Oil-Vinegar form. This equivalent information can now be used to forge signatures.

In this algorithm, the condition that $C_1(\lambda)$ is irreducible is not required at all, and it works as long as we can find a $C_1(\lambda)$ such that $C_1(\bar{W}_{12})$ is not a zero matrix. Step 5 of the algorithm normally requires at most v rounds.

Cryptanalysis of a Toy Example for Odd Characteristic

For this example of an attack on a balanced Oil-Vinegar signature scheme we are only given that $o = v = 3$, that the finite field is $k = GF(7)$ and that the public polynomials for verifying a signature are:

$$\begin{aligned}\bar{f}_1 &= 3x_1^2 + 3x_1x_2 + 2x_1x_3 + 3x_1x_6 + 5x_2^2 + 3x_2x_3 + 6x_2x_4 \\ &\quad + 5x_2x_5 + 5x_2x_6 + x_3x_4 + x_3x_5 + 2x_3x_6 + 3x_4^2 + 3x_4x_5 \\ &\quad + 4x_5^2 + 6x_5x_6 + 2x_6^2,\end{aligned}$$

$$\begin{aligned}\bar{f}_2 &= 5x_1x_2 + 2x_1x_4 + 5x_1x_5 + 6x_1x_6 + 2x_2x_3 + 6x_2x_4 \\ &\quad + 6x_2x_5 + 5x_2x_6 + 3x_3^2 + 6x_3x_4 + 2x_3x_5 + x_3x_6 + x_4^2 \\ &\quad + 6x_4x_5 + 5x_4x_6 + 5x_5^2 + 3x_5x_6 + 6x_6^2,\end{aligned}$$

$$\begin{aligned}\bar{f}_3 &= 6x_1^2 + 2x_1x_2 + 2x_1x_3 + 5x_1x_4 + x_1x_5 + 3x_1x_6 + 5x_2x_3 \\ &\quad + 6x_2x_4 + 4x_2x_5 + 6x_2x_6 + 3x_3^2 + x_3x_4 + 6x_3x_5 + x_3x_6 \\ &\quad + x_4^2 + x_4x_6 + 6x_5^2 + x_5x_6 + x_6^2.\end{aligned}$$

At first glance these quadratic polynomials appear to be of a general nature and it is not obvious that they come from an Oil-Vinegar scheme.

By writing the given public polynomials as bilinear forms, that is $\bar{f}_i = \mathsf{x}^T\bar{Q}_i\mathsf{x}$, we obtain the following three symmetric matrices:

$$\bar{\mathsf{Q}}_i = \begin{pmatrix} 3 & 5 & 1 & 0 & 0 & 5 \\ 5 & 5 & 5 & 3 & 6 & 6 \\ 1 & 5 & 0 & 4 & 4 & 1 \\ 0 & 3 & 4 & 3 & 5 & 0 \\ 0 & 6 & 4 & 5 & 4 & 3 \\ 5 & 6 & 1 & 0 & 3 & 2 \end{pmatrix}$$

$$\bar{\mathsf{Q}}_2 = \begin{pmatrix} 0 & 6 & 0 & 1 & 6 & 3 \\ 6 & 0 & 1 & 3 & 3 & 6 \\ 0 & 1 & 3 & 3 & 1 & 4 \\ 1 & 3 & 3 & 1 & 3 & 6 \\ 6 & 3 & 1 & 3 & 5 & 5 \\ 3 & 6 & 4 & 6 & 5 & 6 \end{pmatrix},$$

$$\bar{\mathsf{Q}}_3 = \begin{pmatrix} 6 & 1 & 1 & 6 & 4 & 5 \\ 1 & 0 & 6 & 3 & 2 & 3 \\ 1 & 6 & 3 & 4 & 3 & 4 \\ 6 & 3 & 4 & 1 & 0 & 4 \\ 4 & 2 & 3 & 0 & 6 & 4 \\ 5 & 3 & 4 & 4 & 4 & 1 \end{pmatrix}.$$

Next we form linear combinations of these matrices until we two non-singular matrices. Since our computer program selects the linear combinations at random we ended up with the following two matrices $\bar{\mathsf{W}}_1 = 6\bar{\mathsf{Q}}_1 + 2\bar{\mathsf{Q}}_2 + 5\bar{\mathsf{Q}}_3$ and $\bar{\mathsf{W}}_2 = 5\bar{\mathsf{Q}}_1 + 5\bar{\mathsf{Q}}_2 + 4\bar{\mathsf{Q}}_3$. These two matrices are then used to compute

$$\bar{\mathsf{W}}_{12} = \bar{\mathsf{W}}_1^{-1}\bar{\mathsf{W}}_2 = \begin{pmatrix} 6 & 4 & 0 & 0 & 4 & 2 \\ 2 & 3 & 3 & 5 & 1 & 2 \\ 4 & 3 & 6 & 0 & 2 & 6 \\ 1 & 6 & 5 & 3 & 4 & 5 \\ 6 & 2 & 2 & 2 & 1 & 3 \\ 1 & 5 & 1 & 1 & 2 & 2 \end{pmatrix}.$$

The characteristic polynomial of this matrix is

$$C(\lambda) = \lambda^6 + 3\lambda^4 + 5\lambda^3 + 4\lambda^2 + 4\lambda + 1 = (\lambda + 1)^2(\lambda^2 + 6\lambda + 6)^2.$$

Since $C(\lambda)$ has only quadratic factors we can use the square root of $C(\lambda)$, that is

$$C_1(\lambda) = (\lambda+1)(\lambda^2+6\lambda+6) = \lambda^3+5\lambda+6$$

and evaluate it at the matrix $\bar{\mathsf{W}}_{12}$. We find

$$C_1(\bar{\mathsf{W}}_{12}) = \begin{pmatrix} 3 & 1 & 6 & 6 & 3 & 6 \\ 1 & 5 & 1 & 2 & 4 & 4 \\ 6 & 5 & 5 & 2 & 3 & 4 \\ 5 & 1 & 2 & 6 & 4 & 6 \\ 4 & 4 & 1 & 3 & 6 & 4 \\ 4 & 0 & 1 & 0 & 3 & 3 \end{pmatrix}$$

and that this matrix has rank 3. A basis for the kernel of this matrix are the three vectors $(1,0,0,1,5,3)^T$, $(0,1,0,1,0,0)^T$ and $(0,0,1,6,4,5)^T$. Extending this basis to the 6 dimensional space we obtain the transformation matrix

$$(\mathsf{L}')^{-1} = \begin{pmatrix} 1 & 0 & 0 & 0 & 0 & 0] \\ 0 & 1 & 0 & 0 & 0 & 0 \\ 0 & 0 & 1 & 0 & 0 & 0 \\ 1 & 1 & 6 & 1 & 0 & 0 \\ 5 & 0 & 4 & 0 & 1 & 0 \\ 3 & 0 & 5 & 0 & 0 & 1 \end{pmatrix}.$$

Let us set $\mathsf{T} = (\mathsf{L}')^{-1}$. When we apply this transformation to the matrices $\bar{\mathsf{Q}}_i$ we find

$$\mathsf{Q}'_1 = \mathsf{T}^T\bar{\mathsf{Q}}_1\mathsf{T} = \begin{pmatrix} 0 & 0 & 0 & 0 & 6 & 5 \\ 0 & 0 & 0 & 6 & 4 & 6 \\ 0 & 0 & 0 & 0 & 2 & 2 \\ 0 & 6 & 0 & 3 & 5 & 0 \\ 6 & 4 & 2 & 5 & 4 & 3 \\ 5 & 6 & 2 & 0 & 3 & 2 \end{pmatrix},$$

$$\mathsf{Q}'_2 = \mathsf{T}^T\bar{\mathsf{Q}}_2\mathsf{T} = \begin{pmatrix} 0 & 0 & 0 & 0 & 0 & 3 \\ 0 & 0 & 0 & 4 & 6 & 5 \\ 0 & 0 & 0 & 2 & 1 & 6 \\ 0 & 4 & 2 & 1 & 3 & 6 \\ 0 & 6 & 1 & 3 & 5 & 5 \\ 3 & 5 & 6 & 6 & 5 & 6 \end{pmatrix},$$

$$\mathsf{Q}_3' = \mathsf{T}^T\bar{\mathsf{Q}}_3\mathsf{T} = \begin{pmatrix} 0 & 0 & 0 & 5 & 4 & 4 \\ 0 & 0 & 0 & 4 & 2 & 0 \\ 0 & 0 & 0 & 2 & 5 & 0 \\ 5 & 4 & 2 & 1 & 0 & 4 \\ 4 & 2 & 5 & 0 & 6 & 4 \\ 4 & 0 & 0 & 4 & 4 & 1 \end{pmatrix}.$$

With $f_i' = \mathsf{x}^T\mathsf{Q}i'\mathsf{x}$ for $i = 1, 2, 3$ we obtain the polynomials:

$$\begin{aligned} f_1' &= 5x_1\check{x}_2 + 3x_1\check{x}_3 + 5x_2\check{x}_1 + x_2\check{x}_2 + 5x_2\check{x}_3 + 4x_3\check{x}_2 + 4x_3\check{x}_3 \\ &\quad + 3\check{x}_1^2 + 3\check{x}_1\check{x}_2 + 4\check{x}_2^2 + 6\check{x}_2\check{x}_3 + 2\check{x}_3^2, \\ f_2' &= 6x_1\check{x}_3 + x_2\check{x}_1 + 5x_2\check{x}_2 + 3x_2\check{x}_3 + 4x_3\check{x}_1 + 2x_3\check{x}_2 + 5x_3\check{x}_3 \\ &\quad + \check{x}_1^2 + 6\check{x}_1\check{x}_2 + 5\check{x}_1\check{x}_3 + 5\check{x}_2^2 + 3\check{x}_2\check{x}_3 + 6\check{x}_3^2, \\ f_3' &= 3x_1\check{x}_1 + x_1\check{x}_2 + x_1\check{x}_3 + x_2\check{x}_1 + 4x_2\check{x}_2 + 4x_3\check{x}_1 + 3x_3\check{x}_2 \\ &\quad + \check{x}_1^2 + \check{x}_1\check{x}_3 + 6\check{x}_2^2 + \check{x}_2\check{x}_3 + \check{x}_3^2. \end{aligned}$$

These polynomials are in the format of the Oil-Vinegar polynomials. We can use them together with the inverse of the transformation T in order to forge a signature to any document in the same way a legitimate user would do with the original set of Oil-Vinegar polynomials and the corresponding transformation L.

The Case of Characteristic Two

In the case of characteristic two, the association of the matrix with a quadratic polynomial is different. Any polynomial can be written as $\sum_{i=1}^n \sum_{j=i}^n a_{ij}x_ix_j$, with $a_{ij} = 0$ for $i > j$, or with a matrix A whose entries are a_{ij}. Since A cannot be symmetric (except when $\mathsf{A} = 0$) then the symmetric matrix associated to it is given as

$$\mathsf{A}^T + \mathsf{A}.$$

This association is unique and all diagonal entries are zero. The basic idea of the attack still works, but $C_1(\bar{\mathsf{W}}_{12})$ is always a zero matrix and it does not matter if $C_1(\lambda)$ is irreducible or not.

Therefore the algorithm has to be modified. We first factorize $C_1(\lambda)$ into products of irreducible factors, and we will look for a distinctive linear factor $(\lambda - \lambda_1)$ of multiplicity one. This should occur with a reasonably high probability. The eigenspace of the corresponding $\bar{\mathsf{W}}_{12}$ is exactly two, and in this eigenspace it must have a vector in the Oil space due to the form of V in (3.2).

This implies that we can find the corresponding eigenspace of $\bar{\mathsf{W}}_{12}$ for the eigenvalue λ_1, which is exactly of dimension two, and it must have one vector in the desired invariant subspace. In order to find this vector in the desired invariant subspace, we try all $q+1$ possible eigenvectors in this eigenspace and use a set of additionally randomly chosen $\bar{W}_{12}$ to act on the chosen vector. One of these vectors will generate the desired invariant subspace, the rest of these vectors will generate subspaces of dimension larger than v. Note that it is not necessary to consider eigenvectors which are proportional to each other, one of them is sufficient.

The explicit algorithm is:

1.) For $i = 1, \ldots, o$, find the symmetric matrix $\bar{Q}_i$ associated with $\bar{f}_i$.

2.) Pick invertible $\bar{W}_1, \bar{W}_2$ in Ω and compute $\bar{W}_{12} = \bar{W}_1^{-1}\bar{W}_2$.

3.) Compute the characteristic polynomial of $\bar{W}_{12}$ which has the form $C(\lambda) = C_1(\lambda)^2$. Factor $C_1(\lambda)$, and if it has a linear factor $\lambda - \lambda_1$ of multiplicity one, go to the next step, otherwise go back to Step 2.

4.) Find a basis $\{A_1, A_2\}$ for the eigenspace of the matrix $C_1(\bar{W}_{12})$ with eigenvalue λ_1. Let the set of all possible eigenvectors be

$$S = \{A_1 + \alpha A_2 \forall \alpha \in k\} \cup \{A_2\}.$$

5.) For each s in S do

   (a) Denote the space spanned by s by T.

   (b) Repeat the following step at most $2o - 1$ times or until the dimension of T is greater than o

   (c) Pick a new random pair of invertible $\bar{W}_1$, $\bar{W}_2$ and compute $\bar{W}_{12} = \bar{W}_1^{-1}\bar{W}_2$. Calculate the image of T under the action of this $\bar{W}_{12}$. Find a basis of the space spanned by T and $\bar{W}_{12}(T)$. Again denote this space T.

   (d) If the dimension of T is o then go to Step 6; otherwise return to Step 2.

6.) Extend the basis of T to a basis for k^n and use it to transform the public polynomials into the Oil-Vinegar form. This equivalent information can now be used to forge signatures.

In computer experiments the algorithm typically produces two and sometimes even more transformations L'. The reason for it is that several different eigenvectors can produce a subspace of dimension o. Not all transformations found by our method will work and it is then necessary to determine by trial and error, which is the correct one.

It is clear that in all the attacks above, we never recover the original keys, but an equivalent one. Consequently, any attacker can use the same fast algorithm as the legitimate signer uses to generate forged signatures for arbitrary messages. Finally, we note that this attack can be modified to work when $v < o$. We shall consider the remaining case of $v > o$ in the next section.

Cryptanalysis of the Toy Example for Characteristic Two

As the attacker we only have access to the public polynomials:

$$\begin{aligned}
\bar{f}_1 &= z_1^2 + \alpha^2 z_1 z_2 + \alpha z_1 z_3 + z_1 z_6 + \alpha z_2^2 + z_2 z_3 + \alpha z_2 z_4 + z_2 z_5 \\
&\quad + \alpha^2 z_2 z_6 + \alpha^2 z_3 z_5 + z_3 z_6, \\
\bar{f}_2 &= z_1^2 + z_1 z_2 + \alpha^2 z_1 z_3 + z_1 z_4 + \alpha z_1 z_6 + z_2^2 + \alpha^2 z_2 z_4 + z_2 z_5 \\
&\quad + \alpha z_2 z_6 + z_3^2 + \alpha^2 z_3 z_4 + z_3 z_6 + \alpha z_4^2 + z_5^2 + z_6^2, \\
\bar{f}_3 &= 7\alpha z_1^2 + \alpha z_1 z_2 + \alpha z_1 z_4 + \alpha^2 z_1 z_5 + z_1 z_6 + z_2^2 + \alpha^2 z_2 z_6 \\
&\quad + \alpha^2 z_3 z_4 + \alpha^2 z_3 z_6 + z_4^2 + z_4 z_6 + \alpha z_5^2 + \alpha z_5 z_6 + \alpha z_6^2.
\end{aligned}$$

From these polynomials we form the associated symmetric matrices

$$\bar{Q}_1 = \begin{pmatrix} 0 & \alpha^2 & \alpha & 0 & 0 & 1 \\ \alpha^2 & 0 & 1 & \alpha & 1 & \alpha^2 \\ \alpha & 1 & 0 & 0 & \alpha^2 & 1 \\ 0 & \alpha & 0 & 0 & 0 & 0 \\ 0 & 1 & \alpha^2 & 0 & 0 & 0 \\ 1 & \alpha^2 & 1 & 0 & 0 & 0 \end{pmatrix},$$

$$\bar{Q}_2 = \begin{pmatrix} 0 & 1 & \alpha^2 & 1 & 0 & \alpha \\ 1 & 0 & 0 & \alpha^2 & 1 & \alpha \\ \alpha^2 & 0 & 0 & \alpha^2 & 0 & 1 \\ 1 & \alpha^2 & \alpha^2 & 0 & 0 & 0 \\ 0 & 1 & 0 & 0 & 0 & 0 \\ \alpha & \alpha & 1 & 0 & 0 & 0 \end{pmatrix},$$

and

$$\bar{Q}_3 = \begin{pmatrix} 0 & \alpha & 0 & \alpha & \alpha^2 & 1 \\ \alpha & 0 & 0 & 0 & 0 & \alpha^2 \\ 0 & 0 & 0 & \alpha^2 & 0 & \alpha^2 \\ \alpha & 0 & \alpha^2 & 0 & 0 & 1 \\ \alpha^2 & 0 & 0 & 0 & 0 & \alpha \\ 1 & \alpha^2 & \alpha^2 & 1 & \alpha & 0 \end{pmatrix}.$$

Next we choose linear combinations until we have two non singular matrices. A possible choice is $\bar{W}_1 = \bar{Q}_1$ and $\bar{W}_2 = \bar{Q}_2 + \bar{Q}_3$. With them we compute

$$\bar{\mathsf{W}}_{12} = \bar{\mathsf{W}}_1^{-1}\bar{\mathsf{W}}_2 = \begin{pmatrix} 0 & 1 & \alpha & 1 & \alpha & 0 \\ \alpha & \alpha & 0 & 0 & 0 & \alpha^2 \\ \alpha & 1 & 0 & 0 & 0 & \alpha \\ \alpha^2 & 1 & \alpha & 0 & 1 & \alpha \\ 1 & 0 & 0 & \alpha & 0 & 1 \\ \alpha & 0 & \alpha^2 & \alpha^2 & \alpha^2 & \alpha \end{pmatrix}.$$

The characteristic polynomial of this matrix is $(x+1)^2(x^2+\alpha x+\alpha)^2$. The eigenvalue $\lambda = 1$ can be used for our purpose. Two eigenvectors spanning the eigenspace are

$$\begin{aligned} A_1 &= (1, 0, \alpha^2, \alpha, 1, \alpha^2)^T, \\ A_2 &= (0, 1, \alpha^2, 1, \alpha^2, 1)^T, \end{aligned}$$

but among the possible eigenvectors from the set

$$\{A_1, A_1 + A_2, A_1 + \alpha A_2, A_1 + \alpha^2 A_2, A_2\}$$

we would like to select the one which is in the desired invariant subspace. For that purpose we compute additional matrices $\bar{\mathsf{W}}_{ij}$. Then we go through this list of eigenvectors and look at the dimension of the space which the vector and its images under the mapping with $\bar{\mathsf{W}}_{ij}$ span. When the dimension of this space exceeds three we know that we do not have the correct eigenvector, and we go to the next vector in our list. With

$$A_1 + \alpha^2 A_2 = (1, \alpha^2, 1, 1, \alpha^2, 0)^T$$

we find the three dimensional invariant subspace. A basis for this space consists of the first three columns of the matrix below, and the last three rows are the extension of this basis to the full dimensional space.

$$\mathsf{L}' = \begin{pmatrix} 1 & 0 & 0 & 0 & 1 & 1 \\ 0 & 1 & 0 & 0 & 0 & 0 \\ 0 & 0 & 1 & 0 & 0 & 0 \\ \alpha & 0 & \alpha^2 & 1 & 0 & 0 \\ 1 & \alpha^2 & 0 & 0 & 1 & 0 \\ 1 & \alpha & 0 & 0 & 0 & 1 \end{pmatrix}$$

With

$$\mathsf{z} = (\mathsf{L}')^{-1}\mathsf{x}$$

we can now transform the given public polynomials into another set of Oil-Vinegar polynomials

$$\begin{aligned} f_1' &= x_1\check{x}_3 + \alpha x_2\check{x}_1 + x_2\check{x}_2 + \alpha^2 x_2\check{x}_3 + \alpha^2 x_3\check{x}_2 + x_3\check{x}_3, \\ f_2' &= x_1\check{x}_1 + \alpha x_1\check{x}_3 + \alpha^2 x_2\check{x}_1 + x_2\check{x}_2 + \alpha x_2\check{x}_3 + \alpha^2 x_3\check{x}_1 + x_3\check{x}_3 \\ &\quad + \alpha\check{x}_1^2 + \check{x}_2^2 + \check{x}_3^2, \\ f_3' &= \alpha^2 x_1\check{x}_1 + x_1\check{x}_2 + x_1\check{x}_3 + \alpha x_2\check{x}_1 + \alpha^2 x_2\check{x}_2 + \alpha x_2\check{x}_3 + \alpha^2 x_3\check{x}_1 \\ &\quad + \check{x}_1^2 + \check{x}_1\check{x}_3 + \alpha\check{x}_2^2 + \alpha\check{x}_2\check{x}_3 + \alpha\check{x}_3^2. \end{aligned}$$

As attackers we can now use these Oil-Vinegar polynomials together with L' to create a forged signature. Assume that we want to create a forged signature for the message $M' = (m_1', m_2', m_3') = (1, \alpha, 1)$ we would proceed in the same way as was done in the previous toy example. With an arbitrary set for the vinegar variables, say $(\check{x}_1, \check{x}_2, \check{x}_3) = (\alpha, 0, 0)$, we find the signature

$$S' = (\alpha^2, \alpha, 0, \alpha^2, \alpha, 0).$$

If the signature-message pair (S', M') is received it must be accepted as legitimate since checking it with the public key gives

$$\bar{F}(\alpha^2, \alpha, 0, \alpha^2, \alpha, 0) = (1, \alpha, 1).$$

Unbalanced Oil-Vinegar

Soon after the attack on balanced Oil-Vinegar, Kipnis, Patarin and Goubin proposed a modified scheme called the unbalanced Oil-Vinegar scheme [Kipnis et al., 1999]. With it they presented the best known attacks and they suggested optimal parameters for practical applications.

Case: $v > o$ and $v \approx o$

The attack in this case is essentially an extension of the attack on the balanced case, but $v - o$ has to be reasonably small. We use the same notation as in the previous section.

With $n = o + v$, let $E : k^n \longrightarrow k^n$ be a linear transformation with matrix $n \times n$ matrix E of the form

$$\mathsf{E} = \begin{pmatrix} 0 & \mathsf{E}_{12} \\ \mathsf{E}_{21} & \mathsf{E}_{22} \end{pmatrix},$$

where $\mathsf{0}$ is the $o \times o$ zero matrix and E_{22} is an $v \times v$ matrix.

Lemma 3.2.3. *Let E be as above. Then*

1.) $E(\mathcal{O})$ is an o-dimensional proper subspace of $\mathcal{V}$;

2.) If E is invertible, then $E^{-1}(\mathcal{V})$ is a v-dimensional subspace of k^n in which $\mathcal{O}$ is a proper subspace.

As in the attack on balanced Oil-Vinegar we seek the space $L^{-1}(\mathcal{O})$, which we will denote $\bar{\mathcal{O}}$. To find this space, we will again use with the matrices $\bar{\mathsf{W}}_i$, whose corresponding matrices W_i have the form of E given above. Let Ω be the span of these matrices. We are looking for a common invariant subspace of the elements of Ω. The following lemma says that such a space exists with high probability.

Lemma 3.2.4. *Let $J : k^l \longrightarrow k^l$ be a randomly chosen invertible k-linear map such that:*

1.) There exist two subspaces A, B in k^l such that the dimension of A is v, and the dimension of B is o and $B \subset A$;

2.) $J(B) \subset A$.

Then the probability that J has a nontrivial invariant subspace in B is no less than q^{o-v}.

Proof. The probability that a nonzero vector is mapped to a nonzero multiple of itself is $\frac{q-1}{q^v-1}$. To get the expected number of such vectors, we multiply by the number of nonzero vectors in the image space to get $\frac{(q-1)(q^o-1)}{q^v-1}$. Since if a vector is mapped to a multiple of itself then the same is true of any multiple of that vector, the expected number of invariant subspaces of dimension 1 is roughly $\frac{q^o-1}{q^v-1} \approx q^{o-v}$. □

Theorem 3.2.1. *Let W_i and W_j be any two randomly chosen matrices in Ω such that W_2^{-1} exists. Then the probability that the associated matrix $\bar{\mathsf{W}}_{ij}$ has a nontrivial invariant subspace (which is also a subspace of $L^{-1}(\mathcal{O})$) is roughly q^{o-v}.*

Proof. This follows directly from the fact that

$$\bar{\mathsf{W}}_{ij} = \mathsf{L}^{-1}\,\mathsf{W}_i^{-1}\,\mathsf{W}_j\,\mathsf{L}.$$

□

The algorithm we will present for finding $L^{-1}(\mathcal{O})$ is a probabilistic algorithm. It looks for an invariant subspace associated with the elements in Ω inside $L^{-1}(\mathcal{O})$. Take a random linear combination of $\bar{\mathsf{W}}_i$ and for some j multiply by $\bar{\mathsf{W}}_j^{-1}$. We then calculate all the so-called minimal invariant subspaces (an invariant subspace which contains no nontrivial invariant subspaces) of this matrix. These subspaces correspond to the irreducible factors of the characteristic polynomial of this matrix, and can be found in probabilistic polynomial time using standard linear algebra techniques. This matrix may or may not have an invariant subspace which is a subspace of $L^{-1}(\mathcal{O})$. The following lemma enables us to distinguish in polynomial time between random subspaces and those that are contained in $L^{-1}(\mathcal{O})$.

Lemma 3.2.5. *If $\mathcal{U}$ is a linear subspace of $L^{-1}(\mathcal{O})$, then for every $x, y \in \mathcal{U}$ and for every i,*

$$\mathsf{x}^T \, \bar{\mathsf{Q}}_i \, \mathsf{y} = 0,$$

where x, y are the column vectors associated with x, y, respectively.

Proof. Let $x', y' \in \mathcal{O}$ such that $x = L^{-1}(x'), y = L^{-1}(y')$. Then we have

$$\begin{aligned} \mathsf{x}^T \, \bar{\mathsf{Q}}_i \, \mathsf{y} &= (\mathsf{L}^{-1} \mathsf{x}')^T \, \bar{\mathsf{Q}}_i \, (\mathsf{L}^{-1} \mathsf{y}) \\ &= (\mathsf{x}')^T \, (\mathsf{L}^{-1})^T \, \bar{\mathsf{Q}}_i \, \mathsf{L}^{-1} \, \mathsf{y}' \\ &= (\mathsf{x}')^T \, \mathsf{Q}_i \, \mathsf{y}' \\ &= 0 \end{aligned}$$

□

We can use this test to decide whether or not the minimal invariant subspace lies in $L^{-1}(\mathcal{O})$. If not, we choose another element of Ω and start over. After roughly q^{v-o-1} tries we are very likely to have at least one vector in $L^{-1}(\mathcal{O})$. We continue this process until we have o linearly independent vectors. We expect this process to have complexity $o^4 q^{v-o-1}$.

Case: $v \geq 2o$

In [Braeken et al., 2005], a security analysis is presented for the case $v \geq 2o$. The attack on unbalanced Oil-Vinegar this time is by the F_4 Gröbner basis algorithm [Faugère, 1999]. Since we know that there are a huge number of legitimate signatures (roughly q^{2v}), if we randomly choose a linear (or affine) space of dimension o then there should be a very good chance (roughly $1/e$) that there is a solution in this subspace. Such a subspace is given by a set of v linear equations, so therefore we can just append a set of v linear equations to the quadratic public key equations and try to solve. The F_4 algorithm implemented in Magma was used and the results of the experiments in [Braeken et al., 2005] show that this attack has a 60% chance of success. More specifically, they investigate the time complexity of the attack for a fixed o and varying $v = 2o, 3o$, in the cases $q = 2, 3, 16$.

From these experiments, we can conclude that o should be greater than 38 for characteristic two, and greater than 24 for characteristic three, both for $v = 2o, 3o$, in order to obtain a security level greater than 2^{64}. However, there was not enough data to predict the behavior in the case $q = 16$. On the other hand, these estimates on the minimum value of o needed for the corresponding level of security are much higher than the bounds proposed in [Kipnis et al., 2003] and later in [Courtois et al., 2002].

Also in [Braeken et al., 2005], a method of linear approximation is also proposed for use in an attack on unbalanced Oil-Vinegar. This attack has a complexity of q^v. The linear approximation method was used by Youssef and Gong [Youssef and Gong, 2001] to attack another Matsumoto-Imai cryptosystem [Imai and Matsumoto, 1985]. In any case, it seems that one must be careful in the case of $v \geq 2o$, and more study is needed to clarify the situation.

Case: $v \simeq o^2/2$ (or $v \geq o^2/2$)

From the above analysis, one may be tempted to conclude that the bigger v is compared to o, the more secure the signature scheme becomes. This is in general false, as we shall see in this section.

Let $\mathcal{S}$ be a random set of n quadratic polynomials in the $o+v$ variables $z_1, \ldots, z_{o+v}$. When $v \simeq o^2/2$ (and more generally when $v \geq o^2/2$), it is not difficult to show [Kipnis et al., 1999] that there is very likely (namely, for most sets $\mathcal{S}$) a linear change of variables such that the resulting set of polynomials $\mathcal{T}$ is a set of Oil-Vinegar polynomials. Therefore such a system is not necessarily secure at all.

Additionally, recall that the document to be signed is a vector in k^o and the signature is a vector in k^{o+v}. This means that the signature is far too large compared to the document length, and therefore the signature scheme is very inefficient for any kind of practical application. Since this is so, we omit the details about how to attack this type of scheme, though the interested reader can find them in [Kipnis et al., 1999].

For the case of characteristic two the situation is more or less the same, except that we must use only the concept of bilinear forms, since all symmetric quadratic forms with zero diagonal entries are trivial as a quadratic form.

From these results we can conclude that indeed we can build useful unbalanced Oil-Vinegar schemes for practical applications. However, one must notice that the document to be signed is a vector in k^o and the signature is a vector in k^{o+v}. This means that the signature is at least twice the size of the document and when $v+o$ is large it does not appear to be efficient. One direction for further research is to see if there is a way to make such a type of system more efficient.

One possibility is the multi-layer unbalanced Oil-Vinegar construction, where one uses the Oil-Vinegar construction multiple times such that in the end the signature will be only slightly longer than the document. It is called the Rainbow signature scheme [Ding and Schmidt, 2005b].

3.3 Rainbow: Multilayer Unbalanced Oil-Vinegar

We first present the general construction of Rainbow and then give an example of its practical implementation.

General Construction

Let S be the set $\{1, 2, 3, \ldots, n\}$. Let $v_1, \ldots, v_u$ be any set of u integers ($u \leq n$) such that $0 < v_1 < v_2 < \cdots < v_u = n$, and define the sets of integers $S_l = \{1, 2, \ldots, v_l\}$ for $l = 1, \ldots, u$. Clearly we have that

$$S_1 \subset S_2 \subset \cdots \subset S_u = S,$$

and that v_i is the number of elements in S_i.

For $i = 1, \ldots, u-1$ let $o_i = v_{i+1} - v_i$ and $O_i = S_{i+1} - S_i$, so that o_i is the number of elements in O_i. With this notation let P_l be the linear space of quadratic polynomials spanned by polynomials of the form

$$\sum_{i \in O_l,\, j \in S_l} \alpha_{ij} x_i x_j + \sum_{i,j \in S_l} \beta_{ij} x_i x_j + \sum_{i \in S_{l+1}} \gamma_i x_i + \eta.$$

We can see that these are Oil-Vinegar-type polynomials where x_i is an Oil variable if $i \in O_l$ and x_j is a Vinegar variable if $j \in S_l$.

More specifically, we say that x_i is an lth layer Oil variable if $i \in O_l$, and that x_j is an lth layer Vinegar variable if $j \in S_l$. A polynomial in P_l is called an lth layer Oil-Vinegar polynomial. It is clear that $P_i \subset P_j$ for $i < j$, and that $\{P_1, \ldots, P_{u-1}\}$ is a set of Oil-Vinegar polynomials. Note that the Vinegar variables at the $(l+1)^{\text{th}}$ layer are all the variables at the lth layer since

$$S_{i+1} = O_i \cup S_i.$$

For $i = 1, \ldots, u-1$, let

$$\tilde{F}_i = (\tilde{F}_{i1}, \ldots, \tilde{F}_{i o_i}),$$

where each $\tilde{F}_{ij}$ is a randomly chosen element from P_i, and then define the map $F : k^n \longrightarrow k^{n-v_1}$ by

$$F = (\tilde{F}_1, \ldots, \tilde{F}_{u-1}).$$

To help simplify the notation, denote the $n - v_1$ polynomial components of F by $F_1, \ldots, F_{n-v_1}$.

From this construction we can see that F has $u-1$ layers of Oil-Vinegar construction. The first layer consists of the o_1 polynomials $F_1, \ldots, F_{o_1}$ where $\{x_i \mid i \in O_1\}$ are the Oil variables and $\{x_j \mid j \in S_1\}$ are the Vinegar variables. The lth layer consists of the o_l polynomials $F_{v_l+1}, \ldots, F_{v_{l+1}}$ where $\{x_i \mid i \in O_l\}$ are the Oil variables and $\{x_j \mid j \in S_l\}$ are the Vinegar variables.

From this we can build a "rainbow" of variables:

$$[x_1, \ldots, x_{v_1}], \{x_{v_1+1}, \ldots, x_{v_2}\};$$
$$[x_1, \ldots, x_{v_1}, x_{v_1+1}, \ldots, x_{v_2}], \{x_{v_2+1}, \ldots, x_{v_3}\};$$
$$[x_1, \ldots, x_{v_1}, x_{v_1+1}, \ldots, x_{v_2}, x_{v_2+1}, \ldots, x_{v_3}], \{x_{v_3+1}, \ldots, x_{v_4}\};$$
$$\vdots$$
$$[x_1, \ldots, \ldots, \ldots, \ldots, \ldots, \ldots, \ldots, \ldots, \ldots, x_{v_{u-1}}], \{x_{v_{u-1}+1}, \ldots, x_n\},$$

where each row of variables represents a layer of the rainbow. For the lth layer, the Vinegar variables are enclosed in square brackets $[\ldots]$, while the Oil variables are enclosed in curly brackets $\{\ldots\}$. We say that F is a Rainbow polynomial map with $u-1$ layers.

Let $L_1 : k^{n-v_1} \longrightarrow k^{n-v_1}$ and $L_2 : k^n \longrightarrow k^n$ be two randomly chosen invertible affine linear maps, and define $\bar{F} : k^n \longrightarrow k^{n-v_1}$ by

$$\bar{F} = L_1 \circ F \circ L_2 = (\bar{F}_1, \ldots, \bar{F}_{n-v_1}).$$

Each of the components $\bar{F}_i$ of $\bar{F}$ is a polynomial in $k[x_1, \ldots, x_n]$.

We now present the Rainbow signature scheme.

Public Key

The public key consists of

1.) The field k, including its additive and multiplicative structure.

2.) The $n - v_1$ polynomial components of $\bar{F}$.

Private Key

The private key consists of the maps L_1, L_2 and F.

Signature Generation

To sign the document $Y' = (y'_1, \ldots, y'_{n-v_1}) \in k^{n-v_1}$, we need to find a solution of the equation

$$\bar{F}(x_1, \ldots, x_n) = L_1 \circ F \circ L_2(x_1, \ldots, x_n) = Y',$$

which is done in the following steps:

1.) Compute

$$\bar{Y}' = L_1^{-1}(Y') = (\bar{y}'_1, \ldots, \bar{y}'_{n-v_1}).$$

2.) Next we need to invert F. In particular, we need to solve the equation

$$F(x_1, \ldots, x_n) = \bar{Y}' = (\bar{y}'_1, \ldots, \bar{y}'_{n-v_1}).$$

To do this we first randomly choose values $\bar{x}'_1, \ldots, \bar{x}'_{v_1}$ for $x_1, \ldots, x_{v_1}$ and substitute them into the first layer of o_1 equations to get

$$\tilde{F}_1(\bar{x}'_1, \ldots, \bar{x}'_{v_1}, x_{v_1+1}, \ldots, x_{v_2}) = (\bar{y}'_1, \ldots, \bar{y}'_{o_1}).$$

This represents a set of o_1 linear equations in the o_1 variables, $x_{v_1+1}, \ldots, x_{v_2}$. These equations can be solved to find the values $\bar{x}'_{v_1+1}, \ldots, \bar{x}'_{v_2}$ for $x_{v_1+1}, \ldots, x_{v_2}$.

3.) We now substitute $\bar{x}'_1, \ldots, \bar{x}'_{v_2}$ into the second layer, which produces o_2 linear equations in the o_2 variables $x_{v_2+1}, \ldots, x_{v_3}$. The solution of this linear system gives values $\bar{x}'_{v_2+1}, \ldots, \bar{x}'_{v_3}$ for $x_{v_2+1}, \ldots, x_{v_3}$. We repeat this procedure for each successive layer until we find the desired solution $\bar{X}' = (\bar{x}'_1, \ldots, \bar{x}'_n)$ of $F(x_1, \ldots, x_n) = \bar{Y}'$.

4.) If at any level the associated linear system does not have a solution, then we need to start over at the first layer after choosing a new set of values $\bar{x}'_1, \ldots, \bar{x}'_{v_1}$ for $x_1, \ldots, x_{v_1}$. From [Patarin, 1997] it is expected that with a very high probability we will eventually succeed if the number of layers is not too large.

5.) Finally, we compute

$$X' = L_2^{-1}(\bar{X}') = (x'_1, \ldots, x'_n),$$

which is the signature of $Y' = (y'_1, \ldots, y'_{n-v_1})$.

In order to sign a large document, we can go through the same procedure as is done in Flash [Patarin et al., 2001] by first applying a hash function and then signing the hash value of the document.

Signature Verification

In order to verify that $X' = (x'_1, \ldots, x'_n)$ is the signature of $Y' = (y'_1, \ldots, y'_{n-v_1})$, we only need to check that

$$\bar{F}(X') = Y'.$$

If this equation is true, then the signature is valid; otherwise we reject the signature.

Choice of Parameters for a Practical Implementation

For a practical implementation we choose the finite field $k = GF(q)$, with $q = 2^8$. Let $n = 37$ so that $S = \{1, 2, 3, \ldots, 37\}$ and let $u = 5$. The parameters associated with the layers of the Rainbow are as follows.

$$\begin{aligned} v_1 &= 10, & o_1 &= 10, \\ v_2 &= 20, & o_2 &= 4, \\ v_3 &= 24, & o_3 &= 3, \\ v_4 &= 27, & o_4 &= 10, \\ v_5 &= 37. \end{aligned}$$

Both maps $\bar{F}$ and F are maps from k^{37} to k^{27}.

The public key consists of 27 quadratic polynomials in 37 variables, each of which has $(38 \times 39)/2$ coefficients. Therefore the public key will require roughly 16 KB of memory.

The private key consists of: 10 polynomials in 27 Vinegar variables and 10 Oil variables, 3 polynomials in 24 Vinegar variables and 3 Oil variables, 4 polynomials in 20 Vinegar variables and 4 Oil variables, 10 polynomials in 10 Vinegar and 10 Oil variables, and two invertible affine transformations $L_1 : k^{27} \longrightarrow k^{27}$ and $L_2 : k^{37} \longrightarrow k^{37}$. Therefore the total size of the private key is roughly 10 KB.

This signature scheme signs a document of size $8 \times 27 = 216$ bits with a signature of size $8 \times 37 = 296$ bits. The parameters given here are slightly different from those presented in [Ding and Schmidt, 2005b].

3.4 Security Analysis of Rainbow

We now present the security analysis of a Rainbow signature scheme with the choice of parameters mentioned in the previous section. There are several possible attacks which we deal with one by one. For those methods where quadratic forms are used, one should recall that the theory of quadratic forms over a finite field with characteristic two is different from that of the case when the characteristic is odd [Dickson, 1909].

Rank Reduction

In [Coppersmith et al., 1997] a method of rank reduction is used to break the birational permutation signature scheme of Shamir. The main reason this attack works is that the space spanned by the polynomial components of Shamir's cipher consists of a flag of spaces:

$$V_1 \subset V_2 \subset \cdots \subset V_t,$$

where V_t is the space spanned by the polynomial components of the cipher, and each V_i is a proper subset of V_{i+1}. Also, the rank of the bilinear form corresponding to any element in $V_{i+1} - V_i$ is strictly larger than the rank of the bilinear form corresponding to every element in V_i, and the difference between the dimension of V_i and the dimension of V_{i+1} is exactly one. Due to these properties, in particular the last one, this flag of spaces is easily found. We proceed by first finding V_{n-1}, then V_{n-2}, and so on by rank reduction.

This attack will not work against Rainbow, even though there also exists a flag of spaces for Rainbow. This flag of spaces will satisfy the following properties:

1.) The number of components is exactly the number of layers, so we have

$$V_1 \subset V_2 \subset \cdots \subset V_{u-1};$$

2.) The difference between the dimensions of the last two spaces V_{u-1} and V_{u-2} is exactly o_{u-1}, which we have specifically chosen to be the relatively large number 11 compared to Shamir's cipher where it is one.

The second property above is the main reason why the attack in [Coppersmith et al., 1997] cannot be applied to Rainbow. The rank reduction method cannot be used here since $o_{u-1} = 10$ and this value is too large for the attack to succeed. In other words, the last layer of Oil is sufficiently "thick" to resist the rank reduction attack.

Attacks on Oil-Vinegar

It is clear that the purpose of L_1 is to mix the polynomial components of F. Therefore, each component of the cipher $\bar{F}$ belongs to the top layer of Oil-Vinegar polynomials, and thus they are all elements in P_4. In particular, they are Oil-Vinegar polynomials with 27 Vinegar variables and 10 Oil variables. An attempt to apply the method in [Kipnis et al., 1999] that was used to attack unbalanced Oil-Vinegar is an attempt to discover the Oil-Vinegar variables of the last layer. According to the cryptanalysis in [Kipnis et al., 1999], the attack complexity of this first step would be $q^{27-10-1} \times 10^4 \gg 2^{100}$. Thus, Rainbow is safe from the attack on unbalanced Oil-Vinegar.

MinRank Attack

There are two distinct ways of using the MinRank attack on Rainbow. The first one is to search for the polynomial whose associated matrix has the lowest rank among all possible choices. This polynomial must be in

P_1, the first layer with 10 Vinegar variables and 10 Oil variables, denoted by $\tilde{F}_1$. To do this, we first associate to each polynomial a bilinear form, which in turn has an associated 37×37 matrix. We then consider linear combinations of the matrices associated with each component of $\bar{F}$ to find a polynomial whose associated matrix has rank 20. Thus, the attack on Rainbow amounts to the problem of finding a rank 20 matrix among a group of 27 matrices each of size 37×37. From the results in [Courtois, 2001] we see that the complexity to find such a matrix is $q^{20} \times 27^3$, which is much larger than 2^{100}.

Another possibility is to search for polynomials in the layer P_3 which are linear combinations of the components of $\tilde{F}_i$ for $i < 4$. In this case, the MinRank method fails because such polynomials have associated matrices of rank 26 in general. One way to proceed is by random search, but because the dimension of P_3 is 17, this becomes a problem of searching for an element in a subspace of dimension 17 inside a larger space of dimension 27. Such a random search is likely to have at least q^{10} failures before finding one success, but then we also need to determine if indeed the rank is less than 27 for each search. For such an attack the total complexity will be at least $q^{10} \times (27 \times 37^2/3) > 2^{90}$. This attack idea is actually related to the method in [Coppersmith et al., 1997], which we have just seen cannot be applied to Rainbow.

From the most recent results in this direction, [Wolf et al., 2004] presents a study of a very general system called STS. Attacks on STS are applicable to Rainbow. However, according to their estimates the security of our system is at least 2^{100}.

Attacks that Exploit the Multilayer Structure of Rainbow

In [Patarin, 1995], Patarin realized that if the cipher is made of several independent parallel branches, then we can perform a separation of variables such that each polynomial in the cipher is derived as a linear combination of polynomials over a group of variables. This property can be used to attack the system. At first glance, one could think that Rainbow's layers resemble branches. Nevertheless, one should realize that Rainbow's layers are in no way independent since each layer is built upon the previous one. In simple terms one can say that all layers stick together and there is no known way to perform any kind of separation of variables. This is made clear by looking at the polynomials in the last layer P_4. Therefore, an attack attempting to make use of the property of the parallel independent branches as in [Patarin, 1995] will not work on Rainbow. Similarly, one can argue that the attack using syzygies will not work here due to the fact that there are no independent branches.

General Methods

Other methods that could be used to attack Rainbow are those that attempt to solve the associated polynomial equations directly, for example with the XL-family of algorithms or with Gröbner bases. However, it is very difficult to solve a set of 27 equations in 37 variables over $GF(2^8)$. Since the system is under-determined there will be too many solutions that the algorithm will have to find. In general, it is much better to solve an equation with only one solution. Still, because of the design of Rainbow, one could try and guess the values for any set of $v_1 = 6$ variables. There would then be a probability of $1/e < 1/2.71828 < 0.37$ of having a unique solution. Now the problem becomes a problem to solve a set of 27 quadratic equations with 27 variables. It is reasonable to think of it as if we have a set of randomly chosen quadratic equations. According to what is commonly believed, the complexity of solving this problem is at least $2^{3 \times 27} > 2^{81}$. By considering each attack so far, we can conclude that the security of Rainbow is at least 2^{80}.

General Security Analysis

There are two natural ways to proceed with an attack on Rainbow. One is from the first layer, the other is from the last layer. The efficiency of an attack from the last layer seems to depend on how effectively the MinRank attack can be used. This attack complexity will be in general $q^{v_2-1}o_{u-1}^3$ if $v_1 > o_1$, or $q^{2v_1}o_{u-1}^3$ if $v_1 \leq o_1$. From this it is clear that $v_2 = o_1 + v_1$ should not be too small. In particular, results from [Wolf et al., 2004] indicate that the security of our system is at least $(n - v_1) \times n^3 \times q^{o_1+v_1} \times u$, which again means that $o_1 + v_1$ should not be too small.

As for the attack from the first layer, the attack on unbalanced Oil-Vinegar shows us that $v_{u-1} - o_{u-1}$ should not be too small. Also o_{u-1} should not be too small in order to avoid a random search attack.

3.5 Comparison with other Multivariate Signature Schemes

In this section we highlight the differences between Rainbow and two similar multivariate cryptosystems: unbalanced Oil-Vinegar and Sflash.

Comparison with Unbalanced Oil-Vinegar

Clearly Rainbow is a generalization of the original Oil-Vinegar construction. More specifically, unbalanced Oil-Vinegar is a single-layer Rainbow scheme with $u = 2$. For the sake of comparison, let us assume that we want to build an unbalanced Oil-Vinegar scheme that has the same document length as our practical example above. In this case, we choose k to be a finite field of size $q = 2^8$, and the number of Oil variables should be 27. Because of the attack on unbalanced Oil-Vinegar schemes [Kipnis et al., 1999], we know that the number of Vinegar variables should be at least $27 + 10 = 37$ in order to have the same level of security as our example of Rainbow.

In this case, the public key consists of 27 polynomials with $37 + 27 = 64$ variables. Each polynomial in the public key then has $(65 \times 66)/2$ coefficients, which amounts to a public key size of roughly 59 KB, about 3.5 times the size of our practical example. This implies that the public computation of verifying the signature should take at least 3 times long.

The private key for the unbalanced Oil-Vinegar scheme consists of one affine linear transformation on k^{64} and a set of 27 Oil-Vinegar polynomials with 27 Oil variables and 37 Vinegar variables. This means that the private key is about 40 KB. This implies that the private calculation to sign the document will take about four times longer compared to our example.

The private key for the unbalanced Oil-Vinegar consists of an Oil-Vinegar map of the same size as the public key, plus the invertible affine transformation $L : k^{65} \longrightarrow k^{65}$. Thus the private key needs roughly 63 KB, and so the signature process will take roughly 6.3 times longer than with the Rainbow example.

The length of the signature is $8 \times 64 = 512$ bits, which is also about twice the size of the signature in our example of Rainbow. From this, we conclude that Rainbow generally compares favorably with unbalanced Oil-Vinegar in terms of both security and efficiency.

Comparison with Sflash

Since Sflash^{v2} is again considered to be secure, we compare Rainbow to Sflash^{v2} and not to the newer version Sflash^{v3} [Courtois et al., 2003b], which has a signature length of 469 bits and a public key of 112 KB. Sflash^{v2} has a signature of length $7 \times 37 = 259$ for a document of $7 \times 26 = 182$ bits. The Rainbow example has a signature of length $8 \times 37 = 296$ for a document of $8 \times 27 = 216$ bits. Clearly, in terms of per-bits efficiency the two are essentially the same.

For a comparison of the running times on a PC, we implemented Sflash^{v_2} as described in [Akkar et al., 2003]. The generation of the signature is about twice as fast for the example with Rainbow as compared to that with Sflash. The time required for signature verification is of course nearly identical. From this, we conclude that Rainbow compares favorably with Sflash in both security and efficiency.

Comparison with TTS

We could also compare Rainbow with the new TTS schemes of [Yang et al., 2004a]. However, the first schemes are broken as was shown in [Ding and Yin, 2004]. We can also see that the Tractable Rational Map Signature scheme presented in [Wang et al., 2005] is very similar to TTS. In fact both TTS and TRMC can be viewed as a very special example of Rainbow, though they are built from different ideas (triangular construction). We will discuss the relations between Rainbow and TRMC further in Chapter 6.

3.6 Optimization and Generalization of Rainbow

There is a great amount of freedom in how we can construct a specific implementation of Rainbow, so naturally there is a question of how to choose an optimal scheme. In the practical example given above, we presented a very simple realization of Rainbow for illustrative purposes only. In this section, we will consider how to optimize the scheme in terms of both key size and computational efficiency, given a fixed security requirement.

Let us assume that we want to build a Rainbow system to sign a document of size $m \times r$ bits in the space k^n, where k is a finite field of size $q = 2^r$. Let us also fix the security requirement to be at 2^θ.

For a document of length m, the length of the signature is $v_u = m+v_1$. (The notation o_i, v_j, and u is from the definition of Rainbow in Section 3.) Security from the MinRank attack requires $2^{3r(v_2-1)} > 2^\theta$. We should choose $v_1 > o_1$ to make the system more efficient, and from this we know that $v_2 = o_1 + v_1$ should be at least $1+\theta/3r$. But if we want to make the signature as short as possible, the private key as small as possible, and the private calculations as easy as possible, we can see that we should choose v_1 and o_1 such that the difference between o_1 and v_1 should be 0 or 1.

Now assume that we have fixed v_2, o_1, and v_1 already. Due to the security requirement, we know that we should make sure that $q^{v_u - v_{u-1} - 1} o_{u-1}^4$ is larger than 2^θ.

Let us assume that we have chosen $v_u - v_{u-1}$. The next choice are the in-between layers. The best choice is $v_{i+1} = v_i + 1$, as it has the shortest secret key, the fastest computation speed and it does not affect at all the security of the system. In this case each $\tilde{F}_i$ has only one polynomial.

We suggest a further improvement of the scheme with an even better choice. For this we set all coefficients of any quadratic term to zero, which mixes the one Oil variable with its Vinegar variables at its layer, and only the coefficient of the linear term of Oil variable is chosen to be a nonzero element. This will ensure that the corresponding linear equation in the signing process always has a solution. It also makes the process faster and does not at all affect the security. We call this type of polynomial a linear Oil-Vinegar polynomial.

If we want to achieve the highest probability for success in finding a signature, even the lowest layer should have the same construction; namely, $v_2 - v_1 = 1$ and the Oil-Vinegar polynomial is chosen in the same way. In this case, the only possible place for the signing process to fail will be the top layer. This type of construction, can be viewed also as a combination of the Oil-Vinegar method with the method first suggested in [Shamir, 1993].

As for the case of an attack from the top, the attack method for unbalanced Oil-Vinegar method tells us that $v_{u-1} - o_{u-1}$ cannot be too small. Also, to avoid a random search attack o_{u-1} should not be too small.

For example, we can improve our practical example for $u = 13$ with the choices $v_1 = 10$, $v_2 = 20$, $v_3 = 21$, $v_4 = 22, \ldots, v_{19} = 27$, $v_{13} = 37$, $o_1 = 10$, $o_2 = 1, \ldots, o_9 = 1$, and $o_{10} = 10$. This now is a 10 layer Rainbow scheme.

Another possibility for optimization is to use sparse polynomials when we choose at random the coefficients of the Oil-Vinegar polynomials. This idea was first proposed in TTS [Yang et al., 2004a]. Nevertheless, this is a very subtle and delicate task, as it opens up the possibility of new, often hidden and unexpected weakness. The use of sparse polynomial in the new TTS scheme caused it to be broken in [Ding and Yin, 2004]. Therefore such a method should be used very carefully. In particular one should show that using special sparse polynomials does not affect the security level of a cryptosystem.

Chapter 4

HIDDEN FIELD EQUATIONS

After the direct generalization of the Matsumoto-Imai cryptosystems using the Plus-Minus method, Patarin pushed his work one step further by looking for ways to replace the map $\tilde{F}$ with something else that could make the cryptosystem more secure. Eventually he invented the Hidden Field Equation cryptosystem (HFE) [Patarin, 1996b], which Patarin believed could be the strongest multivariate scheme at the time he proposed it.

The design of HFE depends on a parameter d which determines the efficiency of the cryptosystem. However, Kipnis and Shamir [Kipnis and Shamir, 1999] found a way of obtaining the secret key with the help of the MinRank method when d is sufficiently small. Later, Courtois improved the Shamir-Kipnis attack and presented two new more efficient attacks [Courtois, 2001].

Attacks also exist that use general methods for solving polynomial equations, and these seem to be closely related to the direct attacks of Kipnis and Shamir. For example, Faugère used his new Gröbner basis algorithm F_5 to break the so-called "HFE Challenge 1" of Patarin in 96 hours on an 833 MHz Alpha workstation with 4 GB of memory [Faugère, 2003]. The experimental data in [Daum and Felke, 2002] shows that the complexity of the Buchberger algorithm applied to HFE depends very strongly on the parameter d associated with the hidden polynomial $\tilde{F}$. This has also been confirmed in [Faugère and Joux, 2003].

The HFE cryptosystem can also be generalized by the methods used to create variants of the Matsumoto-Imai cryptosystem, producing HFE$^-$ (HFE-Minus) or HFE$^\pm$ (HFE-Plus-Minus). However, more attention has been paid to other generalizations of HFE, such as HFEv and HFEv$^-$

(HFEv-Minus). These variants are constructed by combining the idea of Oil-Vinegar and the Minus method.

In this chapter we will focus on the basic HFE construction, the Kipnis-Shamir and Courtois attacks on HFE and HFEv and related attacks. We will leave the more general attacks with the Gröbner basis method to Chapter 7.

4.1 Basic HFE

We will use the same basic notation as in Chapter 2. Let k be a finite field with cardinality q and let K be a degree n extension of k. Unlike Matsumoto-Imai, HFE does not require k to have characteristic two. If $g(x) \in k[x]$ is an irreducible polynomial of degree n, then $K \cong k[x]/g(x)$. Let ϕ be the standard k-linear map that identifies K with k^n; that is, $\phi : K \longrightarrow k^n$, where

$$\phi(a_0 + a_1 x + a_2 x^2 + \cdots + a_{n-1} x^{n-1}) = (a_0, a_1, a_2, \ldots, a_{n-1}).$$

The design of HFE is very similar to that of the Matsumoto-Imai cryptosystem. The main difference is that the Matsumoto-Imai map $\tilde{F} = X^{q^\theta+1}$ is replaced with a new map

$$\tilde{F}(X) = \sum_{i=0}^{r_2-1} \sum_{j=0}^{i} a_{ij} X^{q^i+q^j} + \sum_{i=0}^{r_1-1} b_i X^{q^i} + c, \tag{4.1}$$

where the coefficients $a_{ij}, b_i, c \in K$ are randomly chosen, and r_1, r_2 are chosen so that the degree of $\tilde{F}$ is less than some parameter d. The public key polynomials will be the components of

$$\bar{F} = L_1 \circ F \circ L_2,$$

where $F = \phi \circ \tilde{F} \circ \phi^{-1}$, and L_1, L_2 are secret invertible affine transformations on k^n.

Public Key

The public key includes the following information:

1.) The field k, including its additive and multiplicative structure;

2.) The map $\bar{F}$, or equivalently, its n total degree two components $\bar{f}_1(x_1, \ldots, x_n), \ldots, \bar{f}_n(x_1, \ldots, x_n) \in k[x_1, \ldots, x_n]$.

Private Key

The private key includes the following information:

1.) The map $\tilde{F}$;

2.) The two invertible affine transformations L_1, L_2.

Note that unlike the case of the Matsumoto-Imai cryptosystem, $\tilde{F}$ is virtually impossible to guess.

Encryption

Given a plaintext message $(x'_1, \ldots, x'_n)$, the corresponding ciphertext is:

$$(y'_1, \ldots, y'_n) = \bar{F}(x'_1, \ldots, x'_n),$$

or equivalently

$$y'_i = \bar{f}_i(x'_1, \ldots, x'_n) \quad \text{for } i = 1, \ldots, n.$$

Decryption

Given the ciphertext $(y'_1, \ldots, y'_n)$, decryption includes the following steps.

1.) Compute $(\bar{y}_1, \ldots, \bar{y}_n) = L_1^{-1}(y'_1, \ldots, y'_n)$.

2.) Let $\bar{Y} = \phi^{-1}(\bar{y}_1, \ldots, \bar{y}_n)$. Compute the set

$$\mathcal{Z} = \{Z \in K \mid \tilde{F}(Z) = \bar{Y}\}.$$

To compute $\mathcal{Z}$ we will use a variant of the Berlekamp algorithm suitable for use over the field K. If $d = \deg \tilde{F}(X)$, then the complexity of this step will be $O(nd^2 \log d + d^3)$; or we can use an even more efficient method by first finding the gcd of this polynomial with $X^{q^n} - X$ [Geddes et al., 1992] with slightly lower complexity. From this we see that the degree of $\tilde{F}$ cannot be too large, since otherwise the decryption process is inefficient. Equivalently, we must not choose r_1, r_2 to be too large.

3.) For each element $Z_i \in \mathcal{Z}$, compute

$$(x_{i1}, \ldots, x_{in}) = L_2^{-1} \circ \phi(Z_i).$$

Although we would like that $\tilde{F}$ is a one-to-one map; that is, there is only one element in $\mathcal{Z}$, it is possible that $\mathcal{Z}$ has multiple elements. In this case we can use one of several techniques (hash functions, Plus method, etc.) to detect the plaintext among the solutions.

Toy Example for HFE

We will again use the finite field $k = GF(2^2)$ for which addition and multiplication were given in Table 2.1. In the Matsumoto-Imai cryptosystem, n cannot be a power of two, but for HFE such a restriction does not apply. We have chosen $n = 4$ for this toy example. For the irreducible polynomial $g(x)$ we use:

$$g(x) = x^4 + x^3 + \alpha^2 x^2 + \alpha^2 x + \alpha^2,$$

and we use the following k-linear affine transformations

$$\begin{aligned}
L_1(x_1, x_2, x_3, x_4) &= \begin{pmatrix} \alpha & \alpha & 0 & \alpha \\ 0 & \alpha & 1 & 0 \\ 1 & \alpha & \alpha & \alpha^2 \\ 1 & \alpha & 0 & \alpha^2 \end{pmatrix} \begin{pmatrix} x_1 \\ x_2 \\ x_3 \\ x_4 \end{pmatrix} + \begin{pmatrix} 0 \\ 0 \\ \alpha \\ 0 \end{pmatrix}, \\
L_2(x_1, x_2, x_3, x_4) &= \begin{pmatrix} 1 & 0 & \alpha^2 & 1 \\ \alpha^2 & 1 & 1 & \alpha \\ 1 & \alpha^2 & 1 & \alpha^2 \\ 1 & \alpha & \alpha^2 & 1 \end{pmatrix} \begin{pmatrix} x_1 \\ x_2 \\ x_3 \\ x_4 \end{pmatrix} + \begin{pmatrix} \alpha^2 \\ \alpha^2 \\ 0 \\ 0 \end{pmatrix}.
\end{aligned}$$

For the function in (4.1) we select

$$\tilde{F}(X) = X^{4+4} + \alpha X^{4+1} + X + 1 = X^8 + \alpha X^5 + X + 1,$$

so that the upper limits for the two sums are $r_1 = 1$ and $r_2 = 2$ and $d = 8$. From the composition $\bar{F} = L_1 \circ F \circ L_2$ we find the public key

$$\begin{aligned}
y_1 &= \alpha^2 x_1 x_2 + \alpha x_1 x_3 + \alpha^2 x_1 x_4 + \alpha x_1 + x_2 x_3 + \alpha^2 x_2 x_4 + \alpha x_3^2 \\
&\quad + \alpha x_3 x_4 + x_4^2 + \alpha, \\
y_2 &= x_1^2 + \alpha x_1 x_2 + \alpha^2 x_1 x_4 + \alpha^2 x_1 + x_2 x_3 + \alpha x_2 x_4 + \alpha^2 x_3^2 + x_3 \\
&\quad + \alpha^2 x_4^2 + \alpha^2 x_4, \\
y_3 &= \alpha^2 x_1^2 + x_1 x_2 + \alpha^2 x_1 x_4 + \alpha^2 x_1 + x_2^2 + \alpha x_2 + \alpha x_3^2 + x_3 + \alpha^2 x_4^2, \\
y_4 &= \alpha^2 x_1^2 + x_1 x_2 + \alpha^2 x_1 x_3 + \alpha x_1 + x_2 x_3 + \alpha x_2 x_4 + \alpha x_2 + x_3^2 \\
&\quad + \alpha x_3 x_4 + x_3.
\end{aligned}$$

The encryption is a straightforward evaluation of the public key. For example, if $(x_1', x_2', x_3', x_4') = (0, \alpha^2, 1, \alpha)$ is selected as the plaintext, then $(y_1', y_2', y_3', y_4') = (0, 0, \alpha, \alpha^2)$ is the corresponding ciphertext.

The first step of the decryption process finds

$$L_1^{-1}(0, 0, \alpha, \alpha^2) = (1, 1, \alpha, 0),$$

which corresponds to $1 + x + \alpha x^2$ in K under the transformation ϕ^{-1}.

The next step is to find the solutions of the polynomial equation

$$X^8 + \alpha X^5 + X + 1 = 1 + x + \alpha x^2.$$

In this case, the only solution is $X = \alpha + \alpha x + \alpha x^2$. Therefore

$$\mathcal{Z} = \{\alpha + \alpha x + \alpha x^2\}$$

has only one element. It gets mapped back into k^4 with the help of the map ϕ to give $(\alpha, \alpha, \alpha, 0)$.

In the third step the inverse of the transformation L_2 is applied to find

$$L_2^{-1}(\alpha, \alpha, \alpha, 0) = (0, \alpha^2, 1, \alpha),$$

and the original plaintext is recovered.

When $(x'_1, x'_2, x'_3, x'_4) = (1, \alpha^2, 1, \alpha)$ is selected to be the plaintext, then $(y'_1, y'_2, y'_3, y'_4) = (\alpha^2, \alpha, 0, \alpha)$ is the corresponding ciphertext and $L_1^{-1}(\alpha^2, \alpha, 0, \alpha) = (1, 1, 0, \alpha)$. But this time the polynomial equation

$$X^8 + \alpha X^5 + X + 1 = 1 + x + \alpha x^3$$

has three solutions, so that

$$\mathcal{Z} = \{\alpha^2 + x + \alpha^2 x^2 + x^3, \alpha^2 + x + x^2 + \alpha x^3, 1 + \alpha x^2 + \alpha x^3\}.$$

Applying $L_2^{-1} \circ \phi$ to each value in $\mathcal{Z}$ gives three candidates for the plaintext; that is:

$$\{(1, \alpha^2, 1, \alpha), (\alpha^2, 1, 1, 0), (0, 0, 1, 1)\}.$$

In order to determine which of these three possibilities is the original plaintext, additional information is needed. One suggestion is to add redundant information that can be used in order to decide which is the original plaintext.

Investigating all 256 possible plaintexts in our example we find that in 110 cases there is a unique solution, in 32 cases there are two, in 66 cases there are three, in 40 cases there are five and in 8 cases there are 8 different solutions. These results are of course specific for the function $\tilde{F}(X)$ that we have used. Nevertheless, this indicates that when choosing a non-linear function in K at random it is very unlikely that it will be one-to-one. When HFE is used for encryption this has to be taken into consideration. When HFE is used for signatures, then this is less of a problem since any of the solutions can be selected to create the signature. On the other hand, it is possible that for some documents $(y'_1, \ldots, y'_n)$ the polynomial equation $\tilde{F}(X) = \bar{Y}$ has no solutions, and in this case the document has to be modified slightly so that it can be signed.

4.2 Attacks on HFE

In this section we present the Kipnis-Shamir attack HFE. It is the basis of the new attack on MI presented in Chapter 2, and therefore we refer the reader to Section 2.4 for notation and other necessary results.

The key idea of the Kipnis-Shamir attack is to lift the scheme back to its origins. The construction of the public key of HFE is based on the idea that we look for a map on an extension field, and then reduce it to a map on a vector space over the smaller field. The Kipnis-Shamir attack proceeds by moving the problem back to the extension field, where all the underlying structure can be seen. This is a very natural approach if we intend to exploit the design structure of HFE in the attack. In Section 2.4, we use exactly the same idea to find another attack on MI.

From a general point of view, the first step of the attack is to lift $\bar{F}$ up to a map over K. As with the attack on MI, we may as well assume that this is the map $\phi^{-1} \circ \bar{F} \circ \phi$, since any other lifting map $\psi : K \longrightarrow k^n$ will correspond to equivalent secret keys L_1', L_2' and $\tilde{F}'$. Also, we simplify matters by assuming that the components of $\bar{F}$ are homogeneous of degree two. In this case, we relabel $r_2 = r$.

The key difference between the attacks on MI and HFE is that the $n \times n$ symmetric matrix corresponding to $\tilde{F}$ as in Section 2.4 is of the form

$$\tilde{\mathsf{F}} = \begin{pmatrix} \mathsf{A} & 0 \\ 0 & 0 \end{pmatrix},$$

where A is a randomly chosen $r \times r$ matrix. This means that instead of $\tilde{\mathsf{F}}$ having rank two, we can assume that $\tilde{\mathsf{F}}$ very likely has rank r. Thus,

$$\mathsf{M} = \tilde{\mathsf{F}}' = \mathsf{L}_2^T \, \tilde{\mathsf{F}} \, \mathsf{L}_2$$

also very likely has rank r. Since we also have

$$\mathsf{M} = \bar{\mathsf{F}}'' = \sum_{l=0}^{n-1} L_{1l}^{-1} \bar{\mathsf{F}}_l,$$

we proceed with the MinRank attack to recover L_1^{-1}, and hence L_1. As before, we generate the determinant of each $(r+1) \times (r+1)$ submatrix of M and set these equal to zero. This gives us $\binom{n}{r+1}(\binom{n}{r+1} - 1)/2$ degree $r+1$ equations in the n coefficients of L_1^{-1}. Let $(L_{10}^{-1}, \ldots, L_{1n-1}^{-1})$ be any solution of this MinRank problem.

To finish the attack, we must find L_2. We now know M, so we must consider the equation

$$\mathsf{M} = \mathsf{L}_2^T \, \tilde{\mathsf{F}} \, \mathsf{L}_2.$$

Let $\{u_1, \ldots, u_{n-r}\}$ be a basis for the left kernel of M. Since L_2 is invertible, and for simplicity assumed to be linear, it follows that

$$u_i \, L_2^T \, \tilde{F} = 0.$$

Due to the special form of $\tilde{F}$, it must be that

$$u_i \, L_2^T = (0, \ldots, 0, *, \ldots, *),$$

where the first r coordinates are zero and the last $n - r$ coordinates are arbitrary. Thus, for each $i = 1, \ldots, n-r$, we get r equations of the form

$$\sum_{j=0}^{n-1} u_{ij} \, L_{2\,j-l}^{q^l} = 0,$$

where $l = 0, 1, \ldots, r-1$. Raising this equation to the power q^{n-l} yields

$$\sum_{j=0}^{n-1} u_{ij}^{q^{n-l}} \, L_{2\,j-l} = 0,$$

and we obtain $r(n-r)$ linear equations in the variables $L_{20}, \ldots, L_{2n-1}$. So long as $r(n-r) \geq n$, or equivalently $n \geq r^2/(r-1)$, it is very likely that we can solve this system for the $L_{20}, \ldots, L_{2n-1}$, and finally recover L_2.

The computational complexity of this attack is mainly determined by the computational complexity of the MinRank attack. However, using Courtois' method [Courtois, 2001], the complexity of this step is $O(n^r)$, which is manageable for small r. This method is actually different from, but more efficient than, the original method proposed by Kipnis-Shamir which uses the relinearization method.

4.3 Variants of HFE

The first variants that can be constructed are the Plus and Minus variants, just as in the case of MI. For example, HFE^- can be used as a signature scheme, and $\text{HFE}^\pm$ for encryption schemes.

A more sophisticated variant comes from the combination of basic HFE with the idea of Oil-Vinegar. This variant is called HFEv. Using Lemmas A.0.1–A.0.3 in Appendix A, we will present HFEv in a slightly different though equivalent form from its original presentation. Further extensions of this construction include the Quartz signature scheme, an example of HFEv-Minus [Patarin, 1996b].

The function $\tilde{F}$ is replaced with the function $\tilde{F} : K \times k^v \longrightarrow K$ defined by

$$\tilde{F}(X, \check{x}_1, \ldots, \check{x}_v) = \sum_{i=0}^{r_2-1} \sum_{j=0}^{i} a_{ij} X^{q^i+q^j} + \sum_{i=0}^{r_1} b_i X^{q^i} + \sum_{i=0}^{r_1} c_i \Omega_i(\check{x}_1, \ldots, \check{x}_v) X^{q^i} + \Gamma(\check{x}_1, \ldots, \check{x}_v),$$

where $a_{ij}, b_i, c_j \in K$, and $\Omega_i : k^v \longrightarrow K$ and $\Gamma : k^v \longrightarrow K$ are arbitrary linear and quadratic maps, respectively, in the input variables $(\check{x}_1, \ldots, \check{x}_v)$. The variables associated with X act like Oil variables, while the variables $(\check{x}_1, \ldots, \check{x}_v)$ act like Vinegar variables. If $v = 0$ then HFEv reduces to HFE.

If $\pi : k^v \longrightarrow k^n$ is the embedding $\pi(a_1, \ldots, a_v) = (a_1, \ldots, a_v, 0, \ldots, 0)$, then we can rewrite $\tilde{F}$ as a map from $K \times K$ to K:

$$\tilde{F}(X, V) = \sum_{i=0}^{r_2-1} \sum_{j=0}^{i} a_{ij} X^{q^i+q^j} + \sum_{i=0}^{r_1-1} b_i X^{q^i} + \sum_{i=0}^{r_1} \sum_{j=0}^{n-1} c_{ij} X^{q^i} V^{q^j} + \sum_{i=0}^{n-1} \sum_{j=0}^{i} \alpha_{ij} V^{q^i+q^j} + \sum_{i=0}^{n-1} \beta_i V^{q^i} + \gamma$$

where $V = \phi^{-1} \circ \pi(\check{x}_1, \ldots, \check{x}_v)$, and $a_{ij}, b_i, c_{ij}, \alpha_{ij}, \beta_i, \gamma \in K$. Rewriting $\tilde{F}$ in this way will be key to the attack presented in the next section.

Now define the map $F : k^{n+v} \longrightarrow k^n$ by

$$F(x_1, \ldots, x_n, \check{x}_1, \ldots, \check{x}_v) = \phi \circ \tilde{F} \circ (\phi^{-1} \times \phi^{-1} \circ \pi)(x_1, \ldots, x_n, \check{x}_1, \ldots, \check{x}_v) = (f_1, \ldots, f_n),$$

where $f_1, \ldots, f_n \in k[x_1, \ldots, x_n, \check{x}_1, \ldots, \check{x}_v]$ are each of total degree two. Finally, let $L_1 : k^n \longrightarrow k^n$ and $L_2 : k^{n+v} \longrightarrow k^{n+v}$ be two randomly chosen invertible affine transformations, and define $\bar{F} : k^{n+v} \longrightarrow k^n$ by

$$\bar{F} = L_1 \circ F \circ L_2 = (\bar{f}_1, \ldots, \bar{f}_n),$$

as before. We now summarize the HFEv signature scheme.

Public Key

The public key includes:

1.) The field k, including its additive and multiplicative structure;

2.) The map $\bar{F}$, or equivalently, its components $\bar{f}_1, \ldots, \bar{f}_n$.

Private Key

The private key includes:

1.) The map $\tilde{F}$;

2.) The two invertible affine transformations L_1 and L_2.

Signature Verification

Given a message $M = (m_1, \ldots, m_n)$ and signature $S = (s_1, \ldots, s_{n+v})$, we conclude that S is a valid signature for M only if

$$\bar{F}(S) = M.$$

Signature Generation

To sign the message $M = (m_1, \ldots, m_n)$, we perform the following steps:

1.) Compute $\bar{M} = (\bar{m}_1, \ldots, \bar{m}_n)$ by

$$\bar{M} = L_1^{-1}(M).$$

2.) Randomly choose $(\check{x}'_1, \ldots, \check{x}'_v) \in k^v$, let $V = \phi^{-1} \circ \pi(\check{x}'_1, \ldots, \check{x}'_v)$, and substitute V into $\tilde{F}(X, V)$. The result is a polynomial in X, denoted $\tilde{F}_V(X)$, whose roots can be found efficiently. In other words, it will be easy to find all the elements in the set $\tilde{F}_V^{-1}(Y)$, for a given $Y \in k^n$.

3.) Compute U such that

$$\tilde{F}_V(U) = \phi^{-1}(\bar{M}).$$

If no such U exists, then choose another $(\check{x}'_1, \ldots, \check{x}'_v) \in k^v$ and recompute $\tilde{F}_V(X)$; otherwise go to the next step.

We note that though it is possible that no such U exists (for a given V), this is very unlikely if v is large enough. In fact, we expect that the map $\tilde{F}(X, V)$ is a q^v-to-one map, so we should be able to find U with only a few tries.

4.) Compute a valid signature $S = (s_1, \ldots, s_{n+v})$ for $M = (m_1, \ldots, m_n)$ by first computing

$$\bar{S} = (u_1, \ldots, u_n, \check{x}'_1, \ldots, \check{x}'_v),$$

where

$$\phi(U) = (u_1, \ldots, u_n),$$

and then taking

$$S = L_2^{-1}(\bar{S}).$$

HFEv can be modified with either the Plus or Minus methods. HFEv can be used for encryption as well as for authentication so long as q^v is not too large. In this case we will need to do a little more work in the decryption step in order to detect the plaintext in a set of roughly q^v possible preimages of a given ciphertext.

4.4 Cryptanalysis of HFEv

Case: $v = 1$

In this section we will present the attack on HFEv from [Ding and Schmidt, 2005a], an extension of the Kipnis-Shamir attack on HFE. We begin by assuming that $v = 1$ and that the components of $\bar{F}$ are homogeneous degree two polynomials in $n+1$ variables. With these assumptions, the map $\tilde{F} : K \times K \longrightarrow K$ becomes:

$$\begin{aligned}\tilde{F}(X, V) = &\sum_{i=0}^{r_2-1} \sum_{j=0}^{i} a_{ij} X^{q^i+q^j} + \sum_{i=0}^{r_1-1} b_i X^{q^i} \\ &+ \sum_{i=0}^{r_1-1} c_i X^{q^i} V \\ &+ \alpha V^2 + \beta V + \gamma.\end{aligned}$$

There are no terms with V appearing to a power greater than two since V lies in a subfield of K isomorphic to k.

Now let $\hat{K}$ be the $n+1$ dimensional k-subspace in $K \times K$ such that for any element $\hat{X} = (X_1, X_2)$ we have

$$\phi(X_2) = (x, 0, \ldots, 0)$$

for some $x \in k$. Then the map $\tilde{F}(X, V)$ can be reinterpreted as a map from $\hat{K}$ to K.

Let $\pi : k^n \longrightarrow k$ be the projection $\pi(a_1, \ldots, a_n) = a_1$, and let $\psi = \phi \times (\pi \circ \phi)$ be the standard map from $\hat{K}$ to k^{n+1}. Then F is defined by

$$F = \phi \circ \tilde{F} \circ \psi^{-1},$$

and the public polynomials are given by the components of $\bar{F}$ defined by

$$\bar{F} = L_1 \circ F \circ L_2,$$

where $L_1 : k^n \longrightarrow k^n$, $L_2 : k^{n+1} \longrightarrow k^{n+1}$ are invertible affine transformations.

One way to attack the system is to find L_1 and L_2. Then we can compose $\bar{F}$ on the left with L_1^{-1} and on the right with L_2^{-1} to recover F. As in the Kipnis-Shamir attack, we begin by lifting $\bar{F}$ to $\hat{F} : \hat{K} \longrightarrow K$ by

$$\begin{aligned}\hat{F} &= \phi^{-1} \circ \bar{F} \circ \psi \\ &= \phi^{-1} \circ L_1 \circ F \circ L_2 \circ \psi \\ &= (\phi^{-1} \circ L_1 \circ \phi) \circ \tilde{F} \circ (\psi^{-1} \circ L_2 \circ \psi).\end{aligned}$$

Since we had assumed that the polynomial components of $\bar{F}$ are homogeneous of degree two, $\tilde{F} : K \times K \longrightarrow K$ has the simplified form:

$$\tilde{F}(X, V) = \sum_{i=0}^{r_2-1} \sum_{j=0}^{i} a_{ij} X^{q^i+q^j} + \sum_{i=0}^{r_1-1} c_i X^{q^i} V + \alpha V^2.$$

We know the general form of $\phi^{-1} \circ L_1^{-1} \circ \phi$ from Lemma A.0.1 and of $\psi^{-1} \circ L_2 \circ \psi$ from Lemma A.0.3. In particular, we have

$$\bar{L}_1^{-1}(X) = \phi^{-1} \circ L_1^{-1} \circ \phi(X) = \sum_{i=0}^{n-1} l_{1i} X^{q^i}$$

from Lemma A.0.1 and

$$\begin{aligned}\bar{L}_2(X, V) &= \psi^{-1} \circ L_2 \circ \psi(X, V) \\ &= (l'_{20} V + \sum_{i=0}^{n-1} l_{2i} X^{q^i}, l'_{21} V + \sum_{i=0}^{n-1} l_{2i}^{q^i} X^{q^i})\end{aligned}$$

from Lemma A.0.3. This means that

$$\hat{F}(X, V) = \sum_{i=0}^{n-1} \sum_{j=0}^{n-1} \hat{a}_{ij} X^{q^i+q^j} + \sum_{i=0}^{n-1} \hat{c}_i X^{q^i} V + \hat{\alpha} V^2.$$

Rather than directly solve the problem of finding L_1 and L_2, we will find $\bar{L}_1, \bar{L}_2$, from which we will then be able to recover L_1, L_2. The approach will be the same as that for HFE in the sense that we associate $(n+1) \times (n+1)$ matrices with the quadratic forms belonging to the maps $\hat{F}$ and $\tilde{F}$. Let $\hat{\mathrm{A}}$ and $\tilde{\mathrm{A}}$ be the matrices associated with $\hat{F}$ and $\tilde{F}$, respectively. Then we have

$$\hat{\mathsf{A}} = \begin{pmatrix} 0 & \hat{a}_{01}+\hat{a}_{10} & \cdots & \hat{a}_{0\,n-1}+\hat{a}_{n-1\,0} & \hat{c}_0 \\ \hat{a}_{01}+\hat{a}_{10} & \cdot & \cdots & \hat{a}_{1\,n-1}+\hat{a}_{n-1\,1} & \hat{c}_1 \\ \hat{a}_{02}+\hat{a}_{20} & \cdot & \cdots & \hat{a}_{2\,n-1}+\hat{a}_{n-1\,2} & \hat{c}_2 \\ \cdot & \cdot & \cdots & \cdot & \cdot \\ \cdot & \cdot & \cdots & \cdot & \cdot \\ \hat{a}_{0\,n-1}+\hat{a}_{n-1\,0} & \cdot & \cdots & 0 & \hat{c}_{n-1} \\ \hat{c}_0 & \hat{c}_1 & \cdots & \hat{c}_{n-1} & 0 \end{pmatrix},$$

and with $D = r_2$,

$$\tilde{\mathsf{A}} = \begin{pmatrix} 0 & a_{01}+a_{10} & \cdot & a_{0D}+a_{D0} & 0 & \cdot & 0 & c_0 \\ a_{01}+a_{10} & 0 & \cdot & a_{1D}+a_{D1} & 0 & \cdot & 0 & c_1 \\ a_{02}+a_{20} & \cdot & \cdot & a_{2D}+a_{D2} & 0 & \cdot & 0 & c_2 \\ \cdot & \cdot & \cdot & \cdot & \cdot & \cdot & \cdot & \cdot \\ a_{0D}+a_{D0} & \cdot & \cdot & 0 & 0 & \cdot & 0 & c_D \\ 0 & \cdot & \cdot & \cdot & \cdot & \cdot & 0 & c_{D+1} \\ \cdot & \cdot & \cdot & \cdot & \cdot & \cdot & \cdot & \cdot \\ 0 & \cdot & \cdot & \cdot & \cdot & \cdot & 0 & c_{n-1} \\ c_0 & c_1 & \cdot & \cdot & \cdot & \cdot & c_{n-1} & 0 \end{pmatrix}.$$

As in the attack on HFE, we can show that the matrix M associated with $\tilde{F} \circ \bar{L}_2$ is

$$\mathsf{M} = \mathsf{B}_2^T\, \tilde{\mathsf{A}}\, \mathsf{B}_2,$$

where

$$\mathsf{B}_2 = \begin{pmatrix} l_{20} & l_{21} & \cdots & l_{2\,n-2} & l_{2\,n-1} & l'_{20} \\ l_{2\,n-1}^q & l_{20}^q & \cdots & l_{2\,n-3}^q & l_{2\,n-2}^q & l'^q_{20} \\ l_{2\,n-2}^{q^2} & l_{2\,n-1}^{q^2} & \cdots & l_{2\,n-4}^{q^2} & l_{2\,n-3}^{q^2} & l'^{q^2}_{20} \\ \cdot & \cdot & \cdots & \cdot & \cdot & \cdot \\ l_{21}^{q^{n-1}} & l_{22}^{q^{n-1}} & \cdots & l_{2\,n-4}^{q^{n-1}} & l_{2\,n-3}^{q^{n-1}} & l'^{q^{n-1}}_{20} \\ l_2 & l_2^q & \cdots & l_2^{q^{n-2}} & l_2^{q^{n-1}} & l'_{21} \end{pmatrix}.$$

On the other hand, the matrix M is also associated with $\bar{L}_1^{-1} \circ \hat{F}$, and satisfies

$$\mathsf{M} = l_{10}\hat{\mathsf{A}} + l_{11}\hat{\mathsf{A}}_1 + \cdots + l_{1\,n-1}\hat{\mathsf{A}}_{n-1},$$

where $\hat{\mathsf{A}}_l$ corresponds to the polynomial $\hat{F}^{q^l}$, for $l = 0, \ldots, l-1$. We can see that:

$$[\hat{\mathsf{A}}_l]_{ij} = \begin{cases} [\hat{\mathsf{A}}]^{q^l}_{i-l \bmod n,\, j-l \bmod n} & \text{for } 0 < i, j < n+1; \\ [\hat{\mathsf{A}}]^{q^l}_{n+1,\, j-1 \bmod n} & \text{for } i = n+1,\ j < n+1; \\ [\hat{\mathsf{A}}]^{q^l}_{i-l \bmod n,\, n+1} & \text{for } i < n+1,\ i < n+1; \\ 0 & \text{if } i, j = n+1. \end{cases}$$

We are now in a position to execute the steps of the Kipnis-Shamir attack on HFE using the MinRank method, which was also modified for use against MI. The reader is referred to those sections for the details of how the attack proceeds from this point.

Case: $v > 1$

For the more general case of $v > 1$ the method above may be extended directly, and a rough estimate gives that the attack complexity is approximately $(n + v)^{3(r_2+v)+O(1)}$, though the details of the attack are much more complicated and need to be worked out carefully. This attack complexity depends on n, v, and r_2, and the exponent depends on r_2 and v. It would be much better if we could find some attack such that v is not in the exponent. But from the point view of symmetry, this is impossible. If we consider the case when v is large (greater than n), then the property of the HFEv polynomial should be dominated by the v Vinegar variables and these polynomials are more or less what can be treated as randomly chosen polynomials. From this point of view, we think that this attack complexity must include v in some way in the exponent and this attack method could be very close to what might be achieved in general. In addition, this attack could lead to some new ways of attacking HFEv using the XL family of methods; see [Courtois, 2001].

Quartz

Quartz is a HFEv$^-$ signature scheme; namely, it is a combination of HFEv with the Minus method. The basic idea was first suggested by Patarin [Patarin, 1996b]. In 2004 NESSIE recommended using Quartz for short digital signatures.

The parameters for a Quartz scheme are given as follows:

1.) q, the size of the small field k;

2.) d, the degree of the map $\tilde{F}$;

3.) n, the degree of the extension;

4.) v, the number of vinegar variables;

5.) and r the Minus number.

The parameters for the current version of Quartz are $q = 2$, $d = 129$, $n = 103$, $v = 4$, $r = 3$ [Patarin et al., 2001]. Therefore, we know that the public key gives a map from k^{107} to k^{100}. The signature generation takes about 9 seconds on a 500 MHz PC. Quartz is estimated to be at the security level of 2^{80}, but it requires a large amount of memory [Courtois, 2001].

At this point the best attacks on the the Quartz schemes are general attacks that use Gröbner basis algorithms [Faugère and Joux, 2003; Courtois, 2001; Courtois et al., 2003a].

Related Work

In [Felke, 2005] it is shown how to uncover the affine parts of the secret affine transformation for a certain class of HFE cryptosystems. This work is an extension of the work [Geiselmann et al., 2001]. Furthermore, it is shown that any system built on multi-branches can be decomposed into its individual branches in polynomial time on average.

In [Dobbertin, 2002], Dobbertin studied bijective power functions of higher degree, which can be viewed as a high degree form of MI or HFE. In [Michon et al., 2004], the method of binary decision diagram (BDD) was introduced to analyze the security of HFE. The BDD structure allows one to represent boolean functions with graphs, or more precisely, by trees. However, this method does not seem to be very effective.

Chapter 5

INTERNAL PERTURBATION

With all the variants discussed in the previous chapters, it may seem that all the possible extensions and generalizations of Matsumoto-Imai are exhausted. However, the construction of internal perturbation provides yet another alternative [Ding, 2004b]. The motivation for internal perturbation is to develop new constructions that can resist the algebraic attacks of [Patarin, 1995; Kipnis and Shamir, 1999] and the Gröbner basis attacks [Faugère, 2002; Faugère, 2003] without much sacrifice in the efficiency of the system.

From a very general point of view, the variants discussed in the previous chapters (for example Plus and Minus) can be interpreted as an extension of a commonly used idea in mathematics and physics; namely, perturbation. One way to study a continuous system is to examine the effects of small-scale "perturbations." For example, HFEv can be viewed as a perturbed version of the HFE method created by adding Vinegar variables. This perturbation can in some sense be considered an "external" perturbation, since the perturbation comes in the form of additional (external) variables, as opposed to perturbation that uses the existing variables.

When working with finite fields we must be clear about what we mean by "small-scale perturbations." In the construction of internal perturbation a small dimensional subspace is used to produce the perturbation. It is important to note that this approach does not require any new variables. This idea of internal perturbation is very general and can be applied to other existing multivariate cryptosystems.

In this chapter, we will first present the application of internal perturbation to the Matsumoto-Imai cryptosystem as described in [Ding, 2004b]. Due to the special feature of the Matsumoto-Imai cryptosystems, Fouque, Granboulan and Stern [Fouque et al., 2005] developed a differential attack that manages to "denoise" the cryptosystem, which is then subject to the linearization attack of Patarin. After presenting the differential attack, we show how to prevent the differential attack using the Plus method [Ding and Gower, 2006]. We also present how to add internal perturbation to HFE, and then conclude with some brief comments about using internal perturbation with the Hidden Matrix cryptosystems.

5.1 Internal Perturbation of the MI Cryptosystem

We will use the same notation as in the chapter on Matsumoto-Imai, which we briefly repeat here. Let k be a finite field of characteristic two and cardinality q, and choose an irreducible polynomial $g(x) \in k[x]$ of degree n so that $K = k[x]/g(x)$ is a degree n extension of k. We identify the field K with the k-vector space k^n by the map $\phi : K \longrightarrow k^n$ defined by

$$\phi(a_0 + a_1x + a_2x^2 + \cdots + a_{n-1}x^{n-1}) = (a_0, a_1, a_2, \ldots, a_{n-1}).$$

Let $\tilde{F} : K \longrightarrow K$ be the map defined by

$$\tilde{F}(X) = X^{1+q^\theta},$$

where $\gcd(1 + q^\theta, q^n - 1) = 1$. As we have seen, $\tilde{F}$ is an invertible map with inverse

$$\tilde{F}^{-1}(X) = X^t,$$

where $t(1 + q^\theta) \equiv 1 \bmod (q^n - 1)$.

Let $F : k^n \longrightarrow k^n$ be a map defined by

$$F = \phi \circ \tilde{F} \circ \phi^{-1} = (f_1, \ldots, f_n).$$

Then by choosing two invertible affine transformations L_1, L_2 on k^n, we can construct $\bar{F}$ by the composition

$$\bar{F} = L_1 \circ F \circ L_2 = (\bar{f}_1, \ldots, \bar{f}_n).$$

If $x_1, \ldots, x_n$ are the plaintext variables, then the components $\bar{f}_1, \ldots, \bar{f}_n$ of $\bar{F}$ will be polynomials in the ring $k[x_1, \ldots, x_n]$. These polynomials form the public key of the Matsumoto-Imai public key cryptosystem.

We now describe how to internally perturb Matsumoto-Imai. Let r be a small positive integer and randomly choose r linear functions

$$z_1(x_1, \ldots, x_n) = \sum_{j=1}^{n} \alpha_{j1} x_j + \beta_1$$

$$\vdots$$

$$z_r(x_1, \ldots, x_n) = \sum_{j=1}^{n} \alpha_{jr} x_j + \beta_r$$

such that the $z_i - \beta_i$ are linearly independent. Let $Z : k^n \longrightarrow k^r$ be the map defined by

$$Z(x_1, \ldots, x_n) = (z_1(x_1, \ldots, x_n), \ldots, z_r(x_1, \ldots, x_n)).$$

The map Z will be the source of the internal perturbation for MI. Randomly choose n polynomials $\hat{f}_1, \ldots, \hat{f}_n \in k[z_1, \ldots, z_r]$ of total degree two. Let $\hat{F} : k^r \longrightarrow k^n$ be defined by

$$\hat{F}(z_1, \ldots, z_r) = (\hat{f}_1(z_1, \ldots, z_r), \ldots, \hat{f}_n(z_1, \ldots, z_r)),$$

and define the perturbation map $F^* : k^n \longrightarrow k^n$ by

$$F^*(x_1, \ldots, x_n) = \hat{F} \circ Z(x_1, \ldots, x_n) = (f_1^*, \ldots, f_n^*),$$

where $f_1^*, \ldots, f_n^* \in k[x_1, \ldots, x_n]$.

Let us now replace $\bar{F}$ by

$$\bar{F} = L_1 \circ (F + F^*) \circ L_2 = (\bar{f}_1, \ldots, \bar{f}_n).$$

The components of $\bar{F}$ form the public key of the Perturbed Matsumoto-Imai (PMI) public key cryptosystem. See Figure 5.1 for an illustration.

Before we describe the details of decryption and encryption process for PMI, we note that $\bar{F}$ can be written as

$$\bar{F} = (L_1 \circ F \circ L_2) + (L_1 \circ F^* \circ L_2) - C,$$

where $C \in k^n$ is a constant coming from the affine part of the linear transformation L_1. This observation will become important later when we discuss the cryptanalysis of PMI.

Public Key

The public key includes the following:

1.) The field k including its additive and multiplicative structure;

2.) The n total degree two polynomials $\bar{f}_1, \ldots, \bar{f}_n \in k[x_1, \ldots, x_n]$.

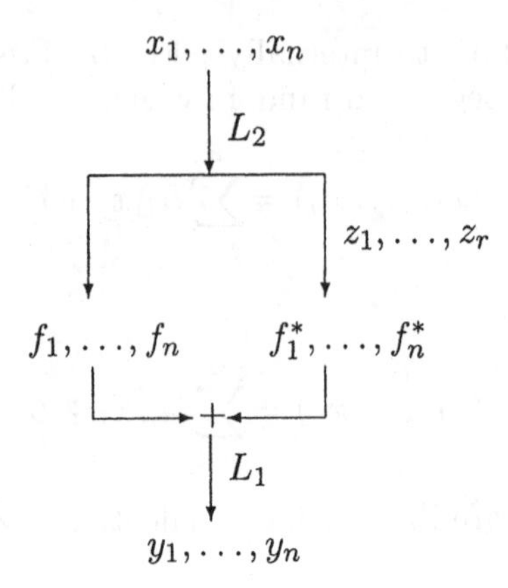

Figure 5.1. Structure of the Perturbed Matsumoto-Imai scheme.

Private Key

The private key includes the following:

1.) The map F;

2.) The set of linear functions $z_1, \ldots, z_r \in k[x_1, \ldots, x_n]$;

3.) The set of points in P defined by

$$P = \{(\mu, \lambda) \mid \hat{F}(\mu) = \lambda\},$$

or equivalently, the set of the polynomials $\hat{f}_1, \ldots, \hat{f}_n \in k[z_1, \ldots, z_r]$;

4.) The two invertible affine transformations L_1, L_2.

We note that P is expected to have about q^r or fewer elements.

Encryption

Given a plaintext message $(x'_1, \ldots, x'_n)$, the corresponding ciphertext $(y'_1, \ldots, y'_n)$ is given by:

$$y'_1 = \bar{f}_1(x'_1, \ldots, x'_n),$$
$$\vdots$$
$$y'_n = \bar{f}_n(x'_1, \ldots, x'_n).$$

Decryption

To decrypt a given ciphertext $(y'_1, \ldots, y'_n)$, we must perform the following steps:

1.) Compute $(\bar{y}_1, \ldots, \bar{y}_n) = L_1^{-1}(y'_1, \ldots, y'_n)$;

2.) For each $(\mu, \lambda) \in P$, compute

$$(y_{\lambda 1}, \ldots, y_{\lambda n}) = F^{-1}\left((\bar{y}_1, \ldots, \bar{y}_n) + \lambda\right),$$

and check if

$$Z(y_{\lambda 1}, \ldots, y_{\lambda n}) = \mu.$$

If not, discard this $(y_{\lambda 1}, \ldots, y_{\lambda n})$; otherwise, go to next step;

3.) Compute $(x_{\lambda 1}, \ldots, x_{\lambda n}) = L_2^{-1}(y_{\lambda 1}, \ldots, y_{\lambda n})$.

If there is only one $(x_{\lambda 1}, \ldots, x_{\lambda n})$, then this must be the plaintext. However, it is very possible that we will have more than one $(x_{\lambda 1}, \ldots, x_{\lambda n})$. In this case we can use the same technique as suggested for HFE (hash functions, Plus method). According to computer experiments, it seems that in general the number of $(x_{\lambda 1}, \ldots, x_{\lambda n})$ seems to be small.

It is evident that this method is very general and can be used to perturb other multivariate cryptosystems. After perturbation, the security should be stronger, though the decryption process is slower by a factor of q^r.

Toy example

We will use the same setup as in the example for the linearization equations in Section 2.3. We use $k = GF(2^2)$ with the field operations defined in Table 2.1. We also have $n = 5$, $\theta = 3$, and $g(x) = x^5 + x^3 + x + \alpha^2$. The linear transformations were not given then and they are displayed here as affine maps

$$L_1(x_1, \ldots, x_5) = \begin{pmatrix} 1 & \alpha & 1 & 0 & \alpha \\ 0 & 1 & 1 & 1 & 1 \\ 1 & 0 & \alpha^2 & 1 & \alpha^2 \\ 1 & 1 & \alpha^2 & 1 & \alpha \\ 0 & \alpha & 0 & 0 & \alpha^2 \end{pmatrix} \begin{pmatrix} x_1 \\ x_2 \\ x_3 \\ x_4 \\ x_5 \end{pmatrix} + \begin{pmatrix} \alpha^2 \\ \alpha^2 \\ 0 \\ 1 \\ 0 \end{pmatrix},$$

$$L_2(x_1, \ldots, x_5) = \begin{pmatrix} \alpha & \alpha^2 & \alpha & \alpha^2 & \alpha^2 \\ 0 & 1 & \alpha & 1 & 0 \\ \alpha & 1 & 0 & \alpha & 1 \\ \alpha & \alpha^2 & 0 & 0 & 1 \\ 1 & 0 & \alpha^2 & \alpha & 1 \end{pmatrix} \begin{pmatrix} x_1 \\ x_2 \\ x_3 \\ x_4 \\ x_5 \end{pmatrix} + \begin{pmatrix} 1 \\ 0 \\ \alpha^2 \\ \alpha^2 \\ \alpha^2 \end{pmatrix}.$$

For the perturbation in this example we use a two dimensional space; that is $r = 2$. The linear transformation $Z : k^5 \longrightarrow k^2$ of rank two is chosen to be

$$\begin{aligned} z_1 &= \alpha^2 x_1 + \alpha x_2 + \alpha^2 x_3, && (5.1)\\ z_2 &= x_1 + \alpha^2 x_2 + \alpha^2 x_3 + \alpha x_4 + \alpha^2 x_5 + \alpha^2. && (5.2) \end{aligned}$$

After the composition $Z \circ L_2$ we have

$$\begin{aligned} z_1 &= \alpha^2 x_2 + \alpha x_3 + x_4 + x_5 + 1, && (5.3)\\ z_2 &= \alpha^2 x_1 + \alpha x_2 + x_3 + x_5 + \alpha^2. && (5.4) \end{aligned}$$

One could select the last map directly at random instead of composing the map in equations (5.1) and (5.2) with L_2 as indicated in Figure 5.1, but the later form is more convenient for decrypting and will speed up that process. The quadratic polynomials $\hat{F}$ used for the perturbation are also selected at random:

$$\begin{aligned} \hat{f}_1 &= z_1^2 + \alpha^2 z_1 z_2 + \alpha z_1 + \alpha z_2^2 + z_2 + \alpha^2, && (5.5)\\ \hat{f}_2 &= z_1^2 + \alpha z_1 z_2 + \alpha^2 z_2^2 + \alpha z_2 + \alpha, && (5.6)\\ \hat{f}_3 &= z_1 z_2 + \alpha z_1 + \alpha z_2 + 1, && (5.7)\\ \hat{f}_4 &= \alpha^2 z_1 z_2 + \alpha^2 z_1 + \alpha z_2^2 + \alpha, && (5.8)\\ \hat{f}_5 &= \alpha z_1 z_2 + z_1 + z_2^2 + \alpha^2 z_2 + 1. && (5.9) \end{aligned}$$

Into these functions we substitute (5.3) and (5.4) to produces new quadratic polynomials, which are added to $F \circ L_2$, before the linear transformation L_1 is applied. The public key reads then

$$\begin{aligned} y_1 &= x_1x_2 + x_1x_3 + \alpha^2 x_1x_4 + x_1x_5 + \alpha x_1 + x_2^2 + \alpha^2 x_2x_3 + \alpha^2 x_2x_4\\ &\quad + x_2 + \alpha^2 x_3^2 + \alpha^2 x_3x_5 + \alpha^2 x_3 + x_4^2 + x_4x_5 + \alpha x_4 + x_5^2 + x_5 + \alpha,\\ y_2 &= \alpha x_1^2 + \alpha^2 x_1x_2 + x_1x_3 + \alpha x_1x_4 + \alpha^2 x_1 + x_2^2 + \alpha^2 x_2x_3 + \alpha^2 x_2x_4\\ &\quad + \alpha x_2x_5 + x_2 + x_3x_5 + \alpha^2 x_3 + x_4x_5 + \alpha x_4 + \alpha x_5^2 + x_5,\\ y_3 &= \alpha x_1^2 + \alpha^2 x_1x_2 + \alpha x_1x_3 + \alpha x_1x_5 + \alpha^2 x_1 + x_2^2 + \alpha x_2x_4 + \alpha^2 x_2x_5\\ &\quad + \alpha^2 x_2 + \alpha^2 x_3x_5 + x_4^2 + x_4x_5 + \alpha^2 x_4 + \alpha^2 x_5^2 + x_5 + \alpha^2\\ y_4 &= x_1^2 + x_1x_2 + x_1x_3 + \alpha x_1x_4 + \alpha x_1x_5 + x_2 + \alpha^2 x_3^2 + x_3x_4\\ &\quad + \alpha^2 x_3x_5 + \alpha^2 x_3 + x_4x_5 + \alpha^2 x_4 + \alpha x_5^2 + \alpha x_5,\\ y_5 &= \alpha^2 x_1^2 + \alpha^2 x_1x_2 + x_1x_3 + \alpha x_1x_4 + x_1x_5 + x_1 + \alpha x_2^2 + \alpha x_2x_3\\ &\quad + \alpha x_2x_4 + \alpha^2 x_2x_5 + \alpha x_2 + x_3^2 + \alpha x_3x_4 + \alpha x_3x_5 + x_3 + \alpha^2 x_4\\ &\quad + x_5^2 + \alpha x_5 + \alpha^2. \end{aligned}$$

If the plaintext is $(0, \alpha, 1, 1, 1)$, then these quadratic polynomials produce the cipher $(y_1', y_2', y_3', y_4', y_5') = (\alpha, 0, 1, \alpha, 0)$.

The decryption process starts by applying L_1^{-1} to the cipher, which gives

$$\bar{y} = (\bar{y}_1, \ldots, \bar{y}_5) = (\alpha, 1, 0, 1, \alpha^2). \tag{5.10}$$

For each value $\mu \in k^2$, we list the corresponding values $\lambda = \hat{F}(\mu) \in k^5$. They are given in the following table, but usually these values will be computed on an as-needed basis from the perturbation polynomials (5.5) to (5.9).

$\mu \in k^2$	$\lambda = \hat{F}(\mu) \in k^5$
$(\alpha, 1)$	$(1, \alpha^2, \alpha, 0, \alpha)$
$(\alpha^2, 1)$	$(1, 0, 1, 0, 1)$
$(0, 1)$	$(0, \alpha^2, \alpha^2, 0, \alpha^2)$
$(1, \alpha)$	$(\alpha, 1, \alpha, 1, 1)$
(α, α)	$(\alpha, 1, \alpha, 0, 0)$
(α^2, α)	$(0, \alpha^2, \alpha, \alpha, \alpha)$
$(0, \alpha)$	$(0, \alpha^2, \alpha, \alpha^2, \alpha^2)$
$(1, \alpha^2)$	$(\alpha, \alpha, 1, 0, 1)$
(α, α^2)	$(0, \alpha^2, \alpha, \alpha^2, 1)$
(α^2, α^2)	$(1, \alpha^2, \alpha^2, \alpha, 1)$
$(0, \alpha^2)$	$(\alpha^2, \alpha, 0, 1, 1)$
$(1, 0)$	$(0, \alpha^2, \alpha^2, 1, 0)$
$(\alpha, 0)$	$(\alpha^2, 1, \alpha, \alpha^2, \alpha^2)$
$(\alpha^2, 0)$	$(0, 0, 0, 0, \alpha)$
$(0, 0)$	$(\alpha^2, \alpha, 1, \alpha, 1)$
$(1, 1)$	$(0, 0, 0, 0, 0)$

No matter which approach is used we must go through the possible values of λ one by one, and find $y_\lambda = F^{-1}(\bar{y} + \lambda)$ with the help of $\tilde{F}^{-1}(X) = X^{362}$, the inverse of the underlying Matsumoto-Imai map. We then check if $\hat{F} \circ Z(y_\lambda)$ matches the μ belonging to λ. In our example it happens only once for $\mu = (\alpha, 0)$, which corresponds to $\lambda = (\alpha^2, 1, \alpha, \alpha^2, \alpha^2)$ and $y_\lambda = (\alpha, 1, \alpha, \alpha^2, \alpha^2)$. Once we have a suitable y_λ, we compute $L_2^{-1}(y_\lambda)$ to find the original plaintext.

It does not suffice to stop when the first matching y_λ is found. Depending on the structure of the perturbation polynomials there can be more than one value of λ. For example, when the perturbation polynomials in (5.5) to (5.9) are all equal to $z_1^2 + \alpha z_2$ then each of the sixteen different μ values is mapped only on one of four different λ values. This is an indication that the perturbation polynomials are not general enough. Nevertheless, in this particular example a unique plaintext will be recovered, but this will no longer be the case when r is closer to n.

A Practical Implementation

In [Ding, 2004b], a 136-bit implementation of PMI is suggested for practical use. In this suggestion, $k = GF(2)$ and K is the degree 136 extension of $GF(2)$ defined by the irreducible polynomial

$$g(x) = 1 + x + x^2 + x^{11} + x^{136}.$$

The dimension of the perturbation space is chosen to be $r = 6$, and the underlying Matsumoto-Imai map $\tilde{F}$ is chosen to be

$$\tilde{F}(X) = X^{2^{5\times 8}+1}.$$

At the time this suggestion was made, this implementation of PMI was expected to have a security level of at least 2^{80} against all known attacks. It was also suggested that n should be at least 96 and r should be at least 5 to maintain this level of security. However, approximately one year later a new attack was found that could defeat this scheme in much less than the supposed 2^{80} security level. We will now present the differential attack by Fouque, Granboulan and Stern [Fouque et al., 2005].

5.2 Differential Attack on PMI

The basic idea of the differential attack is to use the distribution of the rank of the kernel of the differential (or difference) to detect which plaintexts produce noise and which do not. If this can be achieved, then the attacker can "denoise" the system and use Patarin's linearization equation method to break the underlying Matsumoto-Imai scheme.

We begin by establishing the basic notation. For each plaintext message $v \in k^n$, define the differential

$$\bar{L}_v(x) = \bar{F}(x+v) + \bar{F}(x) + \bar{F}(v) + \bar{F}(0),$$

where $\bar{F}$ is a given instance of PMI.

Note that this notion of differential was also used by Patarin, Goubin and Courtois in an attack on MI-Minus [Patarin et al., 1998].

It is straightforward to show that $\bar{L}_v$ is linear in the variable x taking values in k^n, since $\bar{F}$ is quadratic. If $\mathcal{K}$ is the "noise kernel," the kernel of the linear part of the affine transformation $Z \circ L_2$, then the following proposition is easy to prove.

Proposition 5.2.1. *Let $v \in \mathcal{K}$. Then*

$$\dim(\ker(\bar{L}_v)) = \begin{cases} \gcd(\theta, n) & \text{if } v \neq 0; \\ n & \text{if } v = 0. \end{cases}$$

Proof. To prove the proposition we may assume that $\bar{F}$ is simply the underlying Matsumoto-Imai map, since $v \in \mathcal{K}$ means that the perturbation is zero. Furthermore, because L_1 and L_2 are invertible affine transformations, they do not have any effect on the dimension of the kernel of the differential. So, to prove the proposition we need to count the number of solutions (for a fixed $X \in K$) of the following equation:

$$\tilde{F}(X+V)+\tilde{F}(X)+\tilde{F}(V)=0. \tag{5.11}$$

Note that

$$\begin{aligned} 0 &= \tilde{F}(X+V)+\tilde{F}(X)+\tilde{F}(V) \\ &= (X+V)^{q^\theta+1}+X^{q^\theta+1}+V^{q^\theta+1} \\ &= X^{q^\theta}V+XV^{q^\theta}, \end{aligned}$$

so that (5.11) becomes

$$X^{q^\theta}V+XV^{q^\theta}=0. \tag{5.12}$$

It is clear that if $X=0$ then any $V \in K$ will satisfy (5.12). If $X \neq 0$, then by dividing both sides by X we have:

$$X^{q^\theta-1}=V^{q^\theta-1}. \tag{5.13}$$

From Lemma 2.3.3 we know that the number of nonzero solution of (5.13) is exactly $q^{\gcd(\theta,n)}-1$. Adding in the solution $X=0$, the total number of solutions of (5.12) is exactly $q^{\gcd(\theta,n)}$. This completes the proof. □

When $v \notin \mathcal{K}$ then it is likely that $\dim(\ker(L_v)) \neq \gcd(\theta,n)$ due to the randomness of the "noise" provided by the perturbation. Therefore the quantity $\dim(\ker(L_v))$ can be used to detect whether or not a given v is in $\mathcal{K}$. In particular, we can use this quantity to find a basis of $\mathcal{K}$ which can then be used to mount q^r attacks, each attack being against the scheme restricted to one of the q^r affine planes parallel to $\mathcal{K}$. Since these restrictions are just an unperturbed MI, we can use Patarin's linearization attack.

Testing for Membership in $\mathcal{K}$

For each $v \in k^n$, define the function T by

$$T(v)=\begin{cases} 1, & \text{if } \dim(\ker(L_v)) \neq \gcd(\theta,n); \\ 0, & \text{otherwise.} \end{cases}$$

Let $\alpha = P[T(v) = 0]$ and $\beta = P[v \in \mathcal{K}] = q^{-r}$; in other words, α is the probability that $T(v) = 0$, and β is the probability that $v \in \mathcal{K}$. We can use T to devise a test for detecting whether or not a given v is very likely to be in $\mathcal{K}$, assuming the following proposition: If for many different v'_i such that $T(v'_i) = 0$ we have $T(v + v'_i) = 0$, then $v \in \mathcal{K}$ with high probability. Suppose we pick N vectors $v'_1, \ldots, v'_N$ such that $T(v'_i) = 0$. Define $p(v) = P[T(v + v'_i) = 0 \,|\, T(v'_i) = 0]$. If v is chosen at random, then $p(v) = \alpha$; otherwise, $p(v) = \frac{\beta}{\alpha} + \frac{(\alpha-\beta)^2}{\alpha(1-\beta)}$. In this latter case it is not hard to show that $\frac{p(v)}{\alpha} - 1 = \frac{\beta}{1-\beta}(\frac{1}{\alpha} - 1)^2 \approx \beta(\frac{1}{\alpha} - 1)^2$, where $\frac{\beta}{1-\beta} = \beta + \beta^2 + \beta^3 + \cdots \approx \beta$ if β is very small. Thus we have the approximation $p(v) \approx \alpha + \alpha\beta(\frac{1}{\alpha} - 1)^2$ whenever $v \in \mathcal{K}$. Therefore one way to decide whether or not $v \in \mathcal{K}$ is to approximate $p(v)$ and decide whether it is closer to α or $\alpha + \alpha\beta(\frac{1}{\alpha} - 1)^2$.

In [Ding and Gower, 2006], it is suggested instead to consider the function

$$T'(v + v'_i) = \frac{1 - T(v + v'_i)}{\alpha} - 1,$$

which has the expected value

$$E[T'(v + v'_i)] = \frac{p(v)}{\alpha} - 1,$$

and then consider the average $\frac{1}{N}\sum_{i=1}^{N} T'(v + v'_i)$, which is expected to be close to $\frac{p(v)}{\alpha} - 1$, for large enough N by the Central Limit Theorem (see [Feller, 1968]). Then the task would be to determine whether this average is closer to 0 or $\beta(\frac{1}{\alpha} - 1)^2$.

The new function T' is defined as above in terms of T, and is such that

$$T'(v + v'_i) = \begin{cases} \frac{1}{\alpha} - 1, & \text{with probability } p(v); \\ -1, & \text{with probability } 1 - p(v). \end{cases}$$

Also $\mu = E[T'(v + v'_i)] = \frac{p(v)}{\alpha} - 1$ and $\sigma^2 = Var[T'(v + v'_i)] = \frac{p(v)(1-p(v))}{\alpha^2}$. Here E stands for the expected value and Var stands for the variance.

Let X_i be independent and identically distributed random variables with the same distribution as T', and define $S_N = \sum_{i=1}^{N} X_i$. Then the Central Limit Theorem states that

$$P\left[\frac{S_N - N\mu}{\sigma\sqrt{N}} < x\right] \longrightarrow \mathfrak{N}(x) \quad \text{as} \quad N \longrightarrow \infty,$$

where

$$\mathfrak{N}(x) = \frac{1}{\sqrt{2\pi}} \int_{-\infty}^{x} e^{-y^2/2}\, dx$$

is the standard normal distribution function. More informally, the Central Limit Theorem implies that the following approximation is valid for large N:

$$A_N \approx \mu + \frac{\sigma}{\sqrt{N}} \chi,$$

where $A_N = \frac{1}{N} S_N$ and χ is a random variable with standard normal distribution.

Efficiency of the Test

Suppose $v \in K$. In this case $\mu = \frac{p(v)}{\alpha} - 1 = \beta(\frac{1}{\alpha} - 1)^2$, and $\sigma^2 = \frac{p(v)(1-p(v))}{\alpha^2}$, which can be computed in terms of α and β. We also take $N = 1/(\alpha\beta)^2$, as in [Fouque et al., 2005]. We first consider the probability that the question "$A_N > \beta(\frac{1}{\alpha} - 1)^2$" will return true. Equivalently, we consider the probability that

$$\mu + \frac{\sigma}{\sqrt{N}} \chi > \beta \left(\frac{1}{\alpha} - 1 \right)^2 = \mu,$$

which is the probability that $\chi > 0$. But this probability is $1 - \mathfrak{N}(0) = 1 - 0.5 = 0.5$. In other words, the "efficiency" of this test is such that it detects a vector $v \in \mathcal{K}$ (which is actually in $\mathcal{K}$) roughly half of the time. If we are to collect $n - r$ linearly independent vectors in $\mathcal{K}$, then we must perform on average $2(n - r)q^r$ tests.

Reliability of the Test

Let us now compute the probability that an answer to this question returns a false-positive; i.e., the question "$A_N > \beta(\frac{1}{\alpha} - 1)^2$" returns true for $v \notin \mathcal{K}$. Here we must consider the probability that

$$\mu + \frac{\sigma}{\sqrt{N}} \chi > \beta \left(\frac{1}{\alpha} - 1 \right)^2, \tag{5.14}$$

where now $\mu = 0$ and $\sigma^2 = \frac{1-\alpha}{\alpha}$. For example, if we take $\alpha = 0.59$ and $\beta = 2^{-6}$ as in the examples given in [Fouque et al., 2005], then this is the probability that $\chi > 0.9819$, which is $1 - \mathfrak{N}(0.9819) \approx 1 - 0.8369 = 0.1631$. This quantity gives us a measure of the "reliability" of this test in the sense that it tells us that roughly 16% of the $n - r$ vectors that the test leads us to believe are in $\mathcal{K}$ actually are not in $\mathcal{K}$. Though this might seem like a serious problem, it can be remedied by repeating the test a few times, each time with a different set of vectors $v'_1, \ldots, v'_N$. In the example above, by taking $8N$ vectors v'_i, performing the test 8 times with a new set of N vectors each time, and rejecting the vector v if any of the 8 tests fails, the probability that we correctly conclude that $v \in \mathcal{K}$ is approximately $1 - (.1631)^8 \approx 0.9999995$. This in turn means that the probability that there are no false-positives among the final set of $n - r$ vectors is approximately $(1 - (.1631)^8)^{130} \approx 0.9999349$. Therefore, if we perform 8 tests on $\frac{2(n-r)q^r}{0.1631}$ vectors, then the probability that we have $n - r$ vectors in $\mathcal{K}$ is approximately 0.9999349.

We note that the above is a description of a modified version of Technique 1 from [Fouque et al., 2005] for which a much higher degree of reliability is obtained, though the basic idea is the same. In [Fouque et al., 2005] they do not necessarily require such a high level of reliability from Technique 1 since they also use Technique 2 as a filter to find which of the elements from Technique 1 are actually in $\mathcal{K}$.

Verifying Membership in $\mathcal{K}$ Using Graphs

Technique 2 from [Fouque et al., 2005] is essentially another test for membership in $\mathcal{K}$. A graph is defined whose vertices correspond to the elements $v \in k^n$ such that $T(v) = 0$; i.e., elements that may be in $\mathcal{K}$. We put an edge between vertices v, v' if and only if $T(v + v') = 0$. It is clear that all the elements of $\mathcal{K}$ are connected; or in other words, the elements of $\mathcal{K}$ form a large clique within the graph.

In practice we do not need to construct the whole graph. Instead we can construct its restriction to N vertices. Since we are looking for vertices that correspond to $n - r$ independent elements of $\mathcal{K}$, as long as $N > \frac{n}{\beta}$ then it is very likely that the restriction will contain such vertices. The clique containing the elements of $\mathcal{K}$ should contain at least βN vertices. Under the hypothesis that the probability that $T(v+v') = 0$ is independent of the probability that $T(v) = 0$, the graph restricted to N vertices should have αN^2 edges. From the setting of the problem, we believe that such a hypothesis seems to be a very reasonable assumption.

Except for the vertices corresponding to elements of $\mathcal{K}$, we expect that the edges are randomly distributed. Suppose α is the probability that there is an edge between two vertices where not both of them are in $\mathcal{K}$. Then from the general theory of random graphs [Bollobás, 2001], we know that the expected number of vertices in a maximal clique in a random graph of N vertices is given by

$$\frac{\log N^2}{\log \alpha^{-1}} + O(\log \log N).$$

Thus we see that if βN is significantly larger than $\frac{\log N^2}{\log \alpha^{-1}}$, then it is likely that there is an unique large clique from which we can extract a basis of $\mathcal{K}$. When $\beta \ll \alpha$, this condition is equivalent to the condition that

$N \approx \beta^{-1} \log \beta^{-1}$. In this case, the complexity for finding a basis for $\mathcal{K}$ is expected to be $q^{2r} \log^2 q^r$.

It is pointed out in [Fouque et al., 2005] that though this technique seems to be more efficient than Technique 1, there is no max-clique algorithm that uses the fact that we have a random and dense graph with a very large clique. Therefore, for a practical attack against PMI the authors suggest using a mix of Technique 1 and 2.

Cryptanalysis of PMI

Let us now assume that we have correctly derived the kernel $\mathcal{K}$ using some combination of the techniques described above. The rest is straight forward. We find a basis of $\mathcal{K}$ perform an exhaustive search on all the coordinates according to this basis, which will give us r linear equations each time. Then we use Gaussian elimination on these r equations to derive a set of equations, which we can plug into the public polynomials. As we mentioned before, from the point of view algebraic geometry, the internal perturbation becomes just adding constants to the cryptosystem via the public key. We thus can solve the equations derived from any ciphertext using r linearization equations as in [Patarin, 1995]. Then we can use the linear equations to check which answer is the original plaintext.

It is clear that an attack against the practical implementation suggested by [Ding, 2004b] is now easily accomplished. With Technique 1, the complexity of the attack is a computation of order $O(nq^{3r} + n^6 q^r)$, which can be bounded above by 2^{49} with the proposed parameters in [Ding, 2004b]. For the specific PMI implementation suggested above, the complexity is actually $O(n^3 q^r q^{\gcd(\theta,n)})$, which is of the order of 2^{36} binary operations. The conclusion is that there does not exist a practical choice of the parameters which makes PMI secure.

In [Fouque et al., 2005] this new attack is proposed and used to break MI using the original elements in the kernel of the transpose of the differential. They show that there exists a bilinear relation between the ciphertext and the kernel vector. Knowing the kernel allows an attacker to recover the plaintext by solving a linear system of equations.

5.3 Inoculation Against Differential Attacks

It is clear that differential analysis provides a very powerful attack against PMI. However, it is very easy to repair PMI using the Plus method. The resulting scheme, PMI+ [Ding and Gower, 2006], resists the differential attack by adding a few additional polynomials. The MI cryptosystem itself is too "pure" in terms of the rank of the differential as described in Proposition 5.2.1. One way to protect it is to add some additional "noise" to hide or mask this behavior, which is easily accomplished with the Plus method.

PMI+

We will use much of the same notation as before. In particular, let

$$\bar{F} = L_1 \circ (F + F^*) \circ L_2$$

be an instance of PMI, and let $q_1, \ldots, q_a$ be a polynomials of total degree two in the polynomial ring $k[x_1, \ldots, x_n]$. We append these a polynomials to the map $F + F^*$ to form

$$\bar{\bar{F}} = (f_1 + f_1^*, \ldots, f_n + f_n^*, q_1, \ldots, q_a).$$

We choose an invertible affine transformation $L_1 : k^{n+a} \longrightarrow k^{n+a}$, and redefine $\bar{F}$ by

$$\bar{F} = L_1 \circ \bar{\bar{F}} \circ L_2.$$

The public key of the Perturbed Matsumoto-Imai-Plus (PMI+) cryptosystem consists of the $n + a$ total degree two polynomial components $\bar{f}_i$ of $\bar{F}$. Clearly PMI+ is PMI with a additional random quadratic polynomials (externally) mixed into the system by L_1. The full description of PMI+ is the same as PMI except for the decryption process.

Decryption

To decrypt the ciphertext $(y'_1, \ldots, y'_{n+a})$, we must first compute

$$(z_1, \ldots, z_{n+a}) = L_1^{-1}(y'_1, \ldots, y'_{n+a}).$$

We then set aside the last a coordinates that come from the polynomials $q_1, \ldots, q_a$. Now we can continue the decryption process with

$$(z_1, \ldots, z_n)$$

as in the description of PMI. If this produces several candidate plaintexts, we can use the polynomials $q_1, \ldots, q_a$ to help decide which is the true plaintext. This is accomplished by simply evaluating the map

$$Q \circ L_2$$

at each candidate plaintext, where $Q = (q_1, \ldots, q_a)$. In other words, the extra a components serve the additional purpose of helping to determine the true plaintext from among the possibly q^r preimages of the given ciphertext.

Before we examine the effect that the Plus perturbation has on the complexity of the differential attack we give a small example.

Toy Example

We continue with the earlier example illustrating PMI, and use the additional polynomial

$$\begin{aligned} q_1(x) &= \alpha x_1x_2 + \alpha^2 x_1x_3 + \alpha^2 x_1x_4 + \alpha x_1x_5 + x_1 + \alpha^2 x_2^2 + \alpha^2 x_2x_3 + x_2 \\ &\quad + \alpha^2 x_3^2 + \alpha x_3x_4 + \alpha^2 x_3x_5 + \alpha x_4^2 + x_4x_5 + \alpha x_4 + x_5^2 + \alpha x_5 + \alpha. \end{aligned}$$

The affine transformation L_2 from the PMI example can be reused as given and will result in

$$\begin{aligned} q_1 \circ L_2(x) &= \alpha^2 x_1^2 + x_1x_4 + \alpha^2 x_1x_5 + \alpha^2 x_2^2 + x_2x_3 + \alpha x_2x_4 + \alpha^2 x_2x_5 \\ &\quad + \alpha^2 x_2 + \alpha^2 x_3^2 + \alpha x_3x_5 + x_3 + x_4^2 + \alpha^2 x_4x_5 + \alpha^2 x_4 + x_5^2 + \alpha x_5 + \alpha. \end{aligned}$$

On the other hand the cipher will now be six dimensional and we have to select a new affine transformation L_1:

$$L_1(x_1,\ldots,x_6) = \begin{pmatrix} \alpha & \alpha & 1 & 1 & \alpha^2 & 0 \\ 1 & 1 & \alpha & 0 & \alpha & 0 \\ 1 & \alpha & \alpha^2 & \alpha^2 & 0 & 1 \\ 1 & 0 & \alpha & \alpha^2 & 1 & 0 \\ 1 & \alpha & 0 & \alpha^2 & 1 & 0 \\ \alpha & 0 & 1 & 0 & 1 & \alpha^2 \end{pmatrix} \begin{pmatrix} x_1 \\ x_2 \\ x_3 \\ x_4 \\ x_5 \\ x_6 \end{pmatrix} + \begin{pmatrix} 0 \\ \alpha^2 \\ \alpha^2 \\ \alpha^2 \\ \alpha \\ 0 \end{pmatrix}.$$

With these changes the public key becomes

$$\begin{aligned} y_1 &= \alpha x_1^2 + x_1x_2 + \alpha^2 x_1x_3 + \alpha^2 x_1x_5 + \alpha x_2^2 + \alpha^2 x_2x_3 + \alpha^2 x_2x_4 \\ &\quad + x_2x_5 + x_2 + \alpha x_3^2 + \alpha x_3x_4 + \alpha x_3x_5 + \alpha^2 x_3 + x_4^2 + x_4 \\ &\quad + \alpha^2 x_5 + 1, \\ y_2 &= \alpha^2 x_1x_3 + \alpha x_1x_4 + \alpha x_1x_5 + \alpha^2 x_2x_3 + x_2x_4 + x_2x_5 + \alpha x_2 \\ &\quad + \alpha x_3x_5 + \alpha^2 x_3 + x_4^2 + x_4 + x_5^2 + \alpha^2, \\ y_3 &= \alpha x_1^2 + \alpha x_1x_2 + x_1x_4 + \alpha x_1 + x_2^2 + x_2x_3 + \alpha x_3x_4 + x_3x_5 + x_3 \\ &\quad + \alpha^2 x_4^2 + \alpha^2 x_4x_5 + \alpha^2 x_4 + x_5^2 + x_5 + \alpha^2, \\ y_4 &= x_1x_2 + x_1x_3 + x_1x_4 + \alpha^2 x_1 + \alpha^2 x_2^2 + \alpha^2 x_2x_3 + \alpha x_2x_4 + x_2 \\ &\quad + x_3x_4 + \alpha x_3x_5 + \alpha x_3 + \alpha x_4^2 + \alpha x_4x_5 + \alpha x_5^2 + \alpha x_5 + \alpha, \\ y_5 &= \alpha x_1x_2 + \alpha x_1x_4 + \alpha x_1x_5 + \alpha x_1 + \alpha^2 x_2x_3 + \alpha^2 x_2x_4 + \alpha^2 x_2x_5 \\ &\quad + \alpha^2 x_2 + \alpha x_3^2 + x_3x_4 + x_3x_5 + \alpha x_3 + \alpha x_4^2 + x_4x_5 + \alpha x_4 \\ &\quad + \alpha x_5^2 + x_5, \\ y_6 &= x_1^2 + \alpha^2 x_1x_2 + x_1x_4 + x_1x_5 + x_1 + x_2^2 + x_2x_3 + x_2 + \alpha x_3x_5 \\ &\quad + \alpha x_3 + \alpha x_4^2 + x_4x_5 + \alpha^2 x_4 + \alpha^2 x_5 + \alpha. \end{aligned}$$

If the plaintext is again $x = (0, \alpha, 1, 1, 1)$, then this time the public key produces the cipher $(y_1', y_2', y_3', y_4', y_5', y_6') = (\alpha, 1, \alpha^2, 1, \alpha, \alpha)$.

The decryption process starts with computing

$$L_1^{-1}(\alpha, 1, \alpha^2, 1, \alpha, \alpha) = (\alpha, 1, 0, 1, \alpha^2, \alpha^2)$$

and it is seen that the first five components are those of $\bar{y}$ in (5.10). In our example we only changed L_1 to accommodate the additional perturbation polynomial q_1 and thus the decryption continues as given in the example for PMI.

The Effect of Plus Perturbation on $\mathcal{K}$

We begin with the case where gcd $(\theta, n) = 1$, so that dim (ker $(L_v)) = 1$ for every $v \in \mathcal{K}$. The fact that dim (ker $(L_v)) \neq 1$ for many $v \notin \mathcal{K}$ is the very fact that Technique 1 exploits to compute $\mathcal{K}$. So the task is to perturb PMI with Plus so that dim (ker $(L_v)) = 1$ for nearly every $v \notin \mathcal{K}$.

Consider the effect on the linear differential $L_v(x)$ upon adding Plus polynomials. We write $M_{v,a}$ for the matrix associated with the linear differential obtained after adding a Plus polynomials, and in particular $M_{v,0}$ for the matrix associated with the linear differential L_v with no Plus polynomials. Let $R(a)$ be the rank of the matrix $M_{v,a}$. Note that $R(a) < n$, since $M_{v,a}\, v^T = 0$ for any a.

Suppose we add one more Plus polynomial (increase a by one) and consider the probability that $R(a+1) = R(a) + 1$. Note that if $R(a) = n - 1$, then this probability is zero since $R(a) < n$. So we assume $R(a) = n - i$, where $i = 2, 3, \ldots, n-1$. This probability is equivalent to the probability that we choose a new row-vector to be added to form $M_{v,a+1}$ from $M_{v,a}$ which is orthogonal to v and is not in the span of the row-vectors of $M_{v,a}$. The space of vectors orthogonal to v is of dimension $n-1$, and the span of the row-vectors of $M_{v,a}$ is of dimension $n - i$, hence the probability that $R(a+1) = R(a) + 1$ will be $1 - 2^{1-i}$, where $i = 2, 3, \ldots, n-1$. Thus, if $n_{\delta,a}$ is the number of vectors v with dim (ker $(M_{v,a})) = \delta$, for a given a and $\delta = 1, 2, \ldots, n-1$, then we expect:

$$n_{\delta,a+1} = n_{\delta,a} \cdot 2^{1-\delta} + n_{\delta+1,a} \cdot (1 - 2^{-\delta})$$

In order to obtain the distribution for $n_{\delta,a}$ when $a = 0$, and to predict how large we must choose a in order to protect PMI+ from the differential attack, we will use the language of Markov chains [Kemeny and Snell, 1960]. Let $P = [p_{ij}]$ be the $n \times n$ matrix with entries given by:

$$p_{ij} = \begin{cases} 2^{-i+1}, & \text{if } i = j; \\ 1 - 2^{-i+1}, & \text{if } i = j+1; \\ 0, & \text{otherwise.} \end{cases}$$

For a fixed vector $v \in k^n$, p_{ij} gives the one-step transition probability from state s_i to s_j upon appending a randomly chosen row vector to $M_{v,a}$, where state s_i corresponds to nullity$(M_{v,a}) = i$. By nullity$(M_{v,a})$, we mean the dimension of the solution space of the homogeneous linear equations defined by $M_{v,a}$. Here s_1 is an absorbing state, and s_i is a transient state for all other $i \neq 1$.

Let M_v be the matrix associated with MI for a given v. Without loss of generality, assume that L_2 is chosen so that the perturbation Z is a function only of r variables, say $x_1, \ldots, x_r$. Adding the perturbation then is analogous to removing the first r columns of M_v and replacing them with r randomly chosen column vectors. Deleting r columns will increase the nullity to either $r+1$ with probability $\binom{n-1}{r}/\binom{n}{r} = 1 - \frac{r}{n}$, or r with probability $\binom{n-1}{r-1}/\binom{n}{r} = \frac{r}{n}$. If we then add r random column vectors to this matrix one at a time, the nullity will increase in accordance with the r-step transition probability matrix P_r^r, where P_r is the top-left $(r+1) \times (r+1)$ submatrix of P. In particular, if we let $\pi_0 = (0, 0, \ldots, \frac{r}{n}, 1 - \frac{r}{n})$ be the initial state distribution vector, then $\pi_0 P_r^r$ can be used to calculated the probability that nullity$(M_{v,0}) = i$. For example, if $n = 31$ and $r = 6$, then these probabilities are given by:

$$\pi_0 P_6^6 = \begin{pmatrix} 0.350125 \\ 0.539086 \\ 0.106813 \\ 3.94582 \times 10^{-3} \\ 3.01929 \times 10^{-5} \\ 4.67581 \times 10^{-8} \\ 1.17354 \times 10^{-11} \end{pmatrix}.$$

Finally, to obtain the probability that nullity$(M_{v,a}) = i$, we let $\pi' = \pi_0 P_r^r$ and compute $\pi' P^a$.

In [Ding and Gower, 2006] experiments were performed to test the validity of this model. Each experiment was characterized by an instance of PMI defined by the parameters (q, n, r, θ), the number of Plus polynomials a, and κ randomly chosen test vectors. For each test vector v dim$(\ker(M_{v,a}))$. The table below reports the observed values (predicted values in parentheses) of $n_{\delta,a}$ for the experiments performed with parameters $(q, n, r, \theta, \kappa) = (2, 31, 6, 2, 2^{15})$ with $a = 0, 1, 2, \ldots, 11$. The predictions for $a = 0$ are obtained from the matrix $\pi' = \pi_0 P_r^r$, while the predictions for $a > 0$ are obtained by using the observed distribution from $a - 1$ and the 1-step transition matrix P_r.

	$v \notin \mathcal{K}$				$v \in \mathcal{K}$
a	$\delta = 1$	$\delta = 2$	$\delta = 3$	$\delta = 4$	$\delta = 1$
0	19003 (11304)	12182 (17404)	1081 (3448)	19 (127)	483
1	25081 (25094)	6906 (6902)	298 (287)	0 (2)	483
2	28548 (28534)	3660 (3676)	77 (74)	0 (0)	483
3	30366 (30378)	1896 (1888)	23 (19)	0 (0)	483
4	31334 (31314)	944 (965)	7 (6)	0 (0)	483
5	31810 (31806)	473 (477)	2 (2)	0 (0)	483
6	32040 (32046)	244 (238)	1 (0)	0 (0)	483
7	32154 (32162)	130 (123)	1 (0)	0 (0)	483
8	32208 (32219)	77 (66)	0 (0)	0 (0)	483
9	32246 (32246)	39 (38)	0 (0)	0 (0)	483
10	32263 (32266)	22 (20)	0 (0)	0 (0)	483
11	32278 (32274)	7 (11)	0 (0)	0 (0)	483

One may notice that the predictions for $a = 0$ are not as accurate as those for $a > 0$. This is likely due to the fact that the perturbation variables $z_1, \ldots, z_r$ were chosen in a non-random way for convenience with the experiments.

The problem now is to determine how large a must be in order to protect PMI+ against a differential attack. As was previously stated, the effect of adding Plus polynomials is to increase the value of α. In the example given in [Fouque et al., 2005], $\alpha \approx 0.59$ and so the question "$A_N > \beta(\frac{1}{\alpha} - 1)^2$" is answered with a false-positive with the probability that $\chi > 0.9819$, which is 0.1631. Now suppose the attacker is willing to do as much as 2^{2w} work to correctly decide the answer to this test with this same probability. Then inequality (5.14) becomes

$$\chi > \frac{\sqrt{N}}{\sigma}\left[\beta\left(\frac{1}{\alpha} - 1\right)^2 - \mu\right] = 2^{w-r}\left(\frac{1-\alpha}{\alpha}\right)^{3/2}.$$

If we assume that we are using Technique 1 as we have described it above, then the total work will be

$$8N \cdot \frac{n^3}{6} \cdot \frac{2(n-r)q^r}{0.1631} \approx 2^{2w+38.32}.$$

If we want this to be less than 2^{80}, then we must have $w < 20.84$. This implies that we must take $2^{14.84}\left(\frac{1-\alpha}{\alpha}\right)^{3/2} < 0.9819$, or $\alpha > 0.998962$ if we wish to thwart this attack. To compute the value of a necessary to insure $\alpha > 0.998962$, we use the matrix P. In particular, we must compute a so that the first entry of $\pi' P^a$ is greater than 0.998962. If we take $n = 136$, $r = 6$, and $\gcd(\theta, n) = 1$, then we must take $a \geq 10$.

Finally, we consider $g = \gcd(\theta, n) > 1$. If $v \in \mathcal{K}$, then $\text{nullity}(M_{v,0}) = g$; otherwise $\text{nullity}(M_{v,0}) \in \{g-r, \ldots, g+r\}$. We must now add roughly

g Plus polynomials just to get to a situation similar to the $g = 1$ case. Thus, by taking $a \approx g + 10$, we can protect the special case of $g \neq 1$ from the differential attack.

Using Filters with the Differential Attack and Other Security Concerns

It remains to address Technique 2 of [Fouque et al., 2005] as explained above. As we have seen, the basic idea of this technique is to look for a maximal clique in the graph with vertices $v \in k^n$ such that $T(v) = 0$, where two vertices v, v' are connected if and only if $T(v + v') = 0$. The hypothesis underlying a successful use of Technique 2 is that if we look at a big enough subgraph then the maximal clique in this subgraph will consist almost exclusively of vectors from $\mathcal{K}$. However, by increasing the value of α near one, this clique is now very likely to have many elements not in $\mathcal{K}$ (in fact almost every element of k^n is in the clique), and therefore membership in this clique cannot be used as a reliable filter to Technique 1.

Gröbner Basis Attack on PMI+

From the discussion above, it would seem that the larger a is (more Plus perturbation), the more secure the cryptosystem should be. However, we must be careful not to add too many extra polynomials since otherwise we may create a weakness to Gröbner bases attacks [Courtois et al., 2000; Yang et al., 2004b]. Through computer experiments using the implementation of the F_4 algorithm in Magma, it is shown in [Ding et al., 2005] that if we choose $r = 6$ and $n > 83$, then we can expect PMI to have a security of 2^{80} against a Gröbner attack using F_4. Since the analysis above shows that very few Plus polynomials are added to make PMI+ resistant to the differential attack, the results about PMI carry over to PMI+. Therefore, in order to create a secure PMI+ scheme it is suggested that $(q, n, r, \theta) = (2, 84, 6, 4)$ and $a = 14$. Other secure implementations include the now-salvaged scheme $(q, n, r, \theta) = (2, 136, 6, 8)$ with $a = 18$, or any (q, n, r, θ) with $a = 11$, $g = 1$ and $n > 84$. In [Yang and Chen, 2005b], PMI+ was implemented on the IBM 8051 platform, and it was shown to be a very efficient public key encryption scheme.

Other Attacks on PMI+

Of course, it may be possible to attack PMI+ by looking for ways to somehow separate the PMI polynomials from the Plus polynomials. If this were possible then the differential attack could then proceed as with PMI alone. However, this approach has yet to be successfully applied to the MI-Plus-Minus scheme [Patarin et al., 1998], as we have no such method yet to differentiate between MI polynomials and random polynomials. Therefore, at this moment it seems unlikely that such an approach will be successfully applied to PMI+.

5.4 Perturbation of HFE

Internal perturbation has also been applied to the HFE cryptosystem [Ding and Schmidt, 2005a]. We have already discussed HFEv, which can be viewed as a scheme derived from HFE through the Vinegar external perturbation. An attack against HFEv has been demonstrated which can be used to purge the external variables.

We will now present a new variant of HFE using internal perturbation, the so-called internally perturbed HFE cryptosystem, or IPHFE. The idea is to create the perturbation internally using only the plaintext variables without introducing any external variables.

The notation will largely be as in Section 5.1. In particular, let k be a finite field with q elements, and let $g(x) \in k[x]$ be an irreducible polynomial of degree n. For this construction it will not be necessary to assume that k is of characteristic two. Define $K = k[x]/g(x)$ and $\phi : K \longrightarrow k^n$ as before. Let r be a small integer with $r \geq 1$.

Let $Z : K \longrightarrow K$ be a randomly chosen k-linear map

$$Z(X) = \sum_{i=0}^{n-1} z_i X^{q^i},$$

such that the dimension of the image space of Z is r. In other words, chose Z such that the nullity of the k-linear map $\phi \circ Z \circ \phi^{-1}$ from k^n to k^n is $n - r$.

Now define the map $\tilde{F} : K \longrightarrow K$ by

$$\begin{aligned}\tilde{F}(X) = &\sum_{i=0}^{r_2}\sum_{j=0}^{i} a_{ij} X^{q^i+q^j} + \sum_{i=0}^{r_1} b_i X^{q^i} \\ &+ \sum_{i=0}^{r_1}\sum_{j=0}^{n-1} c_{ij} X^{q^i} Z^{q^j} \\ &+ \sum_{i=0}^{n-1}\sum_{j=0}^{n-1} \alpha_{ij} Z^{q^i+q^j} + \sum_{i=0}^{n-1} \beta_i Z^{q^i} + \gamma,\end{aligned}$$

where $a_{ij}, b_i, c_{ij}, \alpha_{ij}, \beta_i, \gamma \in K$, $Z = Z(X)$, and r_2, r_1 are chosen so that $q^i + q^j$ if $j \leq i \leq r_2$ and $q^i < D$ if $i \leq r_1$.

The map $F : k^n \longrightarrow k^n$ is defined by

$$F = \phi \circ \tilde{F} \circ \phi^{-1},$$

and the map $\bar{F} : k^n \longrightarrow k^n$ is defined by

$$\bar{F} = L_1 \circ F \circ L_2,$$

where $L_1, L_2 : k^n \longrightarrow k^n$ are two randomly chosen invertible affine transformations. The polynomial components $\bar{f}_1, \ldots, \bar{f}_n$ form the public key of the IPHFE public key cryptosystem.

Public Key

The public key for IPHFE consists of the following:

1.) The finite field k, including its additive and multiplicative structure;

2.) The map $\bar{F}$, or equivalent, the polynomial components $\bar{f}_1, \ldots, \bar{f}_n$.

Private Key

The private key for IPHFE consists of the following:

1.) The invertible affine transformations L_1, L_2;

2.) The internal perturbation map Z and the map $\tilde{F}$.

Encryption

To encrypt the plaintext message $(x_1', \ldots, x_n')$, we simply compute

$$(y_1', \ldots, y_n') = \bar{F}(x_1', \ldots, x_n').$$

Decryption

To decrypt the ciphertext $(y_1', \ldots, y_n')$, first we compute

$$W = \phi^{-1} \circ L_1^{-1}(y_1', \ldots, y_n').$$

Now for each element Z' in the image space of the mapping Z, we substitute Z' into $\tilde{F}$ to create a new polynomial $\tilde{F}_{Z'}$. We then solve

$$\tilde{F}_{Z'}(X) = W.$$

If this equation has no solution, then we must pick another Z' and try again. There are q^r choices for Z', and for each Z' it is very likely that the equation $\tilde{F}_{Z'} = W$ will have a solution. Thus it is very likely that we can complete this step. Let X' be a solution; we must check that

$Z(X') = Z'$, otherwise we discard X' and start over. This last test will help eliminate most of the unwanted solutions.

Once we have X', we compute the plaintext as

$$(x'_1, \ldots, x'_n) = L_2^{-1} \circ \phi(X').$$

In general, we expect that it is very likely that we have only one solution. However, as with HFE, F is not necessarily injective. Therefore we can add values created by a hash function or use the Plus method to help find the true plaintext.

A Practical Implementation of IPHFE

It is suggested in [Ding and Schmidt, 2005a] that we should take K to be a degree $n = 89$ extension of the finite field $k = GF(2)$ with $q = 2$ elements. The parameters $D = 9$ and $r = 2$ are also suggested. In this case, one can choose the coefficients of $X^{2^3+2^3}$ to be zero. The size of the public key is approximately 50 KB, making this implementation comparable with any existing multivariate cryptosystem. The decryption process requires us to solve an equation of degree 16 over a finite field of size 2^{89} four times, which can be done rather efficiently.

Security Analysis of IPHFE

In [Ding and Schmidt, 2005a], a brief argument is presented to claim that the existing algebraic attacks on previous multivariate cryptosystems cannot be used efficiently against IPHFE, including the method used to attack HFEv. The intuitive reason is that the internal perturbation is fully mixed with the original system and cannot be effectively separated out with the previous techniques.

For example, consider the attack method of [Kipnis and Shamir, 1999; Courtois, 2001] for HFE. From the formula for Z we can see that $\tilde{F}$, when described as a polynomial of X, looks far more complicated than $\tilde{F}$ in the HFE scheme. For IPHFE, $\tilde{F}$ has all possible terms of the form $X^{q^i+q^j}$, and so the corresponding symmetric matrix of the associated bilinear form is expected to have a very high rank in general. In computer simulations in [Ding and Schmidt, 2005a], it turns out that the rank of this matrix is exactly equal to $r+1+\log_q D$. Therefore, it is conjectured that the rank of this matrix is always exactly equal to $r + 1 + \log_q D$, and it might be possible to prove this statement.

Now consider the method of Kipnis and Shamir applied to IPHFE. In the fist step, the MinRank method is used to recover part of the key secret key (namely L_1), and we know that for this step the computational complexity is $89^{3\times 6} > 2^{120}$. Even if we suppose that this could be done and that we have L_1, we have no idea what the matrix corresponding to the original polynomial $\tilde{F}$ is due to the internal perturbation of Z. This situation is far more complicated than it is with HFE, and in particular we have no way of knowing what the null space is like, making recovery of L_2 impossible. Therefore the Kipnis and Shamir method, and in particular the MinRank method, cannot be used to attack IPHFE efficiently. Similarly, the attack used on HFEv cannot succeed due to the internal mixing of the perturbation resulting in an unknown matrix structure.

It would seem then that the only possible efficient attack must use the XL or Gröbner basis algorithms. Our preliminary computer experiments in this direction have shown that this may not be a promising approach, though more extensive experiments must be performed to confirm this. In order to fully understand how IPHFE can resist such attacks, we need to study how the attack complexity changes as r changes for a fixed D. Computer simulations should give us some reasonable way of estimating this behavior, but it is in general a rather time- and memory-consuming task. Recent results in [Diem, 2004] show that the new Gröbner basis algorithm is actually more powerful than the XL method, which means that we can focus on how IPHFE behaves under an attack by Gröbner basis algorithms.

Overall, in accordance with the estimates of the attack complexity for each of the existing attack methods, the security of IPHFE should be at least 2^{80}. In [Ding and Schmidt, 2005a], it is speculated that the security could be even higher, meaning that the best attack method against IPHFE system will be the brute force checking of all possible plaintexts one-by-one.

Efficiency

Another issue we should consider is that of the effect of internal perturbation on the efficiency of the scheme, since it could be argued that we can increase the security of HFE by simply increasing d. However, the problem with increasing d is that the computational complexity in the decryption process increases by a multiple of at least d^2. On the other hand, internal perturbation actually allows us to decrease d (and still maintain the same level of security). In this way we see that IPHFE can actually be much more efficient despite the necessity of the searching process of size q^r.

5.5 Internal Perturbation and Related Work

As mentioned above, internal perturbation is a very general method that can be applied to other MPKCs. For example, in [Wu et al., 2005], internal perturbation was applied to the Hidden Matrix cryptosystem. However, as was pointed out in [Wolf and Preneel, 2005c], the application of internal perturbation may not always improve the cryptosystem if the system is not designed properly. More theoretical research in this area should be pursued.

Recently Patarin [Patarin, 2006] proposed a totally new idea of building MPKCs called probabilistic MPKCs. For example, in a signature scheme, a signature is accepted whenever a substantial number of public equations are satisfied by the document and its signature. The construction of such an example in the paper is also closely related to the idea of internal perturbation.

Also in the recent publications of Tsujii, Tadaki and Fujita [Tsujii et al., 2004; Tsujii et al., 2006], a new family of MPKCs with the name Piece in Hand inspired from the Japanese game *Go* (*Wei-qi* or *Wei-chi* in Chinese) was proposed, and it seems that some of their ideas are in spirit very similar to, or closely related to, internal perturbation.

Chapter 6

TRIANGULAR SCHEMES

Among all the existing constructions of multivariate schemes, the triangular family is a very special one whose origin really belongs to algebraic geometry. The Tame Transformation Method (TTM) cryptosystem was first proposed by T. T. Moh [Moh, 1999a] with a patent in the US in 1998. The origin of this construction can be traced back to the work of Fell and Diffie [Fell and Diffie, 1986], who were unable to find an efficient and secure triangular scheme. A more general form of triangular map called "sequential solution type" was used by Tsujii, Kurosawa, Fujioka, and Matsumoto [Tsujii et al., 1986] to build MPKCs. However, these schemes were defeated by Hasegawa and Kaneoka [Hasegawa and Kaneko, 1987], and by Okamoto and Nakamura [Okamoto and Nakamura, 1986]. A similar form of triangular map was also used by Shamir [Shamir, 1993] to build signature schemes, but not long thereafter it was broken by Coppersmith, Stern and Vaudenay [Coppersmith et al., 1997]. The motivation for this family of schemes is based on the difficulty of decomposing a composition of invertible (nonlinear) polynomial maps. This is closely related to the famous Jacobian Conjecture.

The focus of this chapter is on the cryptanalysis of the TTM cryptosystem and the Tame Transformation Signature (TTS) schemes, which can be viewed as extensions of TTM.

6.1 The Jacobian Conjecture and Tame Transformations

The Jacobian conjecture is a celebrated problem in mathematics about polynomials in several variables. It was first proposed in 1939 by Ott-Heinrich Keller. It was later promoted by Shreeram Abhyankar as an example of a question in algebraic geometry that requires only basic knowledge of calculus to understand. The Jacobian conjecture is also one of Smale's problems [Smale, 1998].

Fix an integer $n > 1$ and a polynomial map $G : \mathbb{C}^n \longrightarrow \mathbb{C}^n$

$$G(x_1, \ldots, x_n) = (g_1(x_1, \ldots, x_n), \ldots, g_n(x_1, \ldots, x_n)).$$

The Jacobian determinant $J(x_1, \ldots, x_n)$ of the map G is the determinant of the $n \times n$ matrix $[g_{ij}]$, where

$$g_{ij} = \frac{\partial g_i}{\partial x_j},$$

is the partial derivative of $g_i(x_1, \ldots, x_n)$ with respect to x_j.

It is relatively straightforward to show that if G is invertible, then $J(x_1, \ldots, x_n)$ must be a nonzero constant. The Jacobian conjecture is the converse for polynomial maps, which states that if $J(x_1, \ldots, x_n)$ is a nonzero constant then G is an invertible map. Several proofs for the two-variable case have been announced, but all were found to have fatal errors. If G is analytical then the statement is not true.

One well-known family of invertible maps is the set of de Jonquières (or triangular) maps. A map $G : k^n \longrightarrow k^n$ is a de Jonquières map if it is of the form:

$$G(x_1, \ldots, x_n) = \begin{pmatrix} x_1 + g_1(x_2, \ldots, x_n) \\ x_2 + g_2(x_3, \ldots, x_n) \\ \vdots \\ x_{n-1} + g_{n-1}(x_n) \\ x_n \end{pmatrix}^T,$$

where k is any field, and $g_i \in k[x_1, \ldots, x_n]$ are arbitrary polynomials.

Due to the very special structure of de Jonquières maps, the inverse can be easily (efficiently) calculated. Furthermore, because these maps are invertible, the closure of this set under composition forms a group. This group is called the group of tame transformations, and any invertible map which is not tame is called a wild transformation. It is a highly nontrivial problem to find an example of a wild transformation, and the famous Nagata problem is such an example [Nagata, 1972].

A more general form of triangular maps was used in [Shamir, 1993]. Such a map $\tilde{G}$ is of the form:

$$\tilde{G}(X) = \begin{pmatrix} x_1 \cdot l_1(x_2, \ldots, x_n) + g_1(x_2, \ldots, x_n) \\ x_2 \cdot l_2(x_3, \ldots, x_n) + g_2(x_3, \ldots, x_n) \\ \vdots \\ x_{n-1} \cdot l_{n-1}(x_n) + g_{n-1}(x_n) \\ x_n \end{pmatrix}^T,$$

where the functions $l_i \in k[x_1, \ldots, x_n]$ are linear (or affine) and the functions $g_i \in k[x_1, \ldots, x_n]$ are quadratic. Such a map $\tilde{G}$ is called a sequentially linearized map. It is clear that if the l_i are chosen to be a non-zero constant, then $\tilde{G}$ is a de Jonquières map. An even more general form of such a map is the case where $l_i \in k[x_1, \ldots, x_n]$ and the $g_i \in k[x_1, \ldots, x_n]$ are rational functions. These maps are called the "sequential solution type," which were used much earlier by Tsujii, Kurosawa, Fujioka, and Matsumoto [Tsujii et al., 1986] to build MPKCs in the 1980s.

6.2 Basic TTM Cryptosystems

Suppose $F : k^n \longrightarrow k^n$ is a composition of l invertible maps $G_1, \ldots, G_l$

$$F(x_1, \ldots, x_n) = G_1 \circ G_2 \circ \cdots \circ G_l(x_1, \ldots, x_n) = (f_1, \ldots, f_n),$$

where each G_i is a map from k^n to k^n and $f \in k[x_1, \ldots, x_n]$. In addition, suppose F has the following properties:

1.) Given $(x'_1, \ldots, x'_n) \in k^n$, $F(x'_1, \ldots, x'_n)$ is efficiently computed;

2.) Given only the polynomial components $f_1, \ldots, f_n$ of F, it is difficult to recover the composition factors $G_1, \ldots, G_l$.

If we further assume that it is computationally difficult to directly solve $F(x_1, \ldots, x_n) = (y'_1, \ldots, y'_n)$ with $(y'_1, \ldots, y'_n) \in k^n$, then we can use F to build a public key cryptosystem. The TTM encryption schemes are built in this way using tame transformations over a finite field.

In some sense all the previous constructions we have considered are built in this way. However, the fundamental difference between TTM and the other constructions we have seen is that TTM uses more than one nonlinear factor in its composition. If we consider public key size and the complexity of public computations, we would like F to be quadratic. Unfortunately this is something that seems to be very difficult to accomplish if we want to maintain a high level of security. The reason is that the degree of a composition of nonlinear tame transformations normally grows very fast as the number of nonlinear composition factors is increased. In this sense, the TTM construction is one attempt at balancing the dual needs of security and efficiency.

In [Moh, 1999a], a quadratic construction is obtained by using the embedding map $\iota : k^n \longrightarrow k^m$ defined by

$$\iota(x_1, \ldots, x_n) = (x_1, \ldots, x_n, 0, \ldots, 0),$$

and then defining $\bar{F} : k^n \longrightarrow k^m$ by

$$\bar{F}(x_1, \ldots, x_n) = F \circ \iota(x_1, \ldots, x_n).$$

Here $F : k^m \longrightarrow k^m$ is defined as the composition

$$F(x_1, \ldots, x_m) = \phi_4 \circ \phi_3 \circ \phi_2 \circ \phi_1(x_1, \ldots, x_m),$$

where ϕ_1, ϕ_4 are invertible linear maps and ϕ_2, ϕ_3 are de Jonquières maps. Additionally, ϕ_2 is of degree two and ϕ_3 is of high degree (say ≥ 8), and ϕ_1 is a map such that

$$\phi_1(x_1, \ldots, x_{n+v}) = (\tilde{\phi}_1(x_1, \ldots, x_n), x_{n+1}, \ldots, x_{n+v}),$$

where $m = n + v$ and $\tilde{\phi}_1$ is an invertible affine transformation on k^n. It is clear that such a map $\bar{F}$ is invertible by construction.

The key component in the construction of TTM is based on a special multivariate polynomial, often denoted by $Q_8(z_1, \ldots, z_l)$, along with a special set of quadratic polynomials $q_i(z_1, \ldots, z_n)$, where $i = 1, \ldots, l$. These polynomials are chosen so that the composition

$$Q_8(q_1(z_1, \ldots, z_l), \ldots, q_l(z_1, \ldots, z_l))$$

is quadratic in terms of the variables z_i, $i = 1, \ldots, l$. These constructions are very interesting from both a theoretical and a practical point of view, and in particular from the point of view of algebraic geometry.

Triangular-Plus-Minus Cryptosystems

The first attacks on the first generation of the TTM cryptosystems use an approach which we have seen before, the MinRank attack . The very first attack on the TTM was presented by Goubin and Courtois [Goubin and Courtois, 2000]. In order to present their attack in a more general way, they first formulated a new family of cryptosystems that include all TTM schemes as a subfamily from the point of view of the MinRank attack.

Triangular-Plus-Minus (TPM) cryptosystems are described by four parameters n, u, r and k, and an implementation with these choices is denoted TPM(n, u, r, k). As usual, k is a given finite field, and n, u, and r are non-negative integers such that $n \geq r$. In what follows, we will use the notation $m = n + u - r$.

We first consider a map F from k^n to $k^m = k^{n+u-r}$ such that

$$F(x_1, \ldots, x_n) = (f_1(x_1, \ldots, x_n), \ldots, f_m(x_1, \ldots, x_n)),$$

where

$$\begin{aligned}
f_1 &= x_1 + g_1(x_{n-r+1}, \ldots, x_n)\\
f_2 &= x_2 + g_2(x_1; x_{n-r+1}, \ldots, x_n)\\
f_3 &= x_3 + g_3(x_1, x_2; x_{n-r+1}, \ldots, x_n)\\
&\vdots\\
f_{n-r} &= x_{n-r} + g_{n-r}(x_1, \ldots, x_{n-r-1}; x_{n-r+1}, \ldots, x_n)\\
f_{n-r+1} &= g_{n-r+1}(x_1, \ldots, x_n)\\
&\vdots\\
f_{n-r+u} &= g_{n-r+u}(x_1, \ldots, x_n)
\end{aligned}$$

and each g_i (for $1 \leq i \leq m$) is a randomly chosen quadratic polynomial.

The public polynomials for an implementation of TPM(n, u, r, k) are given as the components of the map $\bar{F} : k^n \longrightarrow k^m$ defined by

$$\bar{F} = L_1 \circ F \circ L_2 = (\bar{f}_1, \ldots, \bar{f}_m),$$

where $L_1 : k^m \longrightarrow k^m$ and $L_2 : k^n \longrightarrow k^n$ are invertible affine transformations.

The set-up of this cryptosystem is just like what we have seen before. The secret key includes F, L_1 and L_2. However the decryption process requires that we do a search of size q^r (we do a search on $x_{n-r+1}, \ldots, x_n$ variables when we try to "invert" F), where q is the size of k.

Decryption

Given a ciphertext $(y'_1, \ldots, y'_m) \in k^m$, the plaintext can be obtained by executing the following steps.

1.) Compute $(w_1, \ldots, w_m) = L_1^{-1}(y'_1, \ldots, y'_m)$.

2.) For each r-dimensional vector $(z_{n-r+1}, \ldots, z_n) \in k^r$, compute in order the corresponding $z_1, \ldots, z_{n-r}$ to derive the vector $(z_1, \ldots, z_n)$ from the equations

$$z_i = w_i - g_i(z_1, \ldots, z_{i-1}; z_{n-r+1}, \ldots, z_n),$$

for $1 \leq i \leq n - r$, and check if this vector indeed satisfies all the equations

$$g_i(z_1, \ldots, z_n) = w_i,$$

for $n - r + 1 \leq i \leq n - r + u$. If the answer is negative, we discard this vector; if the answer is positive, then we proceed to the next step.

3.) Calculate the plaintext as

$$(x'_1, \ldots, x'_n) = L_2^{-1}(z_1, \ldots, z_n).$$

Clearly the decryption process has a complexity of $O(q^r)$, and therefore a TPM(n, u, r, k) cryptosystem will be practical only if q^r is not too large. If $u > r$ and $u - r$ is large enough, then there is a very high probability that F is collision-free, and thus the cipher $\bar{F}$ can be considered as an injective map from k^n into k^{n+u-r}. This scheme has been considered and attacked by Fell and Diffie [Fell and Diffie, 1986] and by Patarin and Goubin [Patarin and Goubin, 1997]. The decryption process of TTM does not require a search process of size q^r. This explains why TTM is much more efficient than TPM and in this sense TTM is not a subfamily of TPM.

It is not difficult to see that TPM comes from a triangular construction by applying the Minus method (deleting r polynomials) and then the Plus method (adding u polynomials). The reason for this is that if we use only a triangular map, then there is no way that we can hide the one-dimensional linear function in the span of the public key polynomials. This could be used to easily break the cryptosystem. Thus, the Minus and Plus methods are combined to improve the security (using Minus) and ensure the injectivity of the map (using Plus).

The signature scheme proposed in [Shamir, 1993] is nothing but the Minus method applied to the sequential linearized map defined in (6.1) with $u = 0$ and $r = 1$.

First Generation of TTM Cryptosystems

The first set of the TTM cryptosystems was proposed by T.T. Moh in [Moh, 1999a; Moh, 1999b], who is an expert in the research on the Jacobian conjecture. Let $k = GF(256)$ be a finite field with 256 elements.

The map $\phi_2 : k^{n+v} \longrightarrow k^{n+v}$ with

$$\phi_2(x_1, \ldots, x_{n+v}) = (\phi_{2,1}, \ldots, \phi_{2,n+v})$$

is defined by

$$\begin{aligned}
\phi_{2,1} &= x_1 \\
\phi_{2,2} &= x_2 + g_2(x_1) \\
\phi_{2,3} &= x_3 + g_3(x_1, x_2) \\
&\vdots \\
\phi_{2,n} &= x_n + g_n(x_1, \dots, x_{n-1}) \\
\phi_{2,n+1} &= x_{n+1} + g_{n+1}(x_1, \dots, x_n) \\
&\vdots \\
\phi_{2,n+v} &= x_{n+v} + g_{n+v}(x_1, \dots, x_{n+v-1}).
\end{aligned}$$

The g_i are specially chosen quadratic polynomials over k. Now define the map $\phi_3 : k^{n+v} \longrightarrow k^{n+v}$ with

$$\phi_3(x_1, \dots, x_{n+v}) = (\phi_{3,1}, \dots, \phi_{3,n+v}),$$

where

$$\begin{aligned}
\phi_{3,1} &= x_1 + P(x_{n+1}, \dots, x_{n+v}) \\
\phi_{3,2} &= x_2 + Q(x_{n+1}, \dots, x_{n+v}) \\
\phi_{3,3} &= x_3 \\
&\vdots \\
\phi_{3,n+v} &= x_{n+v}.
\end{aligned}$$

The P, Q are two polynomials of degree eight over k. Finally define $\Phi : k^{n+v} \longrightarrow k^{n+v}$ by the composition

$$\Phi_2 = \phi_3 \circ \phi_2 = (\Phi_{2,1}, \dots, \Phi_{2,n+v}),$$

where

$$\begin{aligned}
\Phi_{2,1} &= x_1 \\
&\quad + P(x_{n+1} + g_{n+1}(x_1, \dots, x_n), \dots, x_{n+v} + g_{n+v}(x_1, \dots, x_{n+v-1})) \\
\Phi_{2,2} &= x_2 + g_2(x_1) \\
&\quad + Q(x_{n+1} + g_{n+1}(x_1, \dots, x_n), \dots, x_{n+v} + g_{n+v}(x_1, \dots, x_{n+v-1})) \\
\Phi_{2,3} &= x_3 + g_3(x_1, x_2) \\
&\vdots \\
\Phi_{2,n} &= x_n + g_n(x_1, \dots, x_{n-1}) \\
\Phi_{2,n+1} &= x_{n+1} + g_{n+1}(x_1, \dots, x_n)
\end{aligned}$$

$$\vdots$$

$$\Phi_{2,n+v} = x_{n+v} + g_{n+v}(x_1, \ldots, x_{n+v-1}).$$

Moh found a special way of selecting P, Q and g_i such that $\Phi_{2,1}$ and $\Phi_{2,2}$ both are quadratic in the variables $x_1, \ldots, x_n$ when the last v variables $x_{n+1}, \ldots, x_{n+v}$ are set equal to zero.

Denote j mod 8 by $[j]$, taking $1 \le [j] \le 8$. The de Jonquières map ϕ_2 is defined by

$$\begin{aligned}
\phi_{2,i} &= x_i && \text{for } i = 1,2\\
\phi_{2,i} &= x_i + x_{i-1}x_{i-2} && \text{for } i = 3,\ldots,9\\
\phi_{2,i} &= x_i + x_{[i-1]}^2 + x_{[i]}x_{[i-5]} + x_{[i+1]}x_{i+6} && \text{for } i = 10,\ldots,17\\
\phi_{2,i} &= x_i + x_{[i-1]}x_{[i+1]} + x_{[i]}x_{[i+4]} && \text{for } i = 18,\ldots,25\\
\phi_{2,i} &= x_i + x_{[i-1]}x_{[i+1]} + x_{[i+2]}x_{[i+5]} && \text{for } i = 26,\ldots,30\\
\phi_{2,i} &= x_i + x_{i-10}^2 && \text{for } i = 31,\ldots,60\\
\phi_{2,61} &= x_{61} + x_9^2 && \text{for } i = 61\\
\phi_{2,62} &= x_{62} + x_{61}^2 && \text{for } i = 62\\
\phi_{2,63} &= x_{63} + x_{10}^2 && \text{for } i = 63\\
\phi_{2,64} &= x_{64} + x_{63}^2 && \text{for } i = 64\\
\phi_{2,i} &= x_i + q_{i-64}(x_9, x_{11}, \ldots, x_{16}, x_{51}, x_{52}, \ldots, x_{62}) && \text{for } i = 65,\ldots,92\\
\phi_{2,i} &= x_i + q_{i-92}(x_{10}, x_{17}, \ldots, x_{20}, x_{15}, x_{16}, x_{51}, \ldots, x_{60}, x_{63}, x_{64})\\
& && \text{for } i = 93,\ldots,100,
\end{aligned}$$

where the polynomials $q_i(z_1, \ldots, z_{19})$ are defined by

$$\begin{aligned}
q_1 &= z_1 + z_2 z_6 & q_2 &= z_2^2 + z_3 z_7\\
q_3 &= z_3^2 + z_4 z_{10} & q_4 &= z_3 z_5\\
q_5 &= z_3 z_{11} & q_6 &= z_4 z_7\\
q_7 &= z_4 z_5 & q_8 &= z_7^2 + z_5 z_{11}\\
q_9 &= z_6^2 + z_8 z_9 & q_{10} &= z_8^2 + z_{12} z_{13}\\
q_{11} &= z_9^2 + z_{14} z_{15} & q_{12} &= z_7 z_{10}\\
q_{13} &= z_{10} z_{11} & q_{14} &= z_{12}^2 + z_7 z_8\\
q_{15} &= z_{13}^2 + z_{11} z_{16} & q_{16} &= z_{14}^2 + z_{10} z_{12}\\
q_{17} &= z_{15}^2 + z_{11} z_{17} & q_{18} &= z_{12} z_{16}\\
q_{19} &= z_{11} z_{12} & q_{20} &= z_8 z_{13}\\
q_{21} &= z_7 z_{13} & q_{22} &= z_8 z_{16}
\end{aligned}$$

$$q_{23} = z_{14}z_{17} \qquad q_{24} = z_7z_{11}$$
$$q_{25} = z_{12}z_{15} \qquad q_{26} = z_{10}z_{15}$$
$$q_{27} = z_{12}z_{17} \qquad q_{28} = z_{11}z_{14}$$
$$q_{29} = z_{18} + z_1^2 \qquad q_{30} = z_{19} + z_{18}^2.$$

The other de Jonquières map ϕ_3 is defined by

$$\begin{aligned}
\phi_{3,1} &= x_1 + Q_8(x_{65}, x_{66}, \ldots, x_{92}, x_{61}, x_{62}) \\
\phi_{3,2} &= x_2 + Q_8(x_{93}, \ldots, x_{100}, x_{73}, \ldots, x_{92}, x_{63}, x_{64}) \\
\phi_{3,i} &= x_i \qquad \text{for } i = 3, \ldots, 100,
\end{aligned}$$

where $Q_8 : k^{30} \longrightarrow k$ is given by

$$\begin{aligned}
Q_8(z_1, \ldots, z_{30}) &= z_1^8 \\
&+ \left(z_2^4 + z_3^4 + z_3^2z_8^2 + z_4^2z_5^2 + z_6^2z_{12}^2 + z_7^2z_{13}^2\right) \\
&\quad \times \left(z_9^4 + (z_{10}^2 + z_{14}z_{15} + z_{18}z_{19} + z_{20}z_{21} + z_{22}z_{24}\right) \\
&\quad\quad \times \left(z_{11}^2 + z_{16}z_{17} + z_{23}z_{28} + z_{25}z_{26} + z_{13}z_{27})\right) \\
&+ z_{29}^4 + z_{30}^2.
\end{aligned}$$

The map ϕ_1 is an invertible affine transformation satisfying certain restrictions that in particular imply that ϕ_1 is a trivial extension of an affine linear map $\tilde{\phi}_1 : k^{64} \longrightarrow k^{64}$. The most important consequence of these restrictions for what follows is that

$$\phi_1 \circ \iota = \iota \circ \tilde{\phi}_1,$$

where $\iota : k^{64} \longrightarrow k^{100}$ is the embedding map

$$\iota(x_1, \ldots, x_{64}) = (x_1, \ldots, x_{64}, 0, \ldots, 0).$$

Finally ϕ_4 is an arbitrary invertible affine map. To simplify the notation, we set

$$\begin{aligned}
\hat{\phi}_1 &= \phi_1 \circ \iota \\
\hat{\phi}_{21} &= \phi_2 \circ \phi_1 \circ \iota \\
\hat{\phi}_{321} &= \phi_3 \circ \phi_2 \circ \phi_1 \circ \iota \\
\hat{\phi}_2 &= \phi_2 \circ \iota \\
\hat{\phi}_{32} &= \phi_3 \circ \phi_2 \circ \iota.
\end{aligned}$$

The components of $\hat{\phi}_{32}$ are then given by

$$
\begin{aligned}
&\hat{\phi}_{32,1} = x_1 + x_{62}^2 \\
&\hat{\phi}_{32,2} = x_2 + x_{64}^2 \\
&\hat{\phi}_{32,i} = x_i + x_{i-1}x_{i-2} && \text{for } i = 3, \ldots, 9 \\
&\hat{\phi}_{32,i} = x_i + x_{[i-1]}^2 + x_{[i]}x_{[i-5]} + x_{[i+1]}x_{i+6} && \text{for } i = 10, \ldots, 17 \\
&\hat{\phi}_{32,i} = x_i + x_{[i-1]}x_{[i+1]} + x_{[i]}x_{[i+4]} && \text{for } i = 18, \ldots, 25 \\
&\hat{\phi}_{32,i} = x_i + x_{[i-1]}x_{[i+1]} + x_{[i+2]}x_{[i+5]} && \text{for } i = 26, \ldots, 30 \\
&\hat{\phi}_{32,i} = x_i + x_{i-10}^2 && \text{for } i = 31, \ldots, 60 \\
&\hat{\phi}_{32,61} = x_{61} + x_9^2 \\
&\hat{\phi}_{32,62} = x_{62} + x_{61}^2 \\
&\hat{\phi}_{32,63} = x_{63} + x_{10}^2 \\
&\hat{\phi}_{32,64} = x_{64} + x_{63}^2 \\
&\hat{\phi}_{32,i} = x_i + q_{i-64}(x_9, x_{11}, \ldots, x_{16}, x_{51}, x_{52}, \ldots, x_{62}) \\
& && \text{for } i = 65, \ldots, 92 \\
&\hat{\phi}_{32,i} = x_i + q_{i-92}(x_{10}, x_{17}, \ldots, x_{20}, x_{15}, x_{16}, x_{51}, \ldots, x_{60}, x_{63}, x_{64}) \\
& && \text{for } i = 93, \ldots, 100.
\end{aligned}
$$

This can be checked by direct computation. It is now clear that $\bar{F}$ is indeed quadratic. From the point of view of the MinRank attack the TTM cryptosystems can be viewed as a special case of the general TPM family; that is, they can be viewed as instances of TPM$(64, 38, 2, GF(256))$.

6.3 The MinRank Attack on TPM & TTM

The MinRank problem has already been introduced, including a way of solving it using determinant of certain submatrices. In this section we will present a different way of solving the MinRank problem [Goubin and Courtois, 2000]. Other methods are also presented in [Courtois, 2001]. Among all known methods, it is not possible to say which one is the best due to various considerations, for example the size of the finite field. We will now present the general idea of an attack on TPM(n, u, r, k).

We assume that $m = n+u-r$ is not too large compared with n, since otherwise the problem is easily solved. Also, from the TTM construction, let us assume that $m \leq 2n$. The key idea of the attack is to look at the structure of the polynomials as bilinear forms and to use the rank of the associated matrices.

From the composition $\bar{F} = L_1 \circ F \circ L_2$, we know that L_2 is just a change of variables and that L_1 mixes the polynomials together in such a way that we cannot see the triangular structure. We begin by looking at the structure of F.

Again, for each polynomial component

$$f_i = \sum A_{i,lj} x_l x_j + \sum B_{i,l} x_l + C_i,$$

with $1 \leq i \leq n$, we can derive an $n \times n$ matrix, A_i. Let

$$\mathsf{A}_i = \bar{\mathsf{A}}_i^T + \bar{\mathsf{A}}_i.$$

For the case of odd characteristic, we have that the quadratic part of f_i is actually equal to

$$\mathsf{X}^T \bar{\mathsf{A}}_i \mathsf{X},$$

where $\mathsf{X}^T = (x_1, \ldots, x_n)$. A_i is the symmetric matrix associated with f_i.

Similarly, for each component $\bar{f}_i$ of $\bar{F}$, we associate a symmetric matrix B_i. Because $\bar{F}$ is the public key, the $\bar{\mathsf{B}}_i$ are known.

Let

$$L_1(x_1, \ldots, x_m) = (x_1, \ldots, x_m)\hat{\mathsf{L}}_1 + \mathsf{a}_1$$

and

$$L_2(x_1, \ldots, x_n) = (x_1, \ldots, x_n)\hat{\mathsf{L}}_2 + \mathsf{a}_2,$$

where $\hat{\mathsf{L}}_2$ is an $n \times n$ matrix and $\hat{\mathsf{L}}_1$ is an $m \times m$ matrix. Then we have

$$\mathsf{B}_i = \sum_{j=1}^{m} [\mathsf{L}_1]_{ji} (\hat{\mathsf{L}}_1 \, \mathsf{A}_j \, \hat{\mathsf{L}}_1^T).$$

This also gives us

$$\hat{\mathsf{L}}_1 \, \mathsf{A}_i \, \hat{\mathsf{L}}_1^T = \sum_{j=1}^{m} [\bar{\mathsf{L}}_2]_{ji} \, \mathsf{B}_j,$$

where $\bar{\mathsf{L}}_2 = \hat{\mathsf{L}}_2^{-1}$ and

$$L_2^{-1}(x_1, \ldots, x_n) = (x_1, \ldots, x_n)\bar{\mathsf{L}}_2 + \bar{\mathsf{a}}_2, \tag{6.1}$$

with $\bar{\mathsf{a}}_2 = \mathsf{a}_2 \bar{\mathsf{L}}_2$. From the definition we know that

$$\mathsf{A}_1 =$$

$$\begin{pmatrix}
0 & \cdot & \cdot & 0 & 0 & \cdot & \cdot & 0 \\
0 & \cdot & \cdot & 0 & 0 & \cdot & \cdot & 0 \\
\cdot & \cdot & \cdot & \cdot & 0 & \cdot & \cdot & 0 \\
\cdot & \cdot & \cdot & 0 & [\mathsf{a}_1]_{n-r+1,n-r+2} & \cdot & \cdot & [\mathsf{a}_1]_{n,n-r+1} \\
\cdot & \cdot & \cdot & \cdot & \cdot & \cdot & \cdot & \cdot \\
0 & \cdot & \cdot & \cdot & \cdot & \cdot & \cdot & \cdot \\
0 & \cdot & 0 & [\mathsf{a}_1]_{n,n-r+1} & \cdot & \cdot & \cdot & [\mathsf{a}_1]_{n,m-1} \\
0 & \cdot & 0 & [\mathsf{a}_1]_{n,n-r+1} & \cdot & \cdot & [\mathsf{a}_1]_{n,m-1} & 0
\end{pmatrix},$$

and therefore the rank of A_1 and $\hat{\mathsf{L}}_1 \mathsf{A}_1 \hat{\mathsf{L}}_1^T$ is less than or equal to r. The reason that the diagonal terms are all zero is that the field is of characteristic two.

To find the linear combination of B_i to derive $\hat{\mathsf{L}}_1 \mathsf{A}_1 \hat{\mathsf{L}}_1^T$ is the same situation as in the case of the MinRank attack on HFE. This means that finding $\hat{\mathsf{L}}_1 \mathsf{A}_1 \hat{\mathsf{L}}_1^T$ becomes exactly a MinRank problem. We will now present a different way of solving this problem using the structure of kernels of linear maps.

Alternative MinRank Algorithm

To find a solution of the MinRank problem, we execute the following steps [Goubin and Courtois, 2000].

1.) Let l be the integer defined by $l = \lceil \frac{m}{n} \rceil$. Randomly choose l vectors $X^{[1]}, \ldots, X^{[l]}$. Let c be any nonzero element in k. Since

$$\dim(\ker(c\hat{\mathsf{L}}_1 \mathsf{A}_1 \hat{\mathsf{L}}_1^T)) = n - \text{rank}(c\hat{\mathsf{L}}_1 \mathsf{A}_1 \hat{\mathsf{L}}_1^T) \geq n - r,$$

it is clear that the probability that $X^{[i]} \in \ker(c\hat{\mathsf{L}}_1 \mathsf{A}_1 \hat{\mathsf{L}}_1^T)$ for all $i = 1, \ldots, l$ is q^{-lr}.

2.) Suppose that we have chosen a set of l vectors $X^{[1]}, \ldots, X^{[l]}$ all belonging to the space $\ker(c\hat{\mathsf{L}}_1 \mathsf{A}_1 \hat{\mathsf{L}}_1^T)$. Then we can find a vector $(\lambda_1, \ldots, \lambda_m) \in k^m$ such that

$$(\sum_{i=1}^{m} \lambda_i \mathsf{B}_i) X^{[j]} = (0, \ldots, 0)^T,$$

for $j = 1, \ldots, l$, by solving a set of ln linear equations with m variables $\lambda_1, \ldots, \lambda_m$. Since $l = \lceil \frac{m}{n} \rceil$, the solution space has dimension one with probability nearly 1, and can easily be found using Gaussian elimination.

To make second step work, we will need to do a search of size q^{lr}. Therefore the complexity of the attack is $O(q^{\lceil \frac{m}{n} \rceil r} m^3)$.

Remark 6.3.1. *This is actually a very general method of solving the MinRank problem under the condition that q^r is not too large. In fact, it will be very efficient if $q^{\lceil \frac{m}{n} \rceil r}$ is very small. However, even if r is very small this method will not work well if q is large.*

Due to the triangular structure we can see that if r is odd, then we expect that in general we will find the solution space to have dimension one; namely, we find the space $c\hat{\mathsf{L}}_1 \mathsf{A}_1 \hat{\mathsf{L}}_1^T$, where c is an unknown constant.

If r is even, then we expect that the solution space will have dimension two. This is because any symmetric matrix with entries in a field of characteristic two and odd size will be singular. In this case the solution will give us the space $c_1 \hat{\mathsf{L}}_1 \mathsf{A}_1 \hat{\mathsf{L}}_1^T + c_2 \hat{\mathsf{L}}_1 \mathsf{A}_2 \hat{\mathsf{L}}_1^T$, where c_1, c_2 are unknown constants.

For the first case (r is odd), breaking the cryptosystem becomes very easy. From the matrix $\mathsf{E} = c \hat{\mathsf{L}}_1 \mathsf{A}_1 \hat{\mathsf{L}}_1^T$, we find the row null space of E and a basis for this space, say $v_1, \ldots, v_{n-r}$. We then extend this basis to form a basis $v_1, \ldots, v_n$. Let

$$\mathsf{L} = (v_1, \ldots, v_n)^T.$$

Then we know that

$$\mathsf{L}\,\mathsf{E}\,\mathsf{L}^T =$$

$$\begin{pmatrix}
0 & . & . & 0 & 0 & . & . & 0 \\
0 & . & . & 0 & 0 & . & . & 0 \\
. & . & . & . & 0 & . & . & 0 \\
. & . & . & 0 & \mathsf{E}'_{n-r+1,n-r+2} & . & . & \mathsf{E}'_{n,n-r+1} \\
. & . & . & . & . & . & . & . \\
0 & . & . & . & . & . & . & . \\
0 & . & 0 & \mathsf{E}'_{n,n-r+1} & . & . & . & \mathsf{E}'_{n,m-1} \\
0 & . & 0 & \mathsf{E}'_{n,n-r+1} & . & . & \mathsf{E}'_{n,m-1} & 0
\end{pmatrix},$$

or

$$\mathsf{E} =$$

$$\mathsf{L}^{-1}\begin{pmatrix}
0 & . & . & 0 & 0 & . & . & 0 \\
0 & . & . & 0 & 0 & . & . & 0 \\
. & . & . & . & 0 & . & . & 0 \\
. & . & . & 0 & \mathsf{E}'_{n-r+1,n-r+2} & . & . & \mathsf{E}'_{n,n-r+1} \\
. & . & . & . & . & . & . & . \\
0 & . & . & . & . & . & . & . \\
0 & . & 0 & \mathsf{E}'_{n,n-r+1} & . & . & . & \mathsf{E}'_{n,m-1} \\
0 & . & 0 & \mathsf{E}'_{n,n-r+1} & . & . & \mathsf{E}'_{n,m-1} & 0
\end{pmatrix}(\mathsf{L}^{-1})^T.$$

Let V' be the space spanned by the last r linear function components in $(x_1, \ldots, x_n)\mathsf{L}^{-1}$, which is the same space spanned by the last r components of $(x_1, \ldots, x_n)\hat{\mathsf{L}}_1$ since $\hat{\mathsf{L}}_1$ is invertible. We can easily calculate a basis of V' of the form

$$v_i = x_i + \sum_{1 \le j \le n-r} \alpha_{ij} x_j,$$

for $i = n - r + 1, \ldots, n$.

Now we can proceed to defeat the system. Assume that an attacker has the ciphertext $(y'_1, \ldots, y'_m)$. Then we have m quadratic equations from

$$(\bar{f}_1(x_1, \ldots, x_n), \ldots, \bar{f}_m(x_1, \ldots, x_n)) = (y'_1, \ldots, y'_m).$$

The attacker then, just like a legitimate user, will do a search of size q^r by letting $(v_{n-r+1}, \ldots, v_n)$ range through all possible values in k^r, which gives us a set of r linear equations. Once we have this set of equations, we use them to do a substitution of x_i $(i = n - r + 1, \ldots, n)$ by x_i $(i = 1, \ldots, n - r)$ from this set of equations. This substitution, from the point of algebraic geometry, is equivalent to restricting functions on a subspace, and thus F is equivalent to setting $x_{n-r+1}, \ldots, x_n$ constant.

Now we see that f_1 becomes a linear function. This means that after the substitution, the space spanned by the new set of m equations with $n-r$ variables must contain one nontrivial linear equation corresponding to the equations derived from f_1. Now we can substitute this new linear equation back in, and this time we will get another linear equation since this substitution is equivalent to setting x_1 to be constant. Then we can repeat this process, which in the end gives us n easily solved linearly independent equations.

One important point is that the search for linear equations in each of the steps above needs only be done once, which can then be used again in all the searches. After the first search, we just need to find u equations which are linearly independent from the equation corresponding to each linear equation we found. We then check at the end of each search to see if the solution we derive indeed satisfies these extra u equations. If this is so, then we have found the plaintext.

In the case that r is even, things are only slightly more complicated. In the first step we do the same thing as above, except that we will not search for a linear equation, but instead equations of the form

$$\sum \alpha_i x_i + \sum \beta_i x_i^2 + \gamma = 0.$$

However, we can easily covert these into linear equations due to the form of f_1 and f_2.

From this analysis we see that the most costly steps (in terms of computation) are solving the MinRank problem and searching for linear functions in the space of quadratic functions. In the cases of TTM, we can see that m and n are both relative small (namely, $n = 64$ and $m = 100$), so these steps will not be very expensive and can be done efficiently.

6.4 Another Attack on the First TTM Cryptosystem

Ding and Hodges [Ding and Hodges, 2004] found a very different way to attack the first TTM cryptosystem using the structure of the polynomial Q_8. The strategy of the attack is to find, for any given ciphertext $y' = (y'_1, \ldots, y'_{100})$, a linear subspace $V \subset k^{64}$ such that V contains the unknown plaintext $x' = (x'_1, \ldots, x'_{64})$ and $\bar{F}$ restricted to V coincides with a simpler function $\check{F} : k^{64} \longrightarrow k^{100}$ that is computationally invertible. To do this we find a linear subvariety of k^{64} on which ϕ_3 applied to the image of $\hat{\phi}_{21}$ is essentially linear.

To see what this entails, define

$$\rho_1(x_1, \ldots, x_{100}) = (x_{65}, \ldots, x_{92}, x_{61}, x_{62}),$$
$$\rho_2(x_1, \ldots, x_{100}) = (x_{93}, \ldots, x_{100}, x_{73}, \ldots, x_{92}, x_{63}, x_{64}).$$

Note that

$$(\hat{\phi}_{321})_1(x) = \phi_3(z) = \hat{\phi}_{21}(x)_1 + Q_8 \circ \rho_1 \circ \hat{\phi}_{21}(x)$$
$$(\hat{\phi}_{321})_2(x) = \phi_3(z) = \hat{\phi}_{21}(x)_2 + Q_8 \circ \rho_2 \circ \hat{\phi}_{21}(x)$$
$$(\hat{\phi}_{321})_i(x) = \hat{\phi}_{21}(x)_i \quad \text{for } i = 3, \ldots, 100.$$

Therefore it suffices to find a linear subvariety on which the functions $Q_8 \circ \rho_i \circ \hat{\phi}_{21}(x)$ are constant.

Define $q : k^{19} \longrightarrow k^{30}$, $\pi_i : k^{64} \longrightarrow k^{19}$ and $\rho_i : k^{100} \longrightarrow k^{30}$, for $i = 1, 2$, by

$$q(x_1, \ldots, x_{19}) = (q_1(x_1, \ldots, x_{19}), \ldots, q_{30}(x_1, \ldots, x_{19})),$$
$$\pi_1(x_1, \ldots, x_{64}) = (x_9, x_{11}, \ldots, x_{16}, x_{51}, x_{52}, \ldots, x_{62}),$$
$$\pi_2(x_1, \ldots, x_{64}) = (x_{10}, x_{17}, \ldots, x_{20}, x_{15}, x_{16}, x_{51}, \ldots, x_{60}, x_{63}, x_{64}).$$

Lemma 6.4.1. *For* $i = 1, 2$,

$$\rho_i \circ \hat{\phi}_2 = \rho_i \circ \hat{\phi}_{32} = q \circ \pi_i.$$

Proof. The fact that $\rho_i \circ \hat{\phi}_2 = q \circ \pi_i$ is a routine calculation. The remaining equality follows from the fact that $\rho_i \circ \phi_3 = \rho_i$. □

We now break the polynomial Q_8 into components:

$$Q_8(z_1, \ldots, z_{30}) = z_1^8 + S(z_1, \ldots, z_{30})^2[z_9^4 + T_1(z_1, \ldots, z_{30})T_2(z_1, \ldots, z_{30})] + z_{29}^4 + z_{30}^2,$$

where

$$S(z_1, \ldots, z_{30}) = z_2^2 + z_3^2 + z_3 z_8 + z_4 z_5 + z_6 z_{12} + z_7 z_{13},$$
$$T_1(z_1, \ldots, z_{30}) = z_{10}^2 + z_{14} z_{15} + z_{18} z_{19} + z_{20} z_{21} + z_{22} z_{24},$$
$$T_2(z_1, \ldots, z_{30}) = z_{11}^2 + z_{16} z_{17} + z_{23} z_{28} + z_{25} z_{26} + z_{13} z_{27}.$$

It is easily verified that

$$S(q_1, \ldots, q_{30}) = z_2^4,$$
$$T_1(q_1, \ldots, q_{30}) = z_8^4,$$
$$T_2(q_1, \ldots, q_{30}) = z_9^4,$$

and hence that $Q_8(q_1, \ldots, q_{30}) = z_{19}^2$.

First we find a linear subvariety on which S, T_1 and T_2 are constant. To do this we introduce a little more notation. Let $\mathcal{O}[k^{64}]$ denote the polynomial functions on k^{64} and let $\mathcal{L}$ denote the subspace of linear functions. Set $y_i = \bar{f}(x_1, \ldots, x_n)$. Denote by $\mathcal{S}$ the subspace of $\mathcal{O}[k^{64}]$ generated by the set $\{y_i^{2^6} y_j^{2^6}, y_i^{2^6}, y_i^{2^7}, 1 \mid 1 \leq i, j \leq 100\}$. Consider the space $\mathcal{L} \cap \mathcal{S}$. If $g \in \mathcal{L} \cap \mathcal{S}$, then we may write

$$g = \sum_{i,j} a_{ij} y_i^{2^6} y_j^{2^6} + \sum_l b_l y_l^{2^6} + \sum_l c_l y_l^{2^7} + d,$$

for some $a_{ij}, b_l, c \in k$. For any specific ciphertext $y' \in k^{100}$, we may set

$$g(y') = \sum_{i,j} a_{ij} (y_i')^{2^6} (y_j')^{2^6} + \sum_l b_l (y_l')^{2^6} + \sum_l c_l (y_l')^{2^7} + d.$$

Define the linear subvariety V_1 by

$$V_1 = \mathcal{V}(g - g(y') \mid g \in \mathcal{L} \cap \mathcal{S}).$$

The rationale for considering V_1 is that the functions $S \circ \rho_i \circ \hat{\phi}_2$, $T_1 \circ \rho_i \circ \hat{\phi}_2$ and $T_2 \circ \rho_i \circ \hat{\phi}_2$ are all constant on V_1. This fact is an immediate corollary of the following proposition.

Proposition 6.4.1. *The functions $S^{2^6} \circ \rho_i \circ \hat{\phi}_{21}$, $T_1^{2^6} \circ \rho_i \circ \hat{\phi}_{21}$ and $T_2^{2^6} \circ \rho_i \circ \hat{\phi}_{21}$ all lie in $\mathcal{L} \cap \mathcal{S}$.*

Proof. Notice that

$$S\rho_1 \circ \hat{\phi}_2(x) = S \circ q \circ \pi_1(x) = x_{11}^4,$$

hence $S^{2^6} \circ \rho_i \circ \hat{\phi}_2(x) = (x_{11}^4)^{2^6} = x_{11}$. But $S^{2^6} \circ \rho_i \circ \hat{\phi}_{21} = S^{2^6} \circ \rho_i \circ \hat{\phi}_2 \circ \tilde{\phi}$ and both $S^{2^6} \circ \rho_i \circ \hat{\phi}_2$ and $\tilde{\phi}$ are linear. Hence $S^{2^6} \circ \rho_i \circ \hat{\phi}_{21} \in \mathcal{L}$. Now observe that

$$S^{2^6} \circ \rho_1 \circ \hat{\phi}_{21}(x) = S^{2^6} \circ \rho_1 \circ \hat{\phi}_{321} = S^{2^6} \circ \rho_1 \circ \phi_4^{-1} \circ \bar{F}(x) = S^{2^6} \circ \rho_1 \circ \phi_4^{-1}(y).$$

Since $\rho_1 \circ \phi_4^{-1}$ is linear and

$$S(z)^{2^6} = z_2^{2^7} + z_3^{2^7} + z_3^{2^6} z_8^{2^6} + z_4^{2^6} z_5^{2^6} + z_6^{2^6} z_{12}^{2^6} + z_7^{2^6} z_{13}^{2^6},$$

it is clear that $S^{2^6} \circ \rho_1 \circ \hat{\phi}_{21} = S^{2^6} \circ \rho_1 \circ \phi_4^{-1} \circ \bar{F} \in \mathcal{S}$. An analogous argument works for the other five cases. □

It is easily verified that the six functions described in the above proposition are linearly independent. Thus $\dim(\mathcal{L} \cap \mathcal{S}) \geq 6$. However we will not need this fact in what follows.

We now repeat this procedure to restrict to a smaller linear subvariety. We define $\mathcal{T}$ to be the subspace of $\mathcal{O}[k^{64}]$ generated by the set $\{y_i^4, y_i^2, y_i, 1 \mid 1 \leq i \leq 100\}$. Consider the space

$$\mathcal{L}_{\mathcal{T}} = \{g \in \mathcal{L} \,:\, g|_{V_1} = h|_{V_1} \text{ for some } h \in \mathcal{T}\}.$$

Proposition 6.4.2. *The functions* $Q_8^{2^7} \circ \rho_i \circ \hat{\phi}_{21}$ *belong to* $\mathcal{L}_{\mathcal{T}}$.

Proof. The fact that $Q_8^{2^7} \circ \rho_i \circ \hat{\phi}_{21} \in \mathcal{L}$ follows from the fact that $Q_8(q_1, \ldots, q_{30}) = z_{19}^2$. As in the proof of the previous proposition,

$$Q_8^{2^7} \circ \rho_1 \circ \hat{\phi}_{21}(x) = Q_8^{2^7} \circ \rho_1 \circ \hat{\phi}_{321}(x) = Q_8^{2^7} \circ \rho_1 \circ \phi_4^{-1} \circ F(x) = Q_8^{2^7} \circ \rho_1 \circ \phi_4^{-1}(y)$$

and $\rho_1 \circ \phi_4^{-1}$ is linear. Since

$$\begin{aligned} Q_8(z_1, \ldots, z_{30}) &= z_1^8 \\ &\quad + S(z_1, \ldots, z_{30})^2[z_9^4 + T_1(z_1, \ldots, z_{30})T_2(z_1, \ldots, z_{30})] \\ &\quad + z_{29}^4 + z_{30}^2, \end{aligned}$$

the previous proposition implies that $Q_8^{2^7} \circ \rho_i \circ \hat{\phi}_{21}$ coincides with an element of $\mathcal{T}$ on V_1. □

If $g \in \mathcal{T}$, then we may write

$$g = \sum_i a_i y_i^4 + b_i y_i^2 + c_i y_i + d,$$

for some $a_i, b_i, c_i, d \in k$. For any specific ciphertext $y' \in k^{100}$, again set $g(y') = \sum_i a_i(y_i')^4 + b_i(y_i')^2 + c_i(y_i') + d$. Define the linear subvariety $V \subset V_1$ by

$$V = \mathcal{V}(g - g(y') \mid g \in \mathcal{L}_{\mathcal{T}}).$$

Theorem 6.4.1. *The functions*

$$Q_8 \circ \rho_i \circ \hat{\phi}_{21},$$

for $i = 1, 2$, are constant on the linear subvariety V.

For a given plaintext x', let $a_i = Q_8 \circ \rho_i \circ \hat{\phi}_{21}(x')$. Define ϕ_3' be the function with components

$$\phi_{3,1} = x_1 + a_1$$
$$\phi_{3,2} = x_2 + a_2$$
$$\phi_{3,i} = x_i \quad \text{for } i = 3, \ldots, 100,$$

and let $F' = \phi_4 \circ \phi_3' \circ \phi_2 \circ \phi_1 \circ \iota$. Then $\bar{F}$ and F' coincide on the linear subvariety V containing x'. Since F' is evidently of de Jonquières type, we can invert $\bar{F}|_V$ using the procedure described below, thereby finding the ciphertext x'.

The Attack Procedure and Its Complexity

We perform three steps to derive the plaintext $(x_1', \ldots, x_{64}')$ from the ciphertext $(y_1', \ldots, y_{100}')$. The first step is a common step for any given ciphertext.

Step 1: *Find a basis for the space W_1 of solutions of the equations*

$$\sum_{i,j=1}^{100} a_{ij} y_i^{2^6} y_j^{2^6} + \sum_{k=1}^{100} b_l y_l^{2^6} + c + \sum_{l=1}^{64} d_l x_m = 0.$$

in the unknowns a_{ij}, b_l, c, d_l.

This system of equations involves $100 + 50 \times 99 + 100 + 64 + 1 = 5215$ variables and $1 + 64 + (32 \times 63 + 48) + (48 + 32 \times 63 + (64 \times 63 \times 62/6)) + (48 + 32 \times 63 \times 3 + (64 \times 63 \times 62/6) \times 3 + (64 \times 63 \times 62 \times 61/24) = 812353$ equations. Since the number of equations far exceeds the number of variables, we do not need to use all of the equations to find the solution. In practice we can randomly choose 8000 equations; the probability that we will not find the complete solution is essentially zero. Solving these linear equations involves row operations on an 8000×5215 matrix. However, since we are working over a finite field with only 2^8 elements, the row operations corresponding to each column requires at most $2^8 - 1$ multiplications of any given row. On average, elimination of each variable takes $(2^8 - 1) \times 8000/2$ multiplications. Therefore the solution of these equations requires at most $5215 \times (2^8 - 1) \times 8000/2 \approx 2^{33}$ computations on the finite field k. Moreover this step is independent of the value of the ciphertext y'.

Since we are working over the fixed field k, we can perform the computation of multiplication on k by first finding a generator g of the multiplicative group of k, storing the table of elements of k in the form

g^l, then computing the multiplication by two searches and one addition. This will improve the speed by at least a factor of two. Thus, this preliminary step takes at most 2^{32} computations.

Step 2: *Step 1 yields a set of equations of the form*

$$\sum_{i=1}^{64} d_i x_i = \sum_{i,j=1}^{100} a_{ij} y_i^{2^6} y_j^{2^6} + \sum_{l=1}^{100} b_l y_l^{2^6} + c.$$

For a given ciphertext $(y'_1, \ldots, y'_{100})$, *we substitute these values into the right hand side to derive a set of linear equations in* x_i. *Solving this system by Gaussian elimination enables us to eliminate a certain set of the* x_i, *say* M *of them, by expressing them as linear expressions in the remaining variables. We may then substitute these expressions into the* y_i *to produce a new set of functions* $\tilde{y}_i$, *for* $i = 1, \ldots, 100$, *in the remaining* $64 - M$ *variables.*

This process corresponds to the identification (as a vector space) of the linear subvariety V_1 described in the previous section.

Step 3: *Find a basis for the space of solutions of the system of equations*

$$\sum_{i=1}^{100} \tilde{a}_i \tilde{y}_i^4 + \sum_{i=1}^{100} \tilde{b}_i \tilde{y}_i^2 + \sum_{i=1}^{100} \tilde{c}_j \tilde{y}_j + \tilde{d} + \sum_{i=1}^{64-M} e_i x_{k_i} = 0$$

in the unknowns $\tilde{a}_i, \tilde{b}_i, \tilde{c}_j$ *and* $\tilde{d}$. *We then repeat the procedure of Step 2. Each element of this basis yields an equation of this form into which we can again substitute the ciphertext. This gives a system of linear equations which we can again solve to eliminate further a set of say* $\tilde{M}$ *variables by linear substitution of the solution into* $\tilde{y}_i$, *which we denote as* $\hat{y}_i$.

For the first part, the number of variables is 301 and the number of equations is $3 \times ((64 - M)(64 - M + 3)/2 < 5307$. The computation in this step takes no significant time compared to that of Step 1.

The span of the remaining x_i forms a vector space that we identify naturally with the linear subspace V described above.

Step 4: *We now proceed to "invert"* $\bar{F}|_V$ *by solving the system of polynomial equations* $\hat{y}_i - y'_i = 0$. *Since the map* $\bar{F}|_V$ *is of de Jonquières type, the vector space spanned by the polynomial functions* $\hat{y}_i - y'_i$ *intersects* $\mathcal{L}$ *nontrivially; i.e., it contains a linear function of the* x_i. *This enables us to substitute for one of the* x_i, *thereby reducing the number of variables. The nature of a function of de Jonquières type enables us* *to iterate this process. This elimination process enables us to find the coordinates x'_i of the plaintext corresponding to the variables involved in the $\hat{y}_i$. We use the linear equations derived in the previous steps to find the remaining coordinates.*

Again, this procedure takes no significant time compared to that of Step 1. Thus the four steps together require at most 2^{33} computations. Finally, we note that this attack, as well as the MinRank attack, can also be applied to the new TTM schemes proposed later [Chen et al., 2002].

6.5 Attacks on the New TTM Cryptosystems

Later, new TTM schemes were proposed [Chen and Moh, 2001] that can resist the attacks in [Goubin and Courtois, 2000]. However, Ding and Schmidt [Ding and Schmidt, 2004] show that actually all existing implementation schemes for the TTM cryptosystem at the time have a common defect that make them insecure. For the case of the two new TTM implementation schemes in two different versions of the paper [Chen and Moh, 2001], this defect is used to defeat the schemes. The key idea of this attack comes from an observation that we can extend the linearization equation method of Patarin [Patarin, 1995] to attack all the TTM implementation schemes.

From the construction of the TTM implementation schemes, it is discovered that for all the existing TTM implementation schemes there exist many linearization equations that are satisfied by the quadratic polynomials y_i of the TTM cipher $\bar{F}$. For example, for the new implementation scheme [Chen and Moh, 2001] (the revised version on IACR ePrint archive; the former version has a different implementation scheme), where $m = n+52$, all the linearization equations are computed that the dimension of V is actually 347, where V is the linear space of all the linearization equations satisfied by the quadratic polynomials y_i.

This is the source of the common defect among all the TTM implementation schemes at that time, since the existence of the linearization equations means that for a given a ciphertext $(y'_1, \ldots, y'_m)$, we can immediately produce some linear equations satisfied by the plaintext $(x'_1, \ldots, x'_n)$. For the case of the revised implementation scheme [Chen and Moh, 2001], it is shown that with probability

$$1 - \frac{C_{17}^5}{2^{12\times 8}} > 1 - 2^{-82}$$

the linearization equations will produce 17 linearly independent linear equations satisfied by x_i.

From this case, we can move one step further by performing a substitution of these 17 linear equations into y_i, which makes y_i quadratic polynomials with 17 fewer variables, which we denote by $(x_{v_1}, \ldots, x_{v_{31}})$. Now $\bar{F}$ becomes a new map $\hat{F} : k^{n-17} \longrightarrow k^m$, which can be rewritten as

$$\hat{F} = \hat{\phi}_4 \circ \phi_2 \circ \hat{\Phi}_1,$$

where $\hat{\phi}_4$ (invertible) and $\hat{\Phi}_1$ (injective) are some affine linear maps. The procedure of the substitution of the 17 linear equations eliminates one of the composition factors of the de Jonquières type. Then solving the equations $\hat{F} = (y'_1, \ldots, y'_m)$ for the given ciphertext becomes straightforward due to the triangular form of the de Jonquières type of maps. It is accomplished by an iteration of the procedure of first searching for linear equations by linear combinations and then linear substitution. Finally, the plaintext can be derived by substituting the solution of the values of $(x_{v_1}, \ldots, x_{v_{31}})$ into the original 17 linear equations.

For the practical example $m = 100$ proposed in [Chen and Moh, 2001], it is shown that it takes about 2^{32} computations on a finite field of size 2^8 to defeat the scheme. Similarly, the new TTM scheme in the original version of [Chen and Moh, 2001] can be defeated.

The New Scheme

We will now present the new scheme in the revised version of [Chen and Moh, 2001]. First the finite field k is of size 2^8 and $m = n + 52$. The map F is defined by

$$F = \phi_4 \circ \phi_3 \circ \phi_2 \circ \phi_1(x_1, \ldots, x_{n+52}),$$

which are maps from the $(n + 52)$-dimensional space to itself defined in [Chen and Moh, 2001]. The maps $\phi_1 = (\phi_{1,1}, \ldots, \phi_{1,n+52})$ and $\phi_4 = (\phi_{4,1}, \ldots, \phi_{4,n+52})$ are invertible affine linear maps, and $\phi_{1,i} = x_i$, for $i > n$; ϕ_2 and ϕ_3 are nonlinear maps of de Jonquières type.

The map

$$\begin{aligned} \bar{F}(x_1, \ldots, x_n) &= (y_1, \ldots, y_{n+52}) \\ &= \phi_4 \circ \phi_3 \circ \phi_2 \circ \phi_1(x_1, x_2, \ldots, x_n, 0, \ldots, 0) \\ &= \phi_4 \circ \phi_3 \circ \phi_2 \circ \Phi_1(x_1, \ldots, x_n) \end{aligned}$$

is the cipher, which is public, whereas ϕ_1 and ϕ_4 are private. The map $\Phi_1(x_1, \ldots, x_n) = \phi_1(x_1, x_2, \ldots, x_n, 0, \ldots, 0)$ is an injective map from k^n to k^{n+52}. In the expansion formula, the components y_i of the map $\bar{F}$ are degree two polynomials of variables $(x_1, \ldots, x_n)$.

In [Chen and Moh, 2001], it is proposed that $n = 48$ and $m = 100$ are good parameters for practical applications. It is also claimed that the

MinRank method in [Goubin and Courtois, 2000] has a computational complexity far greater than 2^{84}.

In this scheme, $\phi_2(x_1, \ldots, x_n) = (\phi_{2,1}, \ldots, \phi_{2,100})$ is given as

$$\begin{aligned}
&\phi_{2,1} = x_1, \\
&\phi_{2,i} = x_i + f_i(x_1, \ldots, x_{i-1}), && i = 2, \ldots, 41 \\
&\phi_{2,i} = q_{i-41}(x_{38}, \ldots, x_{48}), && i = 42, \ldots, 48 \\
&\phi_{2,i} = x_i + q_{i-41}(x_{38}, \ldots, x_{48}), && i = 49, \ldots, 76 \\
&\phi_{2,i} = x_i + q_{i-72}(x_{36}, x_{39}, x_{40}, \ldots, x_{45}, x_{37}, x_{47}, x_{48}), && i = 77, \ldots, 84 \\
&\phi_{2,i} = x_i + q_{i-80}(x_{34}, x_{39}, x_{40}, \ldots, x_{45}, x_{35}, x_{47}, x_{48}), && i = 85, \ldots, 92 \\
&\phi_{2,i} = x_i + q_{i-88}(x_{32}, x_{39}, x_{40}, \ldots, x_{45}, x_{33}, x_{47}, x_{48}), && i = 93, \ldots, 100.
\end{aligned}$$

Now define $Q_8 = Q_8(z_1, \ldots, z_{35})$ by

$$\begin{aligned}
Q_8 = &(z_5 z_{13} + z_8 z_{14})(z_{19} z_{32} + z_2(z_{18} + z_{24}))^2 (z_{20} z_{19} + z_{23} z_{18}) \\
&+ (z_{32} z_3 + (z_{18} + z_{24}) z_{21})^2 (z_{22} z_{19} + z_{23} z_{24})(z_9 z_{13} + z_8 z_{15}) \\
&+ a_1^8((z_{25} z_{26} + z_{27} z_{28})(z_6 z_{29} + z_7 z_{16}) \\
&\qquad + (z_{10} z_{30} + z_{11} z_{31})(z_{17} z_1 + z_{18} z_4)) \\
&+ a_1^{12}(z_6 z_{33} + z_{34} z_7 + z_5 z_{35} + z_{14} z_{12}),
\end{aligned}$$

where

$$\begin{aligned}
&q_1(z_1, \ldots, z_{11}) = z_4 z_2 + a_1 z_5, && q_2(z_1, \ldots, z_{11}) = z_3 z_4 + a_1 z_6 \\
&q_3(z_1, \ldots, z_{11}) = z_2 z_5 + a_1 z_7 && q_4(z_1, \ldots, z_{11}) = z_4 z_7 + a_1 z_8 \\
&q_5(z_1, \ldots, z_{11}) = z_1 z_5 + a_1 z_9 && q_6(z_1, \ldots, z_{11}) = z_1 z_2 + a_1 z_{10} \\
&q_7(z_1, \ldots, z_{11}) = z_2 z_9 + a_1 z_{11} && q_8(z_1, \ldots, z_{11}) = z_3 z_9 + a_1 z_1 \\
&q_9(z_1, \ldots, z_{11}) = z_1 z_3 && q_{10}(z_1, \ldots, z_{11}) = z_1 z_7 + a_1 z_9 \\
&q_{11}(z_1, \ldots, z_{11}) = z_4 z_9 + a_1 z_1 && q_{12}(z_1, \ldots, z_{11}) = z_7 z_9 + a_1 z_1 \\
&q_{13}(z_1, \ldots, z_{11}) = z_3 z_{11} + a_1 z_{10} && q_{14}(z_1, \ldots, z_{11}) = z_5 z_{10} + a_1 z_{11} \\
&q_{15}(z_1, \ldots, z_{11}) = z_3 z_{10} && q_{16}(z_1, \ldots, z_{11}) = z_2 z_{10} \\
&q_{17}(z_1, \ldots, z_{11}) = z_7 z_8 + a_1 z_7 && q_{18}(z_1, \ldots, z_{11}) = z_5 z_7 + a_1 z_2 \\
&q_{19}(z_1, \ldots, z_{11}) = z_2 z_3 + a_1 z_7 && q_{20}(z_1, \ldots, z_{11}) = z_5 z_8 + a_1 z_5 \\
&q_{21}(z_1, \ldots, z_{11}) = z_4 z_5 + a_1 z_6 && q_{22}(z_1, \ldots, z_{11}) = z_3 z_8 \\
&q_{23}(z_1, \ldots, z_{11}) = z_3 z_5 + a_1 z_8 && q_{24}(z_1, \ldots, z_{11}) = z_3 z_7 \\
&q_{25}(z_1, \ldots, z_{11}) = z_6 z_8 + a_3 z_5 && q_{26}(z_1, \ldots, z_{11}) = z_2 z_6 \\
&q_{27}(z_1, \ldots, z_{11}) = z_5 z_6 && q_{28}(z_1, \ldots, z_{11}) = z_6 z_7 + a_3 z_2 \\
&q_{29}(z_1, \ldots, z_{11}) = z_2 z_{11} && q_{30}(z_1, \ldots, z_{11}) = z_4 z_{11} + a_1 z_{10} \\
&q_{31}(z_1, \ldots, z_{11}) = z_7 z_{10} + a_1 z_{11} && q_{32}(z_1, \ldots, z_{11}) = z_3 z_6 + z_5 z_6 + a_1 z_4
\end{aligned}$$

$$q_{33}(z_1,\ldots,z_{11}) = z_8z_{11} \qquad q_{34}(z_1,\ldots,z_{11}) = z_8z_{10}$$
$$q_{35}(z_1,\ldots,z_{11}) = z_7z_{11} + a_1z_{10},$$

a_1 and a_3 are any two nonzero elements in the field k, and the quadratic functions $f_i(x_1,\ldots,x_{i-1})$ are chosen at random.

The map $\phi_3(x_1,\ldots,x_n) = (\phi_{3,1},\ldots,\phi_{3,100})$ is given as:

$$\phi_{3,i} = x_i \qquad \text{for } i = 5,\ldots,100$$
$$\phi_{3,i} = x_4 + R_i(x_1,\ldots,x_{100}) \qquad \text{for } i = 1,2,3,4,$$

where the $R_i(x_1,\ldots,x_{100}) = \sum_{j=1}^{4} \beta_{i,j}P_j$ are linearly independent, and the P_i, for $i = 1,2,3$, are given as

$$P_i = Q_8(x_{42},\ldots,x_{45},x_{101-8i},\ldots,x_{108-8i},x_{54},\ldots,x_{76})$$
$$P_4 = Q_8(x_{42},\ldots,x_{76}).$$

Remark 6.5.1. *In the new version of [Chen and Moh, 2001], the polynomials Q_8 and q_i actually have three free parameters a_1, a_2 and a_3. It turns out that in order to make the cipher F have degree two, one must make a_1 equal to a_2. We will assume this condition.*

Because the specific form of ϕ_1, we can write:

$$\phi_1(x_1,x_2,\ldots,x_{48},0,\ldots,0) = \Phi_1(x_1,\ldots,x_{48}) = \pi \circ \hat{\phi}_1(x_1,\ldots,x_{48}),$$

where π is the standard embedding that maps k^{48} into k^{100}

$$\pi(x_1,\ldots,x_{48}) = (x_1,\ldots,x_{48},0,\ldots,0),$$

and $\hat{\phi}_1(x_1,\ldots,x_{48}) = (\hat{\phi}_{1,1}(x_1,\ldots,x_{48}),\ldots,\hat{\phi}_{1,48}(x_1,\ldots,x_{48}))$ is an invertible affine linear transformation from K^{48} to itself.

Let $\phi_3 \circ \phi_2 \circ \pi = \bar{\phi}_{32}$. Then

$$\begin{aligned}\bar{F}(x_1,\ldots,x_{48}) &= \phi_4 \circ \phi_3 \circ \phi_2 \circ \phi_1(x_1,x_2,\ldots,x_{48},0,\ldots,0)\\ &= \phi_4 \circ \phi_3 \circ \phi_2 \circ \pi \circ \hat{\phi}_1(x_1,\ldots,x_{48})\\ &= \phi_4 \circ \bar{\phi}_{32} \circ \hat{\phi}_1(x_1,\ldots,x_{48}).\end{aligned}$$

Let $\bar{\phi}_{32}(x_1,\ldots,x_{48}) = (\bar{\phi}_{32,1},\ldots,\bar{\phi}_{32,100})$. Then

$$\begin{aligned}\bar{\phi}_{32,1} &= x_1 + a_1^{14}\beta_{1,4}(x_{38}x_{48} + x_{47}x_{46})\\ &\quad + a_1^{14}\sum_{1}^{3}\beta_{1,j}(x_{38-2j}x_{48} + x_{39-2j}x_{47})\end{aligned}$$

$$
\begin{aligned}
\bar{\phi}_{32,i} &= x_i + f_i(x_1,\ldots,x_{i-1}) + a_1^{14}\beta_{i,4}(x_{38}x_{48} + x_{37}x_{46}) & \\
&\quad + a_1^{14}\textstyle\sum_1^3 \beta_{i,j}(x_{38-2j}x_{48} + x_{39-2j}x_{47}) & \text{for } i = 2,3,4 \\
\bar{\phi}_{32,i} &= x_i + f_i(x_1,\ldots,x_{i-1}) & \text{for } i = 5,6,\ldots,41 \\
\bar{\phi}_{32,i} &= q_{i-31}(x_{38},\ldots,x_{48}) & \text{for } i = 42,\ldots,48 \\
\bar{\phi}_{32,i} &= q_{i-31}(x_{38},\ldots,x_{48}) & \text{for } i = 49,\ldots,76 \\
\bar{\phi}_{32,i} &= q_{i-72}(x_{36},x_{39},x_{40},\ldots,x_{45},x_{37},x_{47},x_{48}) & \text{for } i = 77,\ldots,84 \\
\bar{\phi}_{32,i} &= q_{i-85}(x_{34},x_{39},x_{40}\ldots,x_{45},x_{35},x_{47},x_{48}) & \text{for } i = 85,\ldots,92 \\
\bar{\phi}_{32,i} &= q_{i-93}(x_{32},x_{39},x_{40},\ldots,x_{45},x_{33},x_{47},x_{48}) & \text{for } i = 93,\ldots,100.
\end{aligned}
$$

The formula above is due to the fact that

$$Q_8(q_1,\ldots,q_{35}) = a_1^{14}(z_9z_{10} + z_1z_{11}),$$

which is the reason why F has degree two.

The Basic Idea of the Cryptanalysis

The new attack starts from the observation that all the q_i are very simple quadratic polynomials having only one quadratic term. In this case, Q_8 has 35 variables and the q_i have 11 variables. For example we have

$$q_9 = z_1z_3, \quad q_{15} = z_3z_{10},$$

which implies that

$$z_{10}q_9 - z_1q_{15} = 0. \tag{6.2}$$

In this implementation scheme, the map $\bar{\phi}_{32}$ has four sets of q_i as its components (with intersections). Since $\bar{F}$ is derived from $\bar{\phi}_{32}$ by composing from both the left side and the right side by an invertible linear map, (6.2) above implies that we must have linearization equations for the y_i, the components of $\bar{F}$. This means there is a possibility of using such linearization equations to attack this scheme, which is the method used by Patarin to defeat the Matsumoto-Imai scheme [Patarin, 1995].

Let V denote the linear space of the linearization equations

$$\sum_{i=1,j=1}^{n,m} a_{ij}x_iy_j(x_1,\ldots,x_n) + \sum_{i=1}^{n} b_ix_i + \sum_{j=1}^{m} c_jy_j(x_1,\ldots,x_n) + d = 0,$$

satisfied by y_i of $\bar{F}$, and let D be its dimension. Let $\bar{V}$ denote the linear space of the linearization equations satisfied by $\bar{\phi}_{32,i}(x_1,\ldots,x_{48})$ of $\bar{\phi}_{32}$; namely,

$$\sum_{i=1,j=1}^{n,m} \bar{a}_{ij}x_i\bar{\phi}_{32,j}(x_1,\ldots,x_{48}) + \sum_{i=1}^{n} \bar{b}_ix_i + \sum_{j=1}^{m} \bar{c}_j\bar{\phi}_{32,j}(x_1,\ldots,x_{48}) + \bar{d} = 0,$$

where $\bar{D}$ is the dimension of $\bar{V}$.

Further define

$$\hat{\phi}_{32}(x_1, \ldots, x_{48}) = (\hat{\phi}_{32,1}, \ldots, \hat{\phi}_{32,100}) = \bar{\phi}_{32} \circ \hat{\phi}_1(x_1, \ldots, x_{48}),$$

and let $\hat{V}$ denote the linear space of the linearization equations satisfied by $\hat{\phi}_{32,i}(x_1, \ldots, x_{48})$ of $\hat{\phi}_{32}$; namely,

$$\sum_{i=1,j=1}^{n,m} \hat{a}_{ij} x_i \hat{\phi}_{32,j}(x_1, \ldots, x_{48}) + \sum_{i=1}^{n} \hat{b}_i x_i + \sum_{j=1}^{m} \hat{c}_j \hat{\phi}_{32,j}(x_1, \ldots, x_{48}) + \hat{d} = 0,$$

where $\hat{D}$ is the dimension of $\hat{V}$.

Now let $\phi_{4,i}$ denote the components of ϕ_4, and let $\hat{\phi}_{1,i}$ denote the components of $\hat{\phi}_1$. Let $(\phi_4^{-1})_i$ denote the components of ϕ_4^{-1}, and let $(\hat{\phi}_1^{-1})_i$ denote the components of $\hat{\phi}_1^{-1}$.

Let M be the map from $\hat{V}$ to V that maps an equation:

$$\sum \hat{a}_{ij} x_i \hat{\phi}_{32,j}(x_1, \ldots, x_{48}) + \sum \hat{b}_i x_i + \sum \hat{c}_j \hat{\phi}_{32,j}(x_1, \ldots, x_{48}) + \hat{d} = 0,$$

to the equation

$$\sum \hat{a}_{ij} x_i (\phi_4^{-1})_j (y_1(x_1, \ldots, x_{48}), \ldots, y_{100}(x_1, \ldots, x_{48})) + \sum \hat{b}_i x_i$$
$$+ \sum \hat{c}_j (\phi_4^{-1})_j (y_1(x_1, \ldots, x_{48}), \ldots, y_{100}(x_1, \ldots, x_{48})) + \hat{d} = 0,$$

and let $\hat{M}$ be the map from $\bar{V}$ to $\hat{V}$ that maps an equation:

$$\sum \bar{a}_{ij} x_i \bar{\phi}_{32,j}(x_1, \ldots, x_{48}) + \sum \bar{b}_i x_i + \sum \bar{c}_j \bar{\phi}_{32,j}(x_1, \ldots, x_{48}) + \bar{d} = 0,$$

to the equation:

$$\sum \bar{a}_{ij} \hat{\phi}_{1,i}(x_1, \ldots, x_{48}) \hat{\phi}_{32,j}(x_1, \ldots, x_{48}) + \sum \bar{b}_i \hat{\phi}_{1,i}(x_1, \ldots, x_{48})$$
$$+ \sum \bar{c}_j \hat{\phi}_{32,j}(x_1, \ldots, x_{48}) + \bar{d} = 0).$$

Theorem 6.5.1. *M and $\hat{M}$ are invertible linear maps and $D = \bar{D} = \hat{D}$.*

The proof follows from the fact that both ϕ_4 are $\bar{\phi}_1$ are invertible affine linear maps. Essentially, the map $\hat{M}$ is a change of basis of x_i and the map M is an affine linear transformation of the substitution of $\hat{\phi}_{32,i}$ by y_i. This means that we only need to find $\bar{D}$ to obtain D, and this can be done by direct computation [Ding and Schmidt, 2004].

To do so a general realization of the field k was chosen; namely $k = GF(2)[x]/(x^8 + x^6 + x^5 + x + 1)$. Because a_1 and a_3 can be any nonzero constants, we choose them both to be equal to 1. Then randomly choose $f_i(x_1, \ldots, x_{i-1})$, for $i = 2, \ldots, 41$, to be quadratic polynomials over k, and the $\beta_{i,j}$ randomly in k (but satisfying the condition that the R_i be linearly independent). In [Ding and Schmidt, 2004], ten different choices of sets of $f_i(x_1, \ldots, x_{48})$ and $\beta_{i,j}$ were computed and for all ten choices we found that:

1.) $\bar{D} = 347$.

2.) All linearization equations are of the form

$$\sum_{i>31}\sum_{j>41} \bar{a}_{ij} x_i \bar{\phi}_{32,j}(x_1, \ldots, x_{48}) + \sum_{i>31} \bar{b}_i x_i + \sum_{j>41} c_j \bar{\phi}_{32,j}(x_1, \ldots, x_{48}) = 0,$$

and the polynomials $\phi_{32,j}(x_1, \ldots, x_{48}), j > 41$ are polynomials of only 17 variables x_i, $i > 31$.

Though we may have such a large number of linearization equations, we are not sure how many linearly independent equations they will produce for a given ciphertext y'_i.

Let $(x'_1, \ldots, x'_{48})$ be an element in k^{48}. Let $y'_i = y_i(x'_1, \ldots, x'_{48})$ and $\hat{\phi}'_{32,i} = \hat{\phi}_{32,i}(x'_1, \ldots, x'_{48})$. Let U be the space of linear equations derived from substitution of y_i by the values y'_i in V. Let $\hat{U}$ be the linear space of linear equations derived from substitution of $\hat{\phi}_{32,i}$ by the values $\hat{\phi}'_{32,i}$ in $\hat{V}$. Let $\bar{U}$ be the linear space of linear equations derived from substitution of $\bar{\phi}_{32,i}$ by the values $\hat{\phi}'_{32,i}$ and x_i by $(\hat{\phi}_1^{-1})_i(x_1, \ldots, x_{48})$ in $\bar{V}$.

For a linear equation $\sum_1^{48} a_i x_i + b = 0$, we define $\tilde{M}$ to be the linear map that maps the equation:

$$\sum_{i=1}^{48} a_i x_i + b = 0$$

to the equation:

$$\sum_{i=1}^{48} a_i (\hat{\phi}_1)_i(x_1, \ldots, x_{48}) + b = 0.$$

Theorem 6.5.2. *The dimension of U is equal to the dimension of $\bar{U}$, the dimension of $\hat{U}$, and the dimension of $\tilde{U}$.*

$$\hat{U} = U = \tilde{M}(\bar{U})$$

Proof. This is proven easily by using the maps M and $\hat{M}$. □

All linearization relations in $\bar{V}$ are expressed in the last 59 components $\bar{\phi}_{32,j}(x_1,\ldots,x_{48})$, $j > 41$ and they are all expressed in terms of the quadratic polynomials q_i. They involve only the last 17 variables x_i, $i = 32,\ldots,48$. In [Ding and Schmidt, 2004], 200 samples were taken of randomly chosen values $\bar{x}'_{32}, \bar{x}'_{33},\ldots,\bar{x}'_{48}$ for $x_{32},\ldots,x_{48}$. The corresponding values of $\bar{\phi}_{32,j}$, $j > 41$ were computed for these $\bar{x}'_{32}, \bar{x}'_{33},\ldots,\bar{x}'_{48}$. Then the values of $\bar{\phi}_{32,j}$, $j > 41$ were substituted into the 347 linearization equations. It was found out that these 347 linearization equations in $\bar{V}$ produce 17 linearly independent equations for x_i, $i > 31$, and by solving these equations the values for $x_i = \bar{x}'_i$, $i = 32,\ldots,48$ are recovered.

Then we can notice that if all the x_i are set to zero, which implies that $\bar{\phi}_{32,i}(0,\ldots,0) = 0$ for any i, and the linearization equations in $\bar{V}$ will not produce 17 linearly independent equations at all. So instead of choosing the values randomly, we chose $(\bar{x}'_{32},\ldots,\bar{x}'_{48})$ with many of them to be zero. In [Ding and Schmidt, 2004] we discovered that as long as at least five of the $x_{32},\ldots,x_{48}$ are not zero, then by substituting the corresponding values of $\bar{\phi}_{32,j}$, $j > 41$ into the 347 linearly independent linearization equations in $\bar{V}$ these 347 linearization equations will produce 17 linearly independent linear equations whose solution gives us the values of $x_i = \bar{x}'_i$ for $i = 32,\ldots,48$. Among all the possible values of x_i, $i = 32,\ldots,48$, the probability that at most five of them among the x_i, $i = 32,\ldots,48$ are nonzero is

$$\frac{C_{17}^5 2^{5\times 8}}{2^{17\times 8}} = \frac{C_{17}^5}{2^{12\times 8}} < 2^{-82}.$$

Therefore we have a probability of

$$1 - \frac{C_{17}^5}{2^{12\times 8}} > 1 - 2^{-82}$$

that the linearization equations will produce 17 linearly independent equations for a given set values of $\bar{\phi}_{32,i}$. Solving these equations will recover the value of $\bar{x}'_{32},\ldots,\bar{x}'_{48}$, if we are given the corresponding values of $\bar{\phi}_{32,j}$ for $j > 41$.

With the previous theorems, we can conclude that with probability

$$1 - \frac{C_{17}^5}{2^{12\times 8}} > 1 - 2^{-82},$$

the linearization equations of y_i in V will produce 17 linearly independent equations satisfied by x_i for a given ciphertext $(y'_1,\ldots,y'_{100})$. This is the first step of the new attack. Here we should emphasize that the statement about the probability to derive 17 linearly independent linear equations from a ciphertext is based on computational experiments not on any theoretical argument. Though it seems possible to actually prove it, this is still an open problem.

Let us assume that we now have 17 linearly independent equations in U derived from a ciphertext $(y'_1, \ldots, y'_{100})$ and its substitution in V. Let $(x'_1, \ldots, x'_{48})$ be the corresponding plaintext. This set of linear equations is not enough to recover the original plaintext. However, we know that if we have 17 linearly independent equations, we can use the Gaussian elimination method to find the sets $A = \{u_1, \ldots, u_{17}\}$, $B = \{v_1, \ldots, v_{31}\}$, with $A \cap B = \emptyset$ and $A \cup B = \{1, \ldots, 48\}$. From this we can derive 17 linearly independent linear equations of the form $x_{u_j} = h_j(x_{v_1}, \ldots, x_{v_{31}})$. Then we substitute these 17 equations into the y_i, which will become quadratic polynomials with only 31 variables. We will call this new set of polynomials $\hat{y}_i$. These can be viewed as components of a map from k^{31} to k^{100}, which will be denoted by $\hat{F}$.

Let ϕ_0 be the map from k^{31} to k^{48}, which is given by

$$\phi_{0,i}(x_{v_1}, \ldots, x_{v_{31}}) = \begin{cases} x_i, & \text{if } i \in B; \\ h_i(x_{v_1}, \ldots, x_{v_{31}}), & \text{otherwise.} \end{cases}$$

Then

$$\hat{F} = \phi_4 \circ \phi_3 \circ \phi_2 \circ \pi \circ \hat{\phi}_1 \circ \phi_0.$$

From the point of view of algebraic geometry, the substitution process is nothing but evaluation of the y_i on the variety defined by the 17 linearly independent linear equations $x_{u_i} = h_i(x_{v_1}, \ldots, v_{31})$, and the existing variables are nothing but the coordinates of this variety. Because for the case of $\bar{\phi}_{32}$, if the dimension of $\bar{U}$ is 17, the variety is defined by $x_i = \bar{x}'_i$ for $i = 32, \ldots, 48$, and $\bar{x}'_i \in K$, with the previous theorems we know that the variety defined by linear equation in U is the same variety defined by $\hat{\phi}_{1,i}(x_1, \ldots, x_{48}) = \hat{\phi}_{1,i}(x'_1, \ldots, x'_{48})$, for $i > 31$. We denote this variety by W. The linear equations in U are nothing but linear combinations of this set of linear equations.

Let

$$\hat{\phi}_{32} = \bar{\phi}_{32} \circ \hat{\phi}_1 \circ \phi_0(x_{v_1}, \ldots, x_{v_{31}}) = (\bar{\phi}_{32,1}, \ldots, \bar{\phi}_{32,100}),$$
$$\phi_{10}(x_{v_1}, \ldots, x_{v_{31}}) = \hat{\phi}_1 \circ \phi_0(x_{v_1}, \ldots, x_{v_{31}}) = (\phi_{10,1}, \ldots, \bar{\phi}_{10,100}).$$

Then using the expansion formula of $\bar{\phi}_{32}$, in particular the formulas for $\bar{\phi}_{32,i}$, we have:

$$\begin{aligned}
\hat{\phi}_{32,1} &= \phi_{10,1} + a_1^{14}\beta_{1,4}(\phi_{10,38}\phi_{10,48} + \phi_{10,47}\phi_{10,46}) \\
&\quad + a_1^{14}\textstyle\sum_1^3 \beta_{1,j}(\phi_{10,38-2j}\phi_{10,48} + \phi_{10,39-2j}\phi_{10,47}) = \phi_{10,1} + R_1' \\
\hat{\phi}_{32,i} &= \phi_{10,i} + f_i(\phi_{10,1}, \ldots, \phi_{10,i-1}) \\
&\qquad + a_1^{14}\beta_{i,4}(\phi_{10,38}\phi_{10,48} + \phi_{10,37}\phi_{10,46}) \\
&\qquad + a_1^{14}\textstyle\sum_1^3 \beta_{i,j}(\phi_{10,38-2j}\phi_{10,48} + \phi_{10,39-2j}\phi_{10,47}) \\
&= \phi_{10,i} + R_i' && \text{for } i = 2,3,4 \\
\hat{\phi}_{32,i} &= \phi_{10,i} + f_i(\phi_{10,1}, \ldots, \phi_{10,i-1}) && \text{for } i = 5,6,\ldots,41 \\
\hat{\phi}_{32,i} &= q_{i-31}(\phi_{10,38}, \ldots, \phi_{10,48}) && \text{for } i = 42,\ldots,48 \\
\hat{\phi}_{32,i} &= q_{i-31}(\phi_{10,38}, \ldots, \phi_{10,48}) && \text{for } i = 49,\ldots,76 \\
\hat{\phi}_{32,i} &= q_{i-72}(\phi_{10,36}, \phi_{10,39}, \phi_{10,40}, \ldots, \phi_{10,45}, \phi_{10,37}, \phi_{10,47}, \phi_{10,48}) \\
& && \text{for } i = 77,\ldots,84 \\
\hat{\phi}_{32,i} &= q_{i-85}(\phi_{10,34}, \phi_{10,39}, \phi_{10,40} \ldots, \phi_{10,45}, \phi_{10,35}, \phi_{10,47}, \phi_{10,48}) \\
& && \text{for } i = 85,\ldots,92 \\
\hat{\phi}_{32,i} &= q_{i-93}(\phi_{10,32}, \phi_{10,39}, \phi_{10,40}, \ldots, \phi_{10,45}, \phi_{10,33}, \phi_{10,47}, \phi_{10,48}) \\
& && \text{for } i = 93,\ldots,100,
\end{aligned}$$

where $R_i' = \sum_1^4 \beta_{i,j} P_i'$ and P_i', for $i = 1,2,3$, are given by

$$\begin{aligned}
P_i' &= \hat{\phi}_{1,31+i+1}(x_1', \ldots, x_{48}')\hat{\phi}_{1,48}(x_1', \ldots, x_{48}') \\
&\quad + \hat{\phi}_{1,31+i}(x_1', \ldots, x_{48}')\hat{\phi}_{1,47}(x_1', \ldots, x_{48}') \\
P_4' &= \hat{\phi}_{1,42}(x_1', \ldots, x_{48}')\hat{\phi}_{1,48}(x_1', \ldots, x_{48}') \\
&\quad + \hat{\phi}_{1,46}(x_1', \ldots, x_{48}')\hat{\phi}_{1,47}(x_1', \ldots, x_{48}'),
\end{aligned}$$

which are constants; namely, the $R_i(\bar{\phi}_{32}(x_1, \ldots, x_{48}))$ are constants on the variety W.

Therefore

$$\hat{F}(x_{v_1}, \ldots, x_{v_{31}})) = (\hat{y}_1, \ldots \hat{y}_{100}) = \phi_4 \circ \hat{\phi}_3 \circ \phi_2 \circ \pi \circ \hat{\phi}_1 \circ \phi_0(x_{v_1}, \ldots, x_{v_{31}}),$$

where $\bar{\phi}_3 = (\bar{\phi}_{3,1}, \ldots, \bar{\phi}_{3,100})$ is given by

$$\begin{aligned}
\hat{\phi}_{3,i} &= x_i \quad \text{for } i = 5, \ldots, 100, \\
\hat{\phi}_{3,4} &= x_4 + R_i' \quad \text{for } i = 1,2,3,4.
\end{aligned}$$

Therefore, ϕ_3 is equivalent to $\hat{\phi}_3$ on the variety W, and is in fact just a linear translation. Also,

$$\hat{F}(x_{v1}, \ldots, x_{v31}) = (\phi_4 \circ \hat{\phi}_3) \circ \phi_2 \circ (\pi \circ \hat{\phi}_1 \circ \phi_0) = \hat{\phi}_4 \circ \phi_2 \circ \hat{\Phi}_1,$$

where $\hat{\phi}_4 = \phi_4 \circ \hat{\phi}_3$ and $\hat{\Phi}_1 = \pi \circ \hat{\phi}_1 \circ \phi_0$; and both $\hat{\phi}_4$ (invertible) and $\hat{\Phi}_1$ (injective) are linear. Then $\hat{F}(x_{v_1}, \ldots, x_{v_{31}}) = (y'_1, \ldots, y'_{100})$ can be easily solved because of the triangular form of ϕ_2; namely, the equation above is equivalent to the equations:

$$\hat{\phi}_4^{-1}(\hat{y}_1, \ldots, \hat{y}_{100}) = \phi_2 \circ \hat{\Phi}_1(x_{v_1}, \ldots, x_{v_{31}}) = \hat{\phi}_4^{-1}(y'_1, \ldots, y_{100'}),$$

whose first nontrivial equation is always a linear equation.

This analysis shows that the equations can be solved by iteration of the procedure of first searching for linear equations using linear combinations of quadratic equations, and then substituting the linear equations into the quadratic equations. Each step of the iteration reduces the number of variables by one. This will require 31 iterations to find the 31 linearly independent linear equations in the triangular form, whose solution gives the values of the 31 variables x_{v_i}. Then we can substitute the values of x_{v_i}, $i = 1, \ldots, 31$ back into the first 17 substitution equations $x_{u_i} = h_i(x_{v_1}, \ldots, v_{31})$, $j = 1, \ldots, 17$, which recovers the complete plaintext.

Overall, the general method is first to search for all linearization equations. Then, for a given ciphertext $(y'_1, \ldots, y'_m)$ corresponding to a plaintext $(x'_1, \ldots, x'_n)$, we use the linearization equations to produce enough (in this case 17) linearly independent linear equations satisfied by the x_i. Then we do a substitution using these linear equations, which essentially makes ϕ_3 linear on the variety defined by the 17 linear equations. The rest of the attack is straightforward.

The Practical Attack Procedure and Its Complexity

There are three steps to derive the plaintext $(x'_1, \ldots, x'_{48})$ from a ciphertext $(y'_1, \ldots, y'_{100})$, and the first step is a common step for any given ciphertext.

Step 1: *We first look for a basis for the space V; namely, the basis of solutions of a_{ij}, b_i, c_j and d for the equations:*

$$\sum_{i=1,j=1}^{n,m} a_{ij} x_i y_j(x_1, \ldots, x_n) + \sum_{i=1}^{n} b_i x_i + \sum_{j=1}^{m} c_j y_j(x_1, \ldots, x_n) + d = 0.$$

For this set of equations, we have $4800 + 48 + 100 + 1 = 4949$ variables and $1 + 48 + (24 \times 47 + 48) + (48 + 24 \times 47 + (8 \times 47 \times 46)) = 19697$ equations. We know that the dimension of the solutions is 347. Though we have 19697 equations, we have only 4949 variables, so we do not need to use all the equations to find the solutions. We can actually randomly choose 6000 equations, with the probability that we will not find the complete solution being essentially zero. To solve these linear equations, we must do row operations on a 6000×4949 matrix. However, because we are working on a finite field with only 2^8 elements, the row operations corresponding to the elimination procedure on each column requires at most $2^8 - 1$ multiplication of a given row. To eliminate each variable, on average, it takes $(2^8 - 1) \times 6000/2$ multiplications. Therefore to solve these equations, it requires at most $4600 \times (2^8 - 1) \times 6000/2 \approx 2^{32}$ computations on the finite field k. This step is also the common step for any attack.

However, because we are working over the fixed field k, we can perform the computation of multiplication on k by finding first a generator g of the multiplicative group of k, and storing the table of elements k as g^l, then computing the multiplication by two searches and one addition. This will improve the speed by at least a factor of two. Therefore, this step takes at most 2^{31} computations.

Step 2: *For a given ciphertext* $(y'_1, \ldots, y'_{100})$, *we substitute the polynomials of* y_i *by* y'_i *into the 347 linearly independent solutions of the linearization equations in* V *and derive 17 linearly independent linear equations of* x_i *by the Gaussian elimination method of the form of* $x_{u_j} = h_j(x_{v_1}, \ldots, x_{v_{31}})$, *where* h_j *is a linear function,* $A = \{u_1, \ldots, u_{17}\}$, $B = \{v_1, \ldots, v_{31}\}$, $A \cap B = \emptyset$ *and* $A \cup B = \{1, \ldots, 48\}$. *We then substitute these into* y_i *yielding polynomials with only 31 variables* $\{v_1, \ldots, v_{31}\}$.

First, the probability that we fail to get 17 linearly independent equations is 2^{-82}, which can be neglected. For the first part, we need to do $347 \times (4800 + 100) \approx 2^{21}$ computations when we substitute y_i by y'_i. Then, to reduce 347 equations to 17 equations for substitution, we must perform $(2^8 - 1) \times 48 \times 347/2 \approx 2^{21}$ computations. Then we perform the substitution of the 17 equations into y_i which takes $100 \times (2 \times 17^2 + 17 \times 31 + 17) \approx 2^{17}$ computations.

For the new 100 polynomials in 31 variables, denoted by $\hat{y}_i$, we will write down the 100 equations $\hat{y}_i - y'_i = \tilde{y}_{1i}(x_{v_1}, \ldots, x_{v_{31}}) = 0$, which are linearly dependent and have dimension only 41.

Step 3: *For the equations* $\tilde{y}_{1i}(x_{v_1}, \ldots, x_{v_{31}}) = 0$, $i = 1, \ldots, 100$, *we will use the Gaussian elimination method, first on the quadratic terms, to derive* $\hat{m} = 41$ *linearly independent equations* $\hat{y}_{1i}(x_{v_1}, \ldots, x_{v_{31}}) = 0$, $i = 1, \ldots, m$, *where the last one is actually linear. Then we take the linear equation out and substitute it back into the remaining* $\hat{m} - 1$ *quadratic equations* $\hat{y}_{1i}(x_{v_1}, \ldots, x_{v_{31}}) = 0$, $i = 1, \ldots, m - 1$. *We denote the new equations* $\tilde{y}_{2i}(x_{v_1}, \ldots, x_{v_i}, x_{v_{i+2}}, \ldots, x_{v_{31}}) = 0$. *Then we repeat the same process on these new equations again and again for a total of 31 times. We then collect all the 31 linear equations derived in this process, a set* *of 31 linearly independent equations in the triangular form. The solution gives us all the values of x_{v_i}, which we can then plug back into the 17 linear equations $x_{u_j} = h_j(x_{v_1}, \ldots, x_{v_{31}})$ in Step 2, which will give us x_{u_i}. This gives us the plaintext.*

For the first part of this step, we need at most $100 \times (31 \times 16 + 31) \times 100/2 \approx 2^{22}$ computations to perform the Gaussian elimination. The substitution takes at most $42 \times 31^2 \approx 2^{15}$ computations. We then need to perform these two procedures at most 31 times. Therefore, it takes at most 2^{25} computations to solve the equations for the 31 variables. We then need to do 2^9 computations to find the values for the other 17 variables. If we add all the three steps together, it takes at most 2^{32} computations, which can be easily checked on a PC.

Cryptanalysis of the Scheme in the First Version of [Chen and Moh, 2001]

In [Ding and Schmidt, 2004] this new attack method was applied to the original version of the scheme in [Chen and Moh, 2001]. The construction of the scheme is similar to the revised case above. We again work on the field k of size 2^8. A map F is composed from ϕ_1, ϕ_2, ϕ_3, ϕ_4, which are maps from the $(n+68)$-dimensional space to itself. The maps ϕ_1, ϕ_4 are invertible affine maps, and ϕ_2 and ϕ_3 are nonlinear of de Jonquières type, though different from that of the previous section.

Again we define the map

$$\begin{aligned} \bar{F}(x_1, \ldots, x_n) &= \phi_4 \circ \phi_3 \circ \phi_2 \circ \phi_1(x_1, x_2, \ldots, x_n, 0, \ldots 0) \\ &= (y_1, \ldots, y_{n+68}) \end{aligned}$$

as the cipher, which is public, though ϕ_1, ϕ_4 are private. To make sure the system is of degree two, another set of polynomials $Q_8(z_1, \ldots, z_{48})$ and $q_i(z_1, \ldots, z_{14})$ are used. The details of ϕ_3, ϕ_2, Q_8, and q_i are omitted here, though they can be found in the appendix of [Ding and Schmidt, 2004].

Through computations and similar arguments as in the section above, it is shown in [Ding and Schmidt, 2004] that:

1.) The dimension of V, the space of linearization equations for the components y_i of F, is 286;

2.) For a given ciphertext $(y_1', \ldots, y_{68+n}')$ the linearization will produce 28 linearly independent linear equations of x_i with a probability of:

$$1 - \left(\frac{C_{14}^4}{2^{10}}\right)^2 \doteq 1 - 2 \times 2^{-60}.$$

3.) For the case of 28 linearly independent equations, we can again do a substitution using these 28 linear equations into y_i to derive new maps $\hat{F}$ from K^{n-28} to K^{n+68}, and $\hat{F} = \hat{\phi}_4 \circ \phi_2 \circ \hat{\Phi}_1$, for some linear maps $\hat{\phi}_4$ (invertible) and $\hat{\Phi}_1$ (injective).

This allows us to use exactly the same attack steps as in the case above. In [Ding and Schmidt, 2004], the parameters $n = 52$ and $m = 120$ are chosen, so it is estimated that the attack takes about 2^{35} computations in k.

Cases for Other Implementation Schemes

In [Ding and Schmidt, 2004], it is noticed that at that time all q_i components are very simple and they never have more than two quadratic monomials. It is easy to see that for the TTM schemes before 2004, the dimension of the linear space of all linearization equations for the components y_i of F is not small. This is the common defect for the implementation schemes, which is undesirable for a secure public key cryptosystem. However, even with these linearization equations, it does not necessarily mean that finding the plaintext from a given ciphertext is easy.

For example, in the case of the first implementation of the TTM cryptosystem [Moh, 1999a] previously presented, it is shown that:

1.) The dimension of V of the space of linearization equations for the components y_i of F is 68;

2.) For a given ciphertext $(y'_1, \ldots, y'_{68+n})$, the linearization will *not* produce enough linearly independent linear equations in the x_i for the subsequent substitution, etc.

This means that what we can achieve is just a reduction of number of variables. Although this could be useful in some way, it is still unclear how we can efficiently obtain the plaintext. In [Ding and Schmidt, 2004], experiments were performed for a search for linearization equations on the components of $\hat{y}_i$. Though there were some successes, there were also some failures as well.

For the case of most of the TTM schemes in the original version of [Moh, 1999a], it is also observed that for the components y_i of the cipher F, a higher order type of linearization equation of the form

$$\sum a_{ij} y_i y_j + \sum b_{ij} y_i x_j + \sum c_k y_k + \sum d_l x_l + e = 0$$

is also satisfied. Finding all the solutions for this case takes more time but is not impossible. In fact, it can also produce nontrivial linear equations in x_i if we are given a ciphertext y'_i. There is a possibility that these linearization equations will produce more linear equations to help defeat the scheme.

The Future of TTM

Though the above work shows that the current implementations of the TTM ideas are not secure, it is not at all clear that in the end such a cryptosystem will not work. The main reason for this is that we do not yet fully understand all the current constructions, and how they work remains in some way very mysterious. Recently, another new TTM cryptosystem was proposed [Moh et al., 2004], and this construction is the same type as the ones before. Its security is still unknown. However, the best solution for the future of the TTM would involve finding some systematic method to establish the TTM cryptosystem. To do so, we need some new knowledge in how to control the degree of the composition of multivariate functions.

6.6 Triangular Signature Schemes

The original TTM schemes were intended for the purpose of public key encryption. Attempts were made to apply a similar but simpler idea for signatures. It was called the TTS (Tamed Transformation Signature) scheme, and it is essentially the result of an application of the Minus method in [Shamir, 1993] to a tame transformation.

The first few TTS schemes were suggested by Chen and Yang in [Chen et al., 2002; Yang and Chen, 2003]. The first generation of the TTS schemes was also closely related to the TPM cryptosystem with $u \leq r$, which can also be used for signatures.

The TPM Signature Protocol (when $u \leq r$): To sign a message, we must execute the following steps:

1.) Given a message M, the user first calculates

$$(y'_1, \ldots, y'_{n+u-r}) = h(M)$$

in k^{n+u-r}, where h is a (collision-free) hash function.

2.) The legitimate user then computes

$$(y_1, \ldots, y_{n+u-r}) = L_1^{-1}(y'_1, \ldots, y'_{n+u-r}).$$

3.) Then the user randomly chooses a vector $(x_{n-r+1}, \ldots, x_n)$ in k^r, until the vector $(x_1, \ldots, x_n)$ obtained by the iteration

$$x_i = y_i - g_i(x_1, \ldots, x_{i-1}; x_{n-r+1}, \ldots, x_n),$$

for $1 \leq i \leq n - r$, satisfies the additional u equations

$$g_i(x_1, \ldots, x_n) = y_i$$

for $n - r + 1 \leq i \leq n - r + u$.

4.) For the last step, the user needs to calculate

$$(x'_1, \ldots, x'_n) = L_2^{-1}(x_1, \ldots, x_n),$$

which is the signature of M.

This signature process has a complexity of $O(q^u)$. Therefore, for any practical use q^u must not be too large. The condition $u \leq r$ ensures that the map $\bar{F}$ is nearly surjective, and we have a high probability of finding an arbitrary document.

The first generation of the TTS schemes are essentially TPM with $u = 0$. The security and efficiency of these schemes were thought to rival those of Sflash^{v2}. The inventors of the TTS schemes later realized in [Yang and Chen, 2004c] that they had not been very careful in its design and they showed that all schemes in [Yang and Chen, 2003] could be defeated easily using the method developed by Coppersmith, Vaudnay, and Stern [Coppersmith et al., 1997].

New schemes were suggested in [Yang and Chen, 2004c] which are considered to have the security and efficiency rivaling those of Sflash. One of them was carefully studied in terms of its practical implementation on low cost smart-cards, and the results were presented at CHES 2004 [Yang et al., 2004a]. The scheme was indeed shown to be very efficient, in particular in the signing process. However, Ding and Yin discovered a way to break this new family of TTS schemes, including the version presented in [Yang et al., 2004a].

First Generation of TTS and its Cryptanalysis

The original TTS schemes [Chen et al., 2002; Yang and Chen, 2003] combine Shamir's idea of Minus [Shamir, 1993] with the basic idea of TTM. This combination was first implicitly pointed out in [Goubin and Courtois, 2000], where it was called a Triangular-Minus system.

For the case of such a TTS scheme [Yang and Chen, 2003; Chen et al., 2002], the public key $\bar{T}$ is made of m quadratic polynomials in n variables over a finite field k,

$$\bar{T}(x_1, \ldots, x_n) = (y_1(x_1, \ldots, x_n), \ldots, y_m(x_1, \ldots, x_n)),$$

where $m < n$. The m polynomials y_i are made public for verifying the authenticity of the signature. In the original construction $k = GF(256)$

was used, but the construction works also for finite fields with odd characteristic. Nevertheless, we will limit the discussion to finite fields with characteristic equal to two.

The map $\bar{T}$ from k^n to k^m is derived as

$$\bar{T} = \bar{L}_1 \circ \mathfrak{J}^- \circ \bar{L}_2,$$

where $\bar{L}_1$ is an invertible affine linear map on a space of dimension m, $\bar{L}_2$ is an invertible affine linear map on a space of dimension n, and $\bar{L}_1$, $\bar{L}_2$ are randomly chosen. The map $\mathfrak{J}^-$ is given as:

$$\begin{aligned}\mathfrak{J}^-(z_1, \ldots, z_n) = (&z_{n-m+1} + g_{n-m+1}(z_1, \ldots, z_{n-m}),\\ &z_{n-m+2} + g_{n-m+2}(z_1, \ldots, z_{n-m+1}),\\ &\ldots,\\ &z_n + g_n(z_1, \ldots, z_{n-1}))\\ = (&\bar{y}_1(z_1, \ldots, z_{n-m+1}), \ldots, \bar{y}_m(z_1, \ldots, z_n)),\end{aligned}$$

which is derived from the upper-triangular de Jonquières map

$$\begin{aligned}\mathfrak{J}(z_1, \ldots, z_n) = (&z_1, z_2 + g_2(z_1), \ldots,\\ &z_{n-m+1} + g_{n-m+1}(z_1, \ldots, z_{n-m}),\\ &z_{n-m+2} + g_{n-m+2}(z_1, \ldots, z_{n-m+1}),\\ &\ldots,\\ &z_n + g_n(z_1, \ldots, z_{n-1}))\end{aligned}$$

by removing (Minus method) the first $n - m$ components.

We can see that

$$\bar{T} = \bar{L}_1 \circ U^{-1} \circ U \circ \mathfrak{J}^- \circ \bar{L}_2,$$

where U is a randomly chosen lower-triangular invertible linear transformation from k^m to k^m such that

$$U(z_1, \ldots, z_m) = (a_{11}z_1, \sum_{j=1}^{2} a_{2j}z_j, \ldots, \sum_{j=1}^{i} a_{ij}z_j, \ldots, \sum_{j=1}^{m} a_{mj}z_j),$$

and $a_{jj} \neq 0$ for $j = 1, \ldots, m$. Then we have

$$U \circ \mathfrak{J}^-(z_1, \ldots, z_n) = (a_{11}\bar{y}_1(z_1, \ldots, z_{n-m+1}), \sum_{j=1}^{2} a_{2j}\bar{y}_j(z_1, \ldots, z_{n-m+j}),$$

$$\ldots, \sum_{j=1}^{i} a_{ij}\bar{y}_j(z_1, \ldots, z_{n-m+j}),$$

$$\ldots, \sum_{j=1}^{m} a_{mj}\bar{y}_j(z_1, \ldots, z_{n-m+j}))$$
$$= (\bar{w}_1(z_1, \ldots, z_{n-m+1}), \ldots, \bar{w}_m(z_1, \ldots, z_n)).$$

Therefore $U \circ \mathfrak{J}^-$ is an equivalent choice for $\mathfrak{J}^-$. In this case, we can associate the standard bilinear form to the quadratic part of $\bar{w}_i$. The rank of these bilinear forms will be in ascending order, although not necessarily strictly ascending. This means that any such scheme cannot work, as the MinRank method [Goubin and Courtois, 2000] can be used due to this property of the ranks. This is a much more efficient attack method than the one of [Yang and Chen, 2004c], which is just an application of [Coppersmith et al., 1997], as discussed in the section on Rainbow signature schemes.

Let

$$V_i = \{v(z_1, \ldots, z_n) \mid v(z_1, \ldots, z_n) = \sum_{j=1}^{i} a_{ij}\bar{y}_j(z_1, \ldots, z_{n-m+j}), a_{ij} \in k\}.$$

It is clear that we have

$$V_1 \subset V_2 \subset V_3 \cdots \subset V_m,$$

and

$$\dim(V_{i+1}) - \dim(V_i) = 1,$$

The V_i form what in mathematics is called a flag.

The highest rank of the quadratic forms of the V_{i+1} is in general one higher than the highest rank for the quadratic forms of V_i, if k is not of characteristic two; otherwise if k is not of characteristic two, the highest rank for the quadratic forms of V_{i+1} is in general either the same (if i is even) or two higher (if i is odd) than the highest rank for the quadratic forms of V_i. This implies that:

1.) For the case where k is not of characteristic two, we can use a linear combination of any two randomly chosen public polynomials to reduce the rank by one;

2.) For the case where k is of characteristic two, we can use a linear combination of any three randomly chosen public polynomials to reduce the rank by two.

Therefore we can use a search to reduce the rank one-by-one or two-by-two to go down to the lowest rank, which solves the MinRank problem. At this point the scheme is broken.

The conclusion is that no matter what parameters we choose, the old TTS schemes given in [Chen et al., 2002; Yang and Chen, 2003] are insecure.

The New TTS Schemes

In [Yang and Chen, 2004c] the authors showed that their previous constructions were insecure. They also suggested some new schemes and studied the security and efficiency of these schemes. The new constructions tries to improve the efficiency (compared with Sflash in particular) by using special sparse polynomials. Their constructions are given in terms of specific formulas, and do not follow from basic general principles. Though there is some explanation how these sparse polynomials are chosen, the formulas themselves are not well understood and we do not yet have any solid justification that such choices do not affect the security. This is a fundamental problem and the reason why there is a new attack method [Ding and Yin, 2004] on the TTS systems in [Yang and Chen, 2004c] using the specific form of the formulas.

The first version of the paper [Yang and Chen, 2004c] suggests four families of formulas. One of these formulas is carefully studied for its practical implementation on low cost smart-cards. This was presented at CHES 2004 [Yang et al., 2004a] and we will now present the attack on this family from [Ding and Yin, 2004]. This new scheme appears on page 373 of [Yang et al., 2004a] and is referred to as TTS(20,28).

According to the claim in [Yang and Chen, 2004c; Yang et al., 2004a] the system is secure with at least a complexity of 2^{80} (a minimum security requirement by NESSIE). This specific construction depends on a map $F(x_0, x_1, \ldots, x_{27}) = (f_1, \ldots, f_{20})$ from k^n to k^m, where k is a finite field of size 2^8, $n = 28$ and $m = 20$. The components of F are defined by

$$\begin{aligned}
f_1 &= x_8 + x_1x_8p_{8,1} + x_2x_9p_{8,2} + x_3x_{10}p_{8,3} + x_4x_{11}p_{8,4} \\
&\quad + x_5x_{12}p_{8,5} + x_6x_{13}p_{8,6} + x_7x_{14}p_{8,7} \\
f_2 &= x_9 + x_1x_9p_{9,1} + x_2x_{10}p_{9,2} + x_3x_{11}p_{9,3} + x_4x_{12}p_{9,4} \\
&\quad + x_5x_{13}p_{9,5} + x_6x_{14}p_{9,6} + x_7x_{15}p_{9,7} \\
f_3 &= x_{10} + x_1x_{10}p_{10,1} + x_2x_{11}p_{10,2} + x_3x_{12}p_{10,3} + x_4x_{13}p_{10,4} \\
&\quad + x_5x_{14}p_{10,5} + x_6x_{15}p_{10,6} + x_7x_{16}p_{10,7} \\
f_4 &= x_{11} + x_1x_{11}p_{11,1} + x_2x_{12}p_{11,2} + x_3x_{13}p_{11,3} + x_4x_{14}p_{11,4} \\
&\quad + x_5x_{15}p_{11,5} + x_6x_{16}p_{11,6} + x_7x_8p_{11,7} \\
f_5 &= x_{12} + x_1x_{12}p_{12,1} + x_2x_{13}p_{12,2} + x_3x_{14}p_{12,3} + x_4x_{15}p_{12,4} \\
&\quad + x_5x_{16}p_{12,5} + x_6x_8p_{12,6} + x_7x_9p_{12,7}
\end{aligned}$$

$$
\begin{aligned}
f_6 &= x_{13} + x_1x_{13}p_{13,1} + x_2x_{14}p_{13,2} + x_3x_{15}p_{13,3} + x_4x_{16}p_{13,4} \\
&\quad + x_5x_8p_{13,5} + x_6x_9p_{13,6} + x_7x_{10}p_{13,7} \\
f_7 &= x_{14} + x_1x_{14}p_{14,1} + x_2x_{15}p_{14,2} + x_3x_{16}p_{14,3} + x_4x_8p_{14,4} \\
&\quad + x_5x_9p_{14,5} + x_6x_{10}p_{14,6} + x_7x_{11}p_{14,7} \\
f_8 &= x_{15} + x_1x_{15}p_{15,1} + x_2x_{16}p_{15,2} + x_3x_8p_{15,3} + x_4x_9p_{15,4} \\
&\quad + x_5x_{10}p_{15,5} + x_6x_{11}p_{15,6} + x_7x_{12}p_{15,7} \\
f_9 &= x_{16} + x_1x_{16}p_{16,1} + x_2x_8p_{16,2} + x_3x_9p_{16,3} + x_4x_{10}p_{16,4} \\
&\quad + x_5x_{11}p_{16,5} + x_6x_{12}p_{16,6} + x_7x_{13}p_{16,7} \\
f_{10} &= x_{17} + x_1x_6p_{17,1} + x_2x_5p_{17,2} + x_3x_4p_{17,3} + x_9x_{16}p_{17,4} \\
&\quad + x_{10}x_{15}p_{17,5} + x_{11}x_{14}p_{17,6} + x_{12}x_{13}p_{17,7} \\
f_{11} &= x_{18} + x_2x_7p_{18,1} + x_3x_6p_{18,2} + x_4x_5p_{18,3} + x_{10}x_{17}p_{18,4} \\
&\quad + x_{11}x_{16}p_{18,5} + x_{12}x_{15}p_{18,6} + x_{13}x_{14}p_{18,7} \\
f_{12} &= x_{19} + x_8x_{10}p_{19,0} + x_0x_{19}p_{19,1} + x_{18}x_{20}p_{19,2} + x_{17}x_{21}p_{19,3} \\
&\quad + x_{16}x_{22}p_{19,4} + x_{15}x_{23}p_{19,5} + x_{14}x_{24}p_{19,6} + x_{13}x_{25}p_{19,7} \\
&\quad + x_{12}x_{26}p_{19,8} + x_{11}x_{27}p_{19,9} \\
f_{13} &= x_{20} + x_9x_{11}p_{20,0} + x_2x_{19}p_{20,1} + x_0x_{20}p_{20,2} + x_{18}x_{21}p_{20,3} \\
&\quad + x_{17}x_{22}p_{20,4} + x_{16}x_{23}p_{20,5} + x_{15}x_{24}p_{20,6} + x_{14}x_{25}p_{20,7} \\
&\quad + x_{13}x_{26}p_{20,8} + x_{12}x_{27}p_{20,9} \\
f_{14} &= x_{21} + x_{10}x_{12}p_{21,0} + x_4x_{19}p_{21,1} + x_2x_{20}p_{21,2} + x_0x_{21}p_{21,3} \\
&\quad + x_{18}x_{22}p_{21,4} + x_{17}x_{23}p_{21,5} + x_{16}x_{24}p_{21,6} + x_{15}x_{25}p_{21,7} \\
&\quad + x_{14}x_{26}p_{21,8} + x_{13}x_{27}p_{21,9} \\
f_{15} &= x_{22} + x_{11}x_{13}p_{22,0} + x_6x_{19}p_{22,1} + x_4x_{20}p_{22,2} + x_2x_{21}p_{22,3} \\
&\quad + x_0x_{22}p_{22,4} + x_{18}x_{23}p_{22,5} + x_{17}x_{24}p_{22,6} + x_{16}x_{25}p_{22,7} \\
&\quad + x_{15}x_{26}p_{22,8} + x_{14}x_{27}p_{22,9} \\
f_{16} &= x_{23} + x_{12}x_{14}p_{23,0} + x_8x_{19}p_{23,1} + x_6x_{20}p_{23,2} + x_4x_{21}p_{23,3} \\
&\quad + x_2x_{22}p_{23,4} + x_0x_{23}p_{23,5} + x_{18}x_{24}p_{23,6} + x_{17}x_{25}p_{23,7} \\
&\quad + x_{16}x_{26}p_{23,8} + x_{15}x_{27}p_{23,9} \\
f_{17} &= x_{24} + x_{13}x_{15}p_{24,0} + x_{10}x_{19}p_{24,1} + x_8x_{20}p_{24,2} + x_6x_{21}p_{24,3} \\
&\quad + x_4x_{22}p_{24,4} + x_2x_{23}p_{24,5} + x_0x_{24}p_{24,6} + x_{18}x_{25}p_{24,7} \\
&\quad + x_{17}x_{26}p_{24,8} + x_{16}x_{27}p_{24,9} \\
f_{18} &= x_{25} + x_{14}x_{16}p_{25,0} + x_{12}x_{19}p_{25,1} + x_{10}x_{20}p_{25,2} + x_8x_{21}p_{25,3} \\
&\quad + x_6x_{22}p_{25,4} + x_4x_{23}p_{25,5} + x_2x_{24}p_{25,6} + x_0x_{25}p_{25,7} \\
&\quad + x_{18}x_{26}p_{25,8} + x_{17}x_{27}p_{25,9} \\
f_{19} &= x_{26} + x_{15}x_{17}p_{26,0} + x_{14}x_{19}p_{26,1} + x_{12}x_{20}p_{26,2} + x_{10}x_{21}p_{26,3}
\end{aligned}
$$

$$
\begin{aligned}
& +x_8x_{22}p_{26,4} + x_6x_{23}p_{26,5} + x_4x_{24}p_{26,6} + x_2x_{25}p_{26,7} \\
& + x_0x_{26}p_{26,8} + x_{18}x_{27}p_{26,9} \\
f_{20} \quad = \quad & x_{27} + x_{16}x_{18}p_{27,0} + x_{16}x_{19}p_{27,1} + x_{14}x_{20}p_{27,2} + x_{12}x_{21}p_{27,3} \\
& +x_{10}x_{22}p_{27,4} + x_8x_{23}p_{27,5} + x_6x_{24}p_{27,6} + x_4x_{25}p_{27,7} \\
& + x_2x_{26}p_{27,8} + x_0x_{27}p_{27,9},
\end{aligned}
$$

where the $p_{i,j}$ are randomly chosen nonzero elements from the field k. We remark that TTS can be viewed as a special case of Rainbow.

The public key for the new TTS system is F, given by

$$\bar{F} = L_1 \circ F \circ L_2,$$

where L_1 is an invertible affine linear transformation over k^{20} and L_2 is an invertible affine linear transformation over k^{28}. These two transformations L_1, L_2 make up the secret key.

In order to sign a document $P = (p_1, \ldots, p_{20})$, an element of k^{20}, we need to find a solution of the equation

$$\bar{F}(x_0, \ldots, x_{27}) = P.$$

We will be able to find a solution due to the triangular-type structure of F.

From the formulas, it is clear that the f_i can be divided into three groups:

$$
\begin{aligned}
(\mathrm{I}) &= \{f_i \mid i = 1, \ldots, 9\} \\
(\mathrm{II}) &= \{f_i \mid i = 10, 11\} \\
(\mathrm{III}) &= \{f_i \mid i = 12, \ldots, 20\}.
\end{aligned}
$$

First, we notice that the quadratic part of Group (I) elements are all of the form

$$\sum_{i=1,\ldots,7;j=8,\ldots,16} a_{ij}x_ix_j. \tag{6.3}$$

If we form any linear combination of these elements, the rank of the associated quadratic form will be 14. Second, the Group (II) elements come from a de Jonquières construction. If we add Group (I) elements to the Group (II) elements then the rank of the corresponding bilinear form increases, though the rank cannot exceed 16. Third, notice that the quadratic parts of Group (III) elements are all of the form

$$\sum_{i=1,\ldots,18;j=19,\ldots,27} a_{ij}x_ix_j + \sum_{i,j=8,\ldots,18} b_{ij}x_ix_j + \sum_{j=19,\ldots,27} c_jx_0x_j. \tag{6.4}$$

When we add any Group (III) elements to any linear combination of Group (I) and (II) elements, the rank of the corresponding bilinear form also increases, and a random linear combination of all f_i will produce a non-degenerate quadratic form.

In order to sign a document P we need to solve the equation

$$F \circ L_2(x_0, \ldots, x_{27}) = L_1^{-1}(P).$$

To do this, we first solve

$$F(x_0, \ldots, x_{27}) = L_1^{-1}(P),$$

and then apply L_2^{-1}.

To solve $F(x_0, \ldots, x_{27}) = (\bar{p}_1, \ldots, \bar{p}_{20})$, because of (6.3), we first randomly fix the values of $x_1, \ldots, x_7$. This allows the polynomials from Group (I) to produce nine linear equations whose solution gives the values of $x_8, \ldots, x_{16}$. Then we plug the values of $x_1, \ldots, x_{16}$ into Group (II) and Group (III). Due to the triangular structure of the de Jonquières type maps, f_{10} produces one linear equation, which gives the value of x_{17}. Then we plug the value of x_{17} into f_{11}, which gives a linear equation, which in turn yields the value of x_{18}. Then we substitute the values of x_{17}, x_{18} into Group (III), and randomly choose a value for x_0. This, due to (6.4), produces again nine linear equations whose solution gives us the values of $x_{19}, \ldots, x_{27}$. Finally, we apply L_2^{-1} to find a solution, which produces a valid signature.

At this point we see that in order to forge a signature, we need to know how to find any solution of the equation $F(x_0, \ldots, x_{27}) = P$.

Cryptanalysis of the New TTS Cryptosystem

The attack method presented here is a combination of searching for invariant subspaces [Kipnis et al., 1999], of MinRank [Goubin and Courtois, 2000], and other general methods for bilinear forms.

Let $L_2(x_0, \ldots, x_{27}) = (L_{2,0}, \ldots, L_{2,27})$ and let

$$\tilde{F}(x_0, x_1, \ldots, x_{27}) = F \circ L_2(x_0, x_1, \ldots, x_{27}) = (\tilde{F}_1, \ldots, \tilde{F}_{20}).$$

We define

$$(\tilde{\mathrm{I}}) = \{\tilde{f}_i \mid i = 1, \ldots, 9\}$$
$$(\widetilde{\mathrm{II}}) = \{\tilde{f}_i \mid i = 10, 11\}$$
$$(\widetilde{\mathrm{III}}) = \{\tilde{f}_i \mid i = 12, \ldots, 20\}.$$

These sets have properties similar to those described in (6.3) and (6.4) corresponding to Groups (I), (II), and (III).

First, we know that for $l = 1, \ldots, 9$,

$$\tilde{f}_l = \sum_{i=1,\ldots,7;j=8,\ldots,16} a_{lij} L_{2,i} L_{2,j}. \tag{6.5}$$

Therefore, if we can find the space of linear combinations of the linear parts (no constant terms) of $L_{2,i}$, $i = 1, \ldots, 7$, then we could do a linear substitution using any linear equation whose linear part is defined by elements from this space. The solution is not unique, but for the attack purpose it suffices to work with a basis for the subspace. According to algebraic geometry, substitution of a linear equation is equivalent to the evaluation on a linear variety. Here, the substitution by linear equations is equivalent to substituting all the $L_{2,i}$, $i = 1, \ldots, 7$ by constants.

The $\bar{f}_i$ and $\tilde{f}_i$ are just linear combinations of each other with additional constant terms due to the invertibility of L_1. Through a search for linear equations by linear combinations, we can find nine linear independent equations whose solution gives the values of $L_{2,j}$, $j = 8, \ldots, 16$. Then, due to the de Jonquières structure of Group ($\widetilde{\text{II}}$), through substitution, the whole system will be reduced to solving a set of equations coming from linear combinations of Group ($\widetilde{\text{III}}$), with all values of $L_{2,j}$, $j = 1, \ldots, 18$ given. This can be handled easily and is the final step of the attack.

The attack finds first the linear span of ($\tilde{\text{I}}$), then the linear span of both ($\tilde{\text{I}}$) and ($\widetilde{\text{II}}$), and finally the linear span of the linear part of $L_{2,i}$, $j = 1, \ldots, 7$.

Step 1: *The Unbalanced Oil-Vinegar Attack Method.*

In preparation for the first proposition below, we make the following definitions. Let S be the set of variables $\{x_0, \ldots, x_{27}\}$, let $O = \{x_1, x_3, x_5, x_7, x_{19}, \ldots, x_{27}\}$ be the set of the "Oil" variables, and let $V = S - O$ be the "Vinegar" variables. Write

$$\mathsf{X} = (x_0, x_1, \ldots, x_{27}) = \sum_{i=0}^{27} x_i E_i,$$

where $E_i = (0, \ldots, 0, 1, 0, \ldots, 0)$ is the vector whose component at position $i + 1$ is 1 and the rest are zero.

Let $\bar{O}$ denote the span of the vectors corresponding to the Oil variables; namely,

$$\bar{O} = \text{Span}(E_1, E_3, E_5, E_7, E_{19}, \ldots, E_{27}),$$

and let $\bar{V}$ denote the span of the vectors corresponding to the Vinegar variables; namely,

$$\bar{V} = \text{Span}(E_0, E_2, E_4, E_6, E_8, \ldots, E_{18}).$$

We must be very careful about the difference between the variables and the corresponding space. The variables are just the coordinates of a vector in terms of the standard basis. They are functions from k^n to k which actually are elements in the dual space of k^n. Therefore these two are the dual of each other.

Now let

$$L_1(x_1, \ldots, x_{20}) = (x_1, \ldots, x_{20}) \times \mathsf{A}_1 + (\alpha_1, \ldots, \alpha_{20}),$$

where A_1 is an $m \times m$ invertible matrix, and let

$$L_1^0(x_1, \ldots, x_{20}) = (x_1, \ldots, x_{20}) \times \mathsf{A}_1 = (L_{1,1}^0, \ldots, L_{1,20}^0),$$

be the linear part of L_1. Let

$$L_2(x_0, \ldots, x_{27}) = (x_0, \ldots, x_{27}) \times \mathsf{A}_2 + (a_0, a_1, \ldots, a_{27}),$$

where A_2 is a $n \times n$ invertible matrix, and let

$$L_2^0(x_0, \ldots, x_{27}) = (x_0, \ldots, x_{27}) \times \mathsf{A}_2,$$

be the linear part of $L_{2,i}$. Then we can see that for any fixed i,

$$L_{2,i} = x_i \circ L_2(x_0, \ldots, x_{27}). \tag{6.6}$$

In other words, $L_{2,i}$ can be derived as a composition of x_i with L_2 from the right.

Let $\tilde{O} = L_2^0(\bar{O})$ be the image of $\bar{O}$ under L_2^0, let ${f^0}_i$ denote the quadratic part of the polynomial f_i, and let

$$F^0(x_0, \ldots, x_{27}) = ({f^0}_1, \ldots, {f^0}_{20}).$$

Then we have

$$F^0 = L_1^0 \circ f^0 \circ L_2^0.$$

For each quadratic polynomial

$${f^0}_l = \sum_{i \geq j} (f_l)_{ij} x_i x_j,$$

we can use the standard method to associate with it an $n \times n$ symmetric matrix m_l such that $[\mathsf{m}_l]_{ii} = 0$ and $[\mathsf{m}_l]_{ij} = [\mathsf{m}_l]_{ji} = (f_l)_{ij}$, if $i > j$. For each m_l, we can associate a bilinear form as $\langle \mathsf{X}, \mathsf{X}' \rangle_l = \mathsf{X}\mathsf{m}_l(\mathsf{X}')^T$, and its quadratic form $\langle \mathsf{X}, \mathsf{X} \rangle_l = \mathsf{X}\mathsf{m}_l\mathsf{X}^T$ is trivial. Here $\mathsf{X}' = (x'_0, \ldots, x'_{27})$. Note that when a field with odd characteristic is used, the definition of these matrices must be modified accordingly.

Similarly, for each quadratic polynomial

$$f^0{}_l = \sum_{i \geq j} (f_l)_{ij} x_i x_j,$$

we can associate an $n \times n$ symmetric matrix M_l. For each M_l, we can also associate a bilinear form as $\langle \mathsf{X}, \mathsf{X}' \rangle^l = \mathsf{X}\mathsf{M}_l(\mathsf{X}')^T$, and its quadratic form $\langle \mathsf{X}, \mathsf{X} \rangle^l = \mathsf{X}\mathsf{M}_l\mathsf{X}^T$ is trivial.

For any fixed l we have that

$$\mathsf{M}_l = \sum_{i=1}^{20} [\mathsf{A}_1]_{il} (\mathsf{A}_2 \mathsf{m}_i \mathsf{A}_2^T), \tag{6.7}$$

where $[\mathsf{A}_1]_{il}$ is the i, l entry of the matrix A_1.

The first observation is that:

Proposition 6.6.1. *For any fixed* l,

$$f^0{}_l(x_0, \ldots, x_{27}) = \sum_{i \in O, j \in V} \alpha_{i,j,l} x_i x_j + \sum_{i,j \in V} \beta_{i,j,l} x_i x_j. \tag{6.8}$$

Let $o = |O|$ and $v = |V|$. In terms of this description, these polynomials are just unbalanced Oil-Vinegar polynomials [Kipnis et al., 1999], and thus all the matrices m_l can be rewritten in the corresponding form if we choose the coordinate system as

$$\bar{\mathsf{X}} = (x_1, x_3, x_5, x_7, x_{19}, \ldots, x_{27}, x_0, x_2, x_4, x_6, x_8, x_9, \ldots, x_{18}).$$

Here we choose the basis of the Oil space as the first o components and the basis of the Vinegar space as the last v components. We also have that $\langle \mathsf{X}, \mathsf{X} \rangle_i = \bar{\mathsf{X}}\bar{\mathsf{m}}_i\bar{\mathsf{X}}^T$, and

$$\bar{\mathsf{m}}_i = \begin{pmatrix} 0 & \mathsf{b}_i \\ \mathsf{b}_i^T & \mathsf{d}_i \end{pmatrix}, \tag{6.9}$$

where b_i is an $o \times v$ matrix and d_i is a symmetric $v \times v$ matrix. This follows directly from (6.8).

Let Z be the 28×28 permutation matrix such that

$$\mathsf{X} = \bar{\mathsf{X}} \times \mathsf{Z}.$$

Since these polynomials are unbalanced Oil-Vinegar polynomials, we can apply the attack method in [Kipnis et al., 1999] to find the hidden Oil space $\tilde{O}$. According to [Kipnis et al., 1999], the complexity is roughly $(2^8)^{v-o-1}o^4 < 2^{23}$.

Now, assume that we have found $\tilde{O}$. Then we can choose a new coordinate system such that the first o components are from $\tilde{O}$ and rewrite the matrix M_i. In terms of matrix notation, we can find an invertible $n \times n$ matrix A_3 such that

$$\mathsf{A}_3\mathsf{M}_i\mathsf{A}_3^T = \bar{\mathsf{M}}_i = \begin{pmatrix} 0 & \mathsf{B}_i \\ \mathsf{B}_i^T & \mathsf{D}_i \end{pmatrix}, \tag{6.10}$$

which follows from (6.9) and (6.8).

Let

$$L_3(x_0, \ldots, x_{27}) = (x_0, \ldots, x_{27}) \times \mathsf{A}_3.$$

Then we know that the subspace $\bar{O}$ is invariant under the linear transformation $L_2^0 \circ L_3$. Equivalently,

$$\mathsf{A}_{32} = \mathsf{A}_3 \times \mathsf{A}_2 = \mathsf{Z}^{-1} \times \begin{pmatrix} \mathsf{Q}_1 & 0 \\ \mathsf{R} & \mathsf{Q}_2 \end{pmatrix} = \mathsf{Z}^{-1} \times Q. \tag{6.11}$$

Remark 6.6.1. *One important thing that we must be careful with is that* A_{32} *preserves the Oil space. However, in terms of coordinate system, it actually preserves the Vinegar coordinates. This is exactly due to the dual relationship mentioned at the beginning of this section.*

Proposition 6.6.2. *Let Q be as defined above. Then*

1.) *The space spanned by E_i, $i = 0, \ldots, 12$, is invariant under the action of Q from the right;*

2.) *The space spanned by $x_i, i = 13, \ldots, 27$, is the same as the space spanned by $x_i \circ Q(x_0, \ldots, x_{27})$, where $Q(x_0, \ldots, x_{27}) = \mathsf{X} \times Q$.*

In other words,

$$\begin{aligned} &\mathrm{Span}\{L_{2,i}^0(x_0, \ldots, x_{27}), \ \ i \in V\} \\ = \ &\mathrm{Span}\{L_{3,i}^-(x_0, \ldots, x_{27}), \ \ i = 13, \ldots, 27\}, \end{aligned}$$

where

$$(L_{3,0}^-(x_0, \ldots, x_{27}), \ldots, L_{3,27}^-(x_0, \ldots, x_{27})) = (x_0, \ldots, x_{27}) \times \mathsf{A}_3^{-1}\mathsf{Z}^{-1}.$$

This can be seen easily from

$$\mathsf{A}_2 = \mathsf{A}_3^{-1}\mathsf{Z}^{-1} \times \begin{pmatrix} \mathsf{Q}_1 & 0 \\ \mathsf{R} & \mathsf{Q}_2 \end{pmatrix} = \mathsf{A}_3^{-1}\mathsf{Z}^{-1}Q.$$

This allows us to find the space spanned by the image of the linear parts of the vinegar variables composed from the right by L_2. This finishes the first step of the attack, which is a simple application of the attack method for the unbalanced Oil-Vinegar scheme.

Step 2: *The MinRank Method.*

Any bilinear form $\langle -, - \rangle_i$ on $k^n \times k^n$ can be restricted to the subspace $\bar{O} \times k^n$, and then has the form

$$\langle \mathsf{X}_o, \mathsf{X}' \rangle_i = \bar{\mathsf{X}}_o (\mathsf{m}_s)_i (\bar{\mathsf{X}}')^T = \mathsf{X}_o \mathsf{m}_i (\bar{\mathsf{X}}')^T = {f^s}_i$$

where

$$\begin{aligned}
\mathsf{X}_o &= (0, x_1, 0, x_3, 0, x_5, 0, x_7, 0, \ldots, 0, x_{19}, \ldots, x_{27}) \\
\bar{\mathsf{X}}' &= (x'_1, x'_3, x'_5, x'_7, x'_{19}, \ldots, x'_{27}, x'_0, x'_2, x'_4, x'_6, x'_8, x'_9, \ldots, x'_{18}) \\
\bar{\mathsf{X}}_o &= (x_1, x_3, x_5, x_7, x_{19}, \ldots, x_{27})
\end{aligned}$$

and

$$(\mathsf{m}_s)_i = \begin{pmatrix} 0 & \mathsf{b}_i \end{pmatrix},$$

where 0 denotes an $o \times o$ matrix and b_i an $o \times v$ matrix.

From this we see that if we restrict the bilinear from $\langle -, - \rangle^i$ to the space $\tilde{O} \times k^n$, then the associated matrix $(\mathsf{M}_s)_i$ under the coordinate system defined by A_3 should be exactly $\begin{pmatrix} 0 & \mathsf{B}_i \end{pmatrix}$, and

$$\mathsf{B}_i = \sum_{j=1,\ldots,20} [\mathsf{A}_1]_{ji} (\mathsf{Q}_1 \mathsf{b}_j Q_2^T). \tag{6.12}$$

This is because

$$\begin{pmatrix} \mathsf{Q}_1 & 0 \\ \mathsf{R} & \mathsf{Q}_2 \end{pmatrix} \times \begin{pmatrix} 0 & \mathsf{b}_j \\ \mathsf{b}_j^T & \mathsf{d}_j \end{pmatrix} \times \begin{pmatrix} \mathsf{Q}_1 & 0 \\ \mathsf{R} & \mathsf{Q}_2 \end{pmatrix}^T$$
$$= \begin{pmatrix} 0 & \mathsf{Q}_1 \mathsf{b}_j \mathsf{Q}_2^T \\ \mathsf{Q}_2 \mathsf{b}_j^T \mathsf{Q}_1^T & \mathsf{Q}_2 \mathsf{b}_j^T \mathsf{R}^T + \mathsf{R} \mathsf{b}_j \mathsf{Q}_2^T + \mathsf{Q}_2 \mathsf{d}_j \mathsf{Q}_2^T \end{pmatrix}.$$

We now take a closer look at the b_i. The key observation is that

$$\begin{aligned}
{f^s}_{10} &= x_1 x'_6 p_{17,1} + x_5 x'_2 p_{17,2} + x_3 x'_4 p_{17,3} \\
{f^s}_{11} &= x_7 x'_2 p_{18,1} + x_3 x'_6 p_{18,2} + x_5 x'_4 p_{17,3},
\end{aligned}$$

where f_i^s denotes the quadratic polynomial derived from restricting the quadratic form to the space $\tilde{O} \times k^n$.

We find that the rank of the corresponding matrices $(\mathsf{m}_s)_i$, or b_i, is exactly three for $i = 10, 11$; and is greater than three for all the rest. We can also see clearly that in the space of all possible linear combinations of the b_i, these two matrices and their constant multiples are the only matrices of the lowest rank three.

In this case, we can use the MinRank method to search for both b_{10} and b_{11} in $\mathsf{Q}_1\mathsf{b}_{10}Q_2^T$, $\mathsf{Q}_1\mathsf{b}_{11}Q_2^T$ through linear combinations of B_i, since A_1 is invertible. We have a total of 20 matrices of size 13×15 and the MinRank is three. From the complexity analysis in [Goubin and Courtois, 2000], we know to find one of them has complexity of $(2^8)^{2\times 3} = 2^{48}$.

Now, let us assume that we have found two rank-three matrices H_i, $i = 10, 11$, and that

$$\mathsf{H}_i = \sum_{j=1}^{20} h_{ij}\mathsf{B}_j. \tag{6.13}$$

Due to the uniqueness of the space of linear combinations of the matrices B_j, we have that

$$\sum_{j=1}^{20} h_{10,j}F^0{}_j = \beta_1 \tilde{f^0}_{11}, \qquad \sum_{j=1}^{20} h_{11,j}f^0{}_j = \beta_2 \tilde{f^0}_{10}, \tag{6.14}$$

or

$$\sum_{j=1}^{20} h_{10,j}f^0{}_j = \beta_1 \tilde{F^0}_{10}, \qquad \sum_{j=1}^{20} h_{11,j}f^0{}_j = \beta_2 \tilde{f^0}_{11}, \tag{6.15}$$

where β_1 and β_2 are nonzero constants in k and

$$(\tilde{f^0}_1, \ldots, \tilde{f^0}_{20}) = f^0 \circ L_2^0 \circ F^0 = (f^0{}_1, \ldots, f^0{}_{20}).$$

The quadratic polynomials $f^0{}_i$ are linearly independent, and so the linear and constant terms are determined by the quadratic terms. This means that we can find constant multiples of both $\tilde{f}_{10}$, $\tilde{f}_{11}$ by applying (6.14) or (6.15); namely,

$$\sum_{j=1}^{20} h_{10,j}f_j = \beta_1 \tilde{f}_{11}, \qquad \sum_{j=1}^{20} h_{11,j}f_j = \beta_2 \tilde{f}_{10} \tag{6.16}$$

or

$$\sum_{j=1}^{20} h_{10,j}f_j = \beta_1 \tilde{f}_{10}, \qquad \sum_{j=1}^{20} h_{11,j}f_j = \beta_2 \tilde{f}_{11}. \tag{6.17}$$

Here $(\tilde{f}_1, \ldots, \tilde{f}_{20}) = F \circ L_2$, and $\tilde{f}_{10}$, $\tilde{f}_{11}$ are essentially f_{10}, f_{11} but with a substitution of variables.

Step 3: *Search for the Null Subspace.*

Now let us take a careful look at both ${f^0}_{11}, {f^0}_{10}$ in terms of their related bilinear forms. Through computation, we know that both $\mathsf{m}_{11}, \mathsf{m}_{10}$ are of rank 14, and therefore the corresponding bilinear form is of rank 14 and the null spaces N_{11}, N_{10} (the space of vectors orthogonal to the whole space) for both bilinear forms have dimension 14. We can see and show by calculation that

$$N_{10} = \mathrm{Span}(E_0, E_7, E_8, E_{17}, E_{18}, E_{19}, \ldots, E_{27})$$
$$N_{11} = \mathrm{Span}(E_0, E_1, E_8, E_9, E_{18}, E_{19}, \ldots, E_{27}).$$

We observe that

$$N_{10} \cap N_{11} \cap \bar{O} = \mathrm{Span}(E_{19}, \ldots, E_{27}) = \bar{O}_1.$$

Because of (6.13), we know that the null space of the bilinear form associated to M_i, $i = 10, 11$ should give us exactly $L_2^0(N_i) = \tilde{N}_i$, $i = 10, 11$. Here $\tilde{N}_i$ denotes the null space for the bilinear form defined by M_i. This can be done by solving a set of n linear equations with n variables:

$$\mathsf{X} \times \mathsf{M}_i = 0.$$

This means that we can find

$$\tilde{O}_1 = L_2^0(\bar{O}_1) = \tilde{N}_{10} \cap \tilde{N}_{11} \cap \tilde{O} = \mathrm{Span}(L_2^0(x_{19}), \ldots, L_2^0(x_{27})).$$

Now, let us assume that we have found $\tilde{O}_1$. Then we can choose a new coordinate system such that the first o_1 components are from $\tilde{O}_1$, where o_1 is the dimension of $\tilde{O}_1$, and rewrite the matrix M_i. In terms of matrix notation, we can find an invertible $n \times n$ matrix A_4 such that

$$\mathsf{A}_4 \mathsf{M}_i \mathsf{A}_4^T = \tilde{\mathsf{M}}_i = \begin{pmatrix} 0 & \tilde{\mathsf{B}}_i \\ \tilde{\mathsf{B}}_i^T & \tilde{\mathsf{D}}_i \end{pmatrix}. \tag{6.18}$$

This follows from the specific formulas for f_i, where there is no $x_i x_j$ term if $19 \leq i, j \leq 27$. The size of the matrix 0 is $o_1 \times o_1$. Let

$$L_4(x_0, \ldots, x_{27}) = (x_0, \ldots, x_{27}) \times \mathsf{A}_4.$$

Then we know that the subspace $\bar{O}_1$ is invariant under the linear transformation $L_2^0 \circ L_4$.

Step 4: *Search for the Subspace of the Linear Span of both* ($\tilde{\mathrm{I}}$) *and* ($\widetilde{\mathrm{II}}$).

In terms of the coordinate system $\tilde{\mathsf{X}} = (x_{19}, \ldots, x_{27}, x_0, x_1, \ldots, x_{18})$, m_i will become a different matrix, say $\tilde{\mathsf{m}}_i$. We observe that

$$\tilde{\mathsf{m}}_i = \begin{pmatrix} 0 & 0 \\ 0 & \mathsf{U}_i \end{pmatrix}, \tag{6.19}$$

for $i = 1, \ldots, 11$, and U_i is of the size $(n - o_1) \times (n - o_1) = 19 \times 19$. This is due to the fact that the polynomials in the Group (I) and (II) contain no $x_i x_j$ for $i \in \{19, \ldots, 27\}$ and $j \in \{0, \ldots, 18\}$. This implies that in the coordinate system defined by A_4, we can find a set of 19 linearly independent matrices $\hat{\mathsf{M}}_j$ which are linear combinations of $\tilde{\mathsf{M}}_i$ such that

$$\hat{\mathsf{M}}_j = \sum_{i=1}^{19} \gamma_{ij} \tilde{\mathsf{M}}_i = \begin{pmatrix} 0 & 0 \\ 0 & \hat{\mathsf{U}}_j \end{pmatrix}. \tag{6.20}$$

This set of matrices can be found easily by solving a small set of linear equations.

Similar to the case of (6.14) and (6.15) we know from formula (6.19) that

$$\text{Span}\{\sum_{i=1}^{19} \gamma_{ij} F_i \mid i = 1, \ldots, 11\} = \text{Span}\{\tilde{F}_i \mid i = 1, \ldots, 11\} \tag{6.21}$$

We will denote this space $\tilde{G}_{12}$.

Step 5: *Search for the Subspace of the Linear Span of Group* ($\tilde{\text{I}}$).

Let us denote the span of the elements in Group ($\tilde{\text{I}}$) by $\tilde{G}_1$. We know that

1.) $\tilde{G}_1$ is a subspace of $\tilde{G}_{12}$, whose dimension is $\dim(\tilde{G}_{12}) - 2$;

2.) If we take any polynomial in $\tilde{G}_{12}$ not in $\tilde{G}_1$, it has the property that the quadratic form corresponding to the quadratic part of this polynomial is of rank 18, and in particular is greater than 14. On the other hand, for any element inside $\tilde{G}_1$, the corresponding rank is exactly 14.

This means that we can find a basis of $\tilde{G}_1$ by choosing three polynomials q_1, q_2 and q_3 from any basis of $\tilde{G}_{12}$, and then searching for all $q_1 + u_1 q_2 + u_2 q_3$ whose corresponding quadratic form is of rank 14, where $u_1, u_2 \in k$. This will definitely produce one element in $\tilde{G}_1$ since a dimension three subspace must non-trivially intersect with a dimension nine subspace in a space of total dimension 11. Using this procedure on the corresponding matrices of the bilinear form for the polynomials of looking for matrix of rank 14, we can find a basis for $\tilde{G}_1$ by searching at most 10 times. The complexity of this step is then less than $(2^8)^2 \times 18^3 \times 10/6 < 2^{30}$.

Step 6: *Reformulation of* $\tilde{G}_1$.

Let $G_{12} = \text{Span}\{f_i \mid i = 1, \ldots, 11\}$ and $G_1 = \text{Span}\{f_i \mid i = 1, \ldots, 9\}$. Let N_i denote the null space for each bilinear form $\langle -, - \rangle_i$. Then we observe and prove by calculation that

$$\bar{N} = \bigcap_{i=1,\ldots,9} N_i = \mathrm{Span}(E_0, E_{17}, E_{18}, E_{19}, \ldots, E_{27}).$$

This implies that we can find a basis of the space consisting of a basis of the subspace of the intersection of all the null spaces of the bilinear forms defined by the quadratic parts of the polynomials in $\tilde{G}_1$. This gives us a matrix A_5 such that

$$\mathsf{A}_5\mathsf{B}\mathsf{A}_5^T = \begin{pmatrix} 0 & \tilde{\mathsf{b}}_i \\ \tilde{\mathsf{b}}_i^T & \tilde{\mathsf{d}}_i \end{pmatrix},$$

where B is any symmetric matrix of the bilinear form corresponding to the quadratic part of any polynomial in $\tilde{G}_1$. This implies that we can define a linear transformation L_5 as

$$L_5(x_0, \ldots, x_{27}) = (x_0, \ldots, x_{27}) \times \mathsf{A}_5,$$

for any $\tilde{f}_i$ in $\tilde{G}_1$, and we have

$$\tilde{f}_i \circ L_5(x_0, \ldots, x_{27}) = \sum_{i \geq j=1}^{16} \tilde{\alpha}_{i,j} x_i x_j + \sum_{i=1}^{16} \tilde{\alpha}_i x_i + \tilde{\alpha}.$$

Therefore, by composing with L_5, all the polynomials in $\tilde{G}_1$ become a set of polynomials with only 16 variables. We will call this new set of polynomials $\widetilde{GL}_1$.

From the above procedure and by solving a set of linear equations, we find an affine linear transformation L_6 on k^{16} such that the space $\widetilde{GL}_1$ is derived from composition of the elements in G_1 from the right by L_6. Now we treat all elements in G_1 and $\widetilde{GL}_1$ as polynomials of only 16 variables and ignore the other variables. Again, we associate the quadratic part of each G_1 with a bilinear form and we can see that all these forms are exactly of rank 14. We randomly pick nine linearly independent polynomials $\hat{f}_i$ from $\tilde{G}L_1$. Let $\langle -, - \rangle_i^s$ denote the bilinear form corresponding to the quadratic part of $\hat{f}_i$ over k^{16}, and let ${N^s}_i$ denote the null space for each bilinear form $\langle -, - \rangle_i^s$.

Through observation and computational simulations, we find

$$\mathrm{Span}(N_i^s, i = 1, \ldots, 9) = \mathrm{Span}(E_i^s, i = 8, \ldots, 16).$$

Using the same argument from Remark 6.6.1, we can find the image of the space spanned by the the image of $L_{6,i}(x_1, \ldots, x_{16})$, $i = 1, \ldots, 7$, where

$$L_6(x_1, \ldots, x_{16}) = (L_{6,1}(x_1, \ldots, x_{16}), \ldots, L_{6,16}(x_1, \ldots, x_{16})).$$

In this way we find the image of the linear parts of the seven variables $\{x_1, \ldots, x_7\}$ composed by L_6. Again following the same argument of Remark 6.6.1 and by combining L_5 and L_6, for any basis of the space spanned by $L_{6,i}(x_1, \ldots, x_{16})$, $i = 1, \ldots, 7$, if we compose each by L_5^{-1} from the right, this gives us a basis of the image space of the span of the linear parts of seven variables $\{x_1, \ldots, x_7\}$ composed by L_2. We will denote a basis we find for this space by $k_i(x_0, \ldots, x_{27})$, $i = 1, \ldots, 7$.

Step 7: *Completing the Attack.*

Assume we have a message P to be signed. We first randomly choose r_i and solve the equation $k_1(x_0, \ldots, x_{27}) = r_i$ by Gaussian elimination, and then substitute the final results into the polynomial equations coming from a basis of $\tilde{G}_1$ found in Step 5. From the point of view of algebraic geometry, this is equivalent to giving specific values to $x_1, \ldots, x_7$ for f_i. This should produce nine linearly independent equations, which we again solve by Gaussian elimination. This is equivalent to solving the polynomials from Group (I).

Next we substitute it into the remaining two polynomial equations from $\tilde{G}_{12}$, whose linear combination would produce one linear equation. Again, we then substitute again, and the remaining equations should produce another linear equation. This solves the polynomials from Group (II).

When we substitute again, we will only have nine nonlinear equations left from (6.6.0). These all come from linear combinations of polynomials from Group (III), but with all $x_1, \ldots, x_{18}$ replaced by given values, and the variables $x_0, x_{19}, \ldots, x_{27}$ have undergone an invertible affine linear transformation.

Let us choose a random set of values v_i, choose $x_1 = v_1, \ldots, x_{18} = v_{18}$, and let

$$F^e{}_i(x_0, x_{19}, \ldots, x_{27}) = f_i(x_0, v_1, \ldots, v_{18}, x_{19}, \ldots, x_{27}),$$

for $i = 12, \ldots, 20$. Let

$$F^e(x_0, x_{19}, \ldots, x_{27}) = (\bar{f}^e_{12}(x_0, x_{19}, \ldots, x_{27}), \ldots, \bar{f}^e_{20}(x_0, x_{19}, \ldots, x_{27})),$$

and let

$$\bar{F}^e(x_0, x_{19}, \ldots, x_{27}) = L_1^e \circ f^e \circ L_2^e(x_0, x_{19}, \ldots, x_{27}),$$

where L_1^e and L_2^e are invertible affine linear transformations. The problem now is to solve a set of equations of the form:

$$\bar{F}^e(x_0, x_{19}, \ldots, x_{27}) = P_e,$$

where P_e belongs to k^9. To do this, the only thing we need to know is how to find the image of the linear part of x_0 under the composition from the right by L_2^e, which is a linear combination of other variables. The observation is that all quadratic parts of the f_i^e are of the form $x_0 \times x_j$ with no other quadratic terms, and the corresponding quadratic form has rank two.

Let f_a^e and f_b^e be two linearly independent elements in the space spanned by f_i^e, and let N_a^e and N_b^e denote the null space for each bilinear form derived from the quadratic part of f_a^e and f_b^e. Through computer simulations and direct proof, we have

$$\text{Span}(N_a^e, N_b^e) = \text{Span}(E_i^e, i = 1, \ldots, 9),$$

where $E_i^e = (0, 0, \ldots, 1, \ldots, 0)$ is the standard basis in k^{10}. Using the same argument from Remark 6.6.1, this implies we could find the image of the space spanned by $L_2^e(x_0, \ldots, x_{27})$, where

$$L_2^e(x_0, x_{19}, \ldots, x_{27}) = (L_{2,0}^e(x_0, x_{19}, \ldots, x_{27}), \ldots, L_{2,9}^e(x_0, x_{19}, \ldots, x_{27})).$$

This is done by finding the corresponding dimension two space of the invariant variables for both f_a^e and f_b^e as described in Remark 6.6.1. The intersection of the two spaces has exactly dimension one, and it is proportional to the linear part of $L_{2,0}^e(x_0, \ldots, x_{27})$.

We choose a random value for $L_{2,0}^e(x_0, \ldots, x_{27})$ and then substitute it into the nonlinear equations, which is equivalent to the case of giving x_0 a specific value in addition to $x_1, \ldots, x_{27}$ in all the f_i. This will produce again 9 linear independent equations. Finally, we collect all the linear independent equations whose solution will give a forgery of a signature.

On the other hand, as was pointed out in [Yang and Chen, 2005a], the result from Step 1 can be used to attack the system directly by assigning values to the Vinegar variables right away, but this approach has a much higher complexity.

The Complexity

In all steps above, we analyze the computational complexity, except the cases where we only have to solve some simple linear equations, which is of very small complexity. We can easily see that the complexity of the attack procedure is dominated by Step 2. Therefore we conclude that the attack complexity is less than 2^{50}.

Further Comments about TTS

One of the interesting ideas from the first TTS schemes, something that it inherited from the original design of the TTM cryptosystem, is the use of sparse polynomials. In other words, whenever we are supposed to choose random polynomials, we instead choose some special sparse polynomials with randomly chosen coefficients. The purpose of this is to make the secret (decryption) computation, or the signing process, much faster. This is a very good idea, and much more work is needed to be done to justify the specific choice of the sparse polynomials. The new TTS schemes later proposed another version, which also uses sparse polynomials and it seems that these schemes could resist all existing attacks [Yang and Chen, 2005a; Yang and Chen, 2004c]. However due to the use of sparse polynomials, the security of these new TTS cryptosystems remains an open problem.

We note in passing that all the TTS and Tractable Rational Map Signature schemes [Wang et al., 2005] can be viewed as special examples of the Rainbow signature scheme. Also the so-called tractable Rational maps [Wang et al., 2005] are nothing but the "sequential solution type" maps used by Tsujii, Kurosawa, Fujioka, and Matsumoto [Tsujii et al., 1986].

6.7 Further Generalizations of Triangular Maps

In [Wolf et al., 2004], a further generalization of the triangular map is presented, called step-wise triangular maps. The key map

$$S(x_1, \ldots, x_n) = (s_1(x_1, \ldots, x_n), \ldots, s_{Lr}(x_1, \ldots, x_n))$$

is given as

$$\begin{aligned}
s_1(x_1, \ldots, x_n) &= p_1(x_1, \ldots, x_r) \\
&\vdots \\
s_r(x_1, \ldots, x_n) &= p_r(x_1, \ldots, x_r) \\
&\vdots \\
s_{(l-1)r+1}(x_1, \ldots, x_n) &= p_{(l-1)r+1}(x_1, \ldots, x_r, \ldots, x_{lr}) \\
&\vdots \\
s_{lr}(x_1, \ldots, x_n) &= p_{lr}(x_1, \ldots, x_r, \ldots, x_{lr}) \\
&\vdots \\
s_{(L-1)r+1}(x_1, \ldots, x_n) &= p_{(L-1)r+1}(x_1, \ldots, x_r, \ldots, x_{lr}, \ldots, x_n) \\
&\vdots \\
s_{Lr}(x_1, \ldots, x_n) &= p_{Lr}(x_1, \ldots, x_r, \ldots, x_{lr}, \ldots, x_n).
\end{aligned}$$

Clearly TTS, Tractable Rational Map, and the Rainbow signature scheme are covered under such a generalization. Cryptanalysis of this general scheme is presented in [Wolf et al., 2004], making use of similar methods to those presented above, and in particular the rank analysis either from the top or the bottom. Using this formulation, the authors defeat the RSE(2) or RSSE(2) schemes from [Kasahara and Sakai, 2004a; Kasahara and Sakai, 2004b].

6.8 Other Related Work

The more general form of triangular maps is used in [Tsujii et al., 1987; Tsujii et al., 1989; Tsujii et al., 2004]. For example, in [Tsujii et al., 1989], a degree four MPKC on a relative large field with five variables is given. Its security needs a more careful analysis. In [Yang and Chen, 2005a], an improved MinRank method is proposed. This algorithm uses a new concept of interlinked kernels among the kernels for the corresponding quadratic form with the lowest rank. Essentially, the efficiency is improved by a factor of the number of interlinked kernels. The basic idea of TTM is also closely related to the Feistel family of symmetric cryptosystems like DES.

Chapter 7

DIRECT ATTACKS

The cipher of a multivariate cryptosystem is typically a set of m quadratic polynomials in n variables over a finite field

$$G(x_1, \ldots, x_n) = (g_1(x_1, \ldots, x_n), \ldots, g_m(x_1, \ldots, x_n)).$$

This set of polynomials is the public key and thus publicly accessible. Since we assume that the communications are through an open channel, an attacker will also have access to the ciphertext $(y'_1, \ldots, y'_m)$ in the case of encryption. It could also be a signature in case the attacker plans to forge the signature to a document. Therefore, the attacker automatically has the set of equations

$$\begin{aligned} g_1(x_1, \ldots, x_n) &= y'_1, \\ &\vdots \\ g_m(x_1, \ldots, x_n) &= y'_m, \end{aligned}$$

which is the same as the set of equations

$$\begin{aligned} g_1(x_1, \ldots, x_n) - y'_1 &= 0, \\ &\vdots \\ g_m(x_1, \ldots, x_n) - y'_m &= 0. \end{aligned}$$

We call this a set of public equations for the ciphertext $(y'_1, \ldots, y'_m)$. Therefore, any general method to solve a set of multivariate polynomial equations can be used for attacking an MPKC. Thus it is very important to know how efficient general methods are in solving these polynomial equations for the unknowns and thus breaking the cryptographic system.

If the set of equations are over the field of the real numbers, where certain numerical approximations are allowed, then it suffices to find approximate solutions. In this case numerical methods are used and they are often based on the ideas of Newton.

When equations over a finite field are considered, numerical methods do not appear to be applicable. Exact solutions are desired and so methods of algebraic geometry come into play. Only a limited number of distinct methods exist for solving systems of polynomial equations over a finite field and they can be grouped as follows:

- Gröbner bases method,
- XL method,
- Zhuang-Zi method.

The first one is the most important method of the three. It was introduced by Buchberger in the 1970s and it has been refined since then. Given a set of polynomial equations

$$\begin{aligned} f_1(x_1, \ldots, x_n) &= 0, \\ &\vdots \\ f_m(x_1, \ldots, x_n) &= 0, \end{aligned} \tag{7.1}$$

the problem is to compute a basis for the ideal generated by the set $F = \{f_1, \ldots, f_m\}$. Once a basis has been obtained, it is possible to investigate what kind of solutions exist and if possible, to compute them. Furthermore, this basis can be used to determine if an element is in the ideal or not.

When the coefficients of (7.1) come from a finite field, say $k = GF(q)$, the solutions $x_1, \ldots, x_n$ are to be found in the same field. The method of Gröbner bases does not take advantage of this fact, and it is necessary to augment (7.1) with

$$\begin{aligned} f_{m+1} &= x_1^q - x_1 = 0, \\ &\vdots \\ f_{m+n} &= x_n^q - x_n = 0, \end{aligned}$$

which are used to eliminate the spurious solutions in an extension field of the finite field k, but not in k.

Another method for solving systems of nonlinear equations attempts to convert the nonlinear equations (7.1) into linear ones by introducing additional variables which represent the nonlinear terms. The nonlinear terms are derived through the method of generating the corresponding ideal by multiplying monomials. This method goes by the name XL, (eXtended Linearization) and variations thereof. From the name it can be seen that its origin lies in the linearization equation method introduced by Patarin [Patarin, 1995] to attack the Matsumoto-Imai system.

The Zhuang-Zi method attempts to take advantage of the fact that the solutions are in a finite field. The set of equations and the variables are lifted into a polynomial ring with coefficients in an extension field, $K[X]$, so that solving (7.1) is the same as finding the roots of a single polynomial in the variable X. As usual, nothing comes for free and the resulting polynomial can have a very high degree. The Zhuang-Zi algorithm provides a method to convert this polynomial to an equivalent one with a lower degree. When the degree of the resulting polynomial is low enough, its roots are found by one of the well known methods for a single variable polynomial in a finite field [von zur Gathen and Gerhard, 2003; Geddes et al., 1992; Knuth, 1981]. It is possible that spurious roots are introduced by this approach and it is necessary to check that each root is actually a solution of (7.1), eliminating those that do not. This method is very much inspired by the ideas of Matsumoto-Imai [Matsumoto and Imai, 1988; Patarin, 1996b; Kipnis and Shamir, 1998].

It is clear that the topics for this chapter deserve their own book. The purpose of this chapter is to give a reasonable introduction to the basic ideas in the related research areas. We will start by introducing some basic notions of algebraic geometry which we will use. Then we will discuss the three algorithms and the connection between the first two algorithms.

7.1 Basic Results from Algebraic Geometry

For this chapter we would like to introduce some basic notation and theorems from algebraic geometry without any proofs. This will make it much easier to explain the basic ideas and concepts for the XL algorithm in particular. The proofs can be found in any standard text book [Cox et al., 2005]. Let $k[x_1, \ldots, x_n]$ be the space of functions over k^n as previously defined.

Definition 7.1.1. *An ideal I in the space of functions $k[x_1, \ldots, x_n]$ is a subring in $k[x_1, \ldots, x_n]$ such that for any element $a \in k[x_1, \ldots, x_n]$ and any element $b \in I$, we have $ab \in I$.*

Theorem 7.1.1. *Let $G = \{g_1, \ldots, g_s\}$ be a set of polynomial functions. The smallest ideal which contains this set consists of elements in the* *following form:*

$$\sum_{i=1}^{s} h_i g_i,$$

for all possible $h_i \in k[x_1, \ldots, x_n]$. *This ideal is also called the ideal generated by* $G = \{g_1, \ldots, g_l\}$ *and is denoted by* $\langle G \rangle$ *or* $\langle g_1, \ldots, g_l \rangle$.

Theorem 7.1.2. *There is a bijection between the space of all sets in* k^n *and the set of all ideals in* $k[x_1, \ldots, x_n]$. *The correspondence is given by*

$$C(I) = \{(a_1, \ldots, a_n) \in k^n \mid f(a_1, \ldots, a_n) = 0 \quad \text{for any} \quad f \in I\}.$$

Theorem 7.1.3. *If a set of polynomial equations*

$$g_1(x_1, \ldots, x_n) = 0,$$
$$\vdots$$
$$g_m(x_1, \ldots, x_n) = 0,$$

has a unique solution

$$x_1 = a_1, \ \ldots, \ x_n = a_n,$$

then the ideal generated by $g_1, \ldots, g_m$ *contains the elements* $x_1 - a_1, \ldots, x_n - a_n$, *and in fact is* $\langle x_1 - a_1, \ldots, x_n - a_n \rangle$.

Theorem 7.1.4. *If a set of polynomial equations*

$$g_1(x_1, \ldots, x_n) = 0,$$
$$\vdots$$
$$g_m(x_1, \ldots, x_n) = 0,$$

has multiple solutions such that there does not exist a proper affine subset which contains all these solutions, then no linear (affine) function exists in the ideal generated by $g_1, \ldots, g_m$.

7.2 Gröbner Bases

Gröbner bases have been studied intensively for more than thirty years and many books have been written on this topic. Our presentation is only superficial and is only a small part of what can be found in books like [Adams and Loustaunau, 1994; Becker and Weispfenning, 1991; Kreuzer and Robbiano, 2000] and others.

Gröbner bases provide a very general method to solve systems of multivariate polynomial equations and all practical multivariate cryptosystems must be able to resist this attack. Since the exact complexity for computing a Gröbner basis is typical not known, it is common to experiment with small examples to see how the time and space requirements increase as the size of the example grows. With this data, some heuristic statements can be given about how well a particular system can resist a brute force attack via Gröbner bases.

Gröbner bases are typically discussed for polynomial rings over the integers or rational numbers. In these cases it has been shown that the problem is NP-hard. Unfortunately, or fortunately for cryptography, the complexity of Gröbner bases does not change when polynomials are considered over finite fields. Therefore, the discussion that follows is of more general nature and follows other presentations of this topic.

The concept of Gröbner bases was introduced by Bruno Buchberger in his thesis [Buchberger, 1965] when he tried to do algorithmic computations in residue classes of polynomial rings. The theory of Gröbner bases is centered around ideals generated by finite sets of multivariate polynomials. The original motivation for studying polynomials comes from the relationship between algebra and geometry. Today, multivariate polynomials arise in many areas of pure and applied mathematics. Solving the polynomial equations is often an essential step for understanding the given problem.

The name *Gröbner basis* was introduced by Buchberger in 1975, in honor of his thesis advisor Professor Wolfgang Gröbner who had suggested to Buchberger the use of the S-polynomial. This led Buchberger to a powerful algorithm, and he also proposed at once several improvements to this algorithm in order to make it useful in computations [Buchberger, 1979]. Over the years, Buchberger's algorithm has been studied extensively, improved, and refined. It also has been implemented in most computer algebra systems, in many cases in a version proposed by Gebauer and Möller [Gebauer and Möller, 1988], and thus is known as the *Gebauer-Möller Installation.*

More recently Faugère [Faugère, 1999] proposed additional refinements, in particular one which simplifies in parts the programming of the algorithm and thus leads to a more efficient implementation. His versions for computing Gröbner bases are known as the F_4 and F_5 algorithms. Alan Steel has implemented F_4 for the computer algebra system Magma [Computational Algebra Group, 2005]. At the time of the writing of this book, it is one of the best and fastest methods for computing a Gröbner basis. Nevertheless, due to the exponential increase in time and space requirements for the method, many polynomial equations remain unsolvable for all practical purposes.

Term Ordering

The Gaussian algorithm is often used to solve systems of linear equations efficiently. When Buchberger's method of Gröbner bases is used on linear systems of equations, it is similar to Gaussian elimination. Therefore, the method of Gröbner bases can be considered as an extension of Gaussian elimination to nonlinear polynomial equations. In the Gaussian algorithm, the order in which variables are eliminated is not very important and it usually follows from the order in which the coefficient matrix is written. On the other hand, for Buchberger's algorithm the ordering of variables and of terms is very important. There exists many examples where the algorithm will succeed in finding a basis with one ordering, but will fail to do so in another ordering due to time or memory constraints.

We will denote a polynomial ring over a field k by $k[x]$, where $x = (x_1, x_2, \ldots, x_n)$ represents an (ordered) set of variables. We will use the following definitions.

Definition 7.2.1.

1.) A polynomial of the form $f = x_1^{i_1} x_2^{i_2} \cdots x_n^{i_n}$ with $i_1, i_2, \ldots, i_n \in \mathbb{N}$ is called a **monomial**.

2.) The monomial $x_1^0 x_2^0 \cdots x_n^0$ is usually written as 1.

3.) The set of all possible monomials in $k[x]$ is denoted by T^n.

4.) For $c \in k$, $f = cx_1^{i_1} \cdots x_n^{i_n}$ is called a **term**.

5.) Every polynomial $f \in k[x]$ can be written as a sum of terms:

$$f = c_1 t_1 + \cdots + c_m t_m,$$

with $m \in \mathbb{N}$, $c_i \neq 0$ and all monomials t_i are distinct from each other.

6.) A constant polynomial is given by $f = c$. When $c = 0$ it is called the null-polynomial. The null-polynomial must sometimes be treated differently from other polynomials.

7.) The set $\mathrm{Supp}(f) = \{t_1, \ldots, t_n\}$ *is called the* **support** *of f. When a subset $F \subset k[x]$ of polynomials is considered, then the definition of support is extended accordingly to* $\mathrm{Supp}(F) = \cup_{f \in F} \mathrm{Supp}(f)$.

8.) For a monomial $t = x_1^{i_1} \cdots x_n^{i_n}$ with $i_1, \ldots, i_n \in \mathbb{N}$, we call the number $\deg(t) = i_1 + i_2 + \cdots + i_n$ *the* **degree** *of the monomial t.*

9.) For an arbitrary polynomial, the degree of f is defined as:

$$\deg(p) = \max\{\deg(t) \mid t \in \mathrm{Supp}(f)\}.$$

A constant polynomial has degree 0, but the null-polynomial by definition is given any degree in $\mathbb{N}$.

The notations for monomials and terms are often interchanged in the literature. When a coefficient is unity then it does not matter, but in other cases it does. We follow the convention that seems to be in common use in recent years and also has been adopted by the computer algebra system Magma [Computational Algebra Group, 2005]. With this convention, a polynomial is a sum of terms and the factors of each term are a coefficient and a monomial.

Some computer algebra programs also do not distinguish between a constant polynomial and the null-polynomial. Those programs that do make a distinction often assign a negative integer or some other special value to the degree of the null-polynomial.

Buchberger's algorithm requires that the set of monomials

$$T^n = \{x_1^{i_1} \cdots x_n^{i_n} \mid i_1, \ldots, i_n \in \mathbb{N}\}$$

obey an admissible ordering as given in the following definition.

Definition 7.2.2. *Let $\sigma \subseteq T^n \times T^n$ be a complete relationship defined on T^n. Instead of $(t_1, t_2) \in \sigma$ we write $t_1 \leq_\sigma t_2$, and we will also use the other common symbols indicating an order. An ordering is called admissible if for all $t_1, t_2, t_3 \in T^n$ the following holds:*

1.) $1 \leq_\sigma t_1$,

2.) $t_1 \leq_\sigma t_2 \implies t_1 t_3 \leq_\sigma t_2 t_3$.

A variety of admissible orderings is possible, but only three are in common use in computer algebra systems, since they turned out to be the most useful ones in practice. For writing down the definitions it is helpful to define $\log : T^n \longrightarrow \mathbb{N}^n$, the **logarithm** of a monomial by:

$$\log\left(x_1^{i_1} \cdots x_n^{i_n}\right) = (i_1, \ldots, i_n).$$

Definition 7.2.3. *In the pure* **lexicographic** *(lex) ordering we have:*

$$t_1 <_{lex} t_2$$

whenever the first non-zero component in $\log(t_2) - \log(t_1)$ *is positive.*

Note that by specifying the polynomial ring as $k[x_1, x_2, \ldots, x_n]$, the precedence of the variables

$$x_1 >_{lex} x_2 >_{lex} \cdots >_{lex} x_n$$

is implied. The ordering is called lexicographic since the terms appear in the order as they would be given in a dictionary.

Definition 7.2.4. *In the* **graded lexicographic** *(glex) ordering we have:*

$$t_1 <_{glex} t_2$$

if either $deg(t_1) < deg(t_2)$, or both $deg(t_1) = deg(t_2)$ and $t_1 <_{lex} t_2$.

The ordering is called graded lexicographic since it first grades the monomials by total degree and then uses the lexicographic ordering for monomials with the same total degree. This ordering is rarely used in practical applications as the next ordering often produces Gröbner bases with fewer terms.

Definition 7.2.5. *In the* **graded reverse lexicographic** *(grevlex) ordering we have:*

$$t_1 <_{grevlex} t_2$$

if either $deg(t_1) < deg(t_2)$, or both $deg(t_1) = deg(t_2)$ and the last non-zero component of $\log(t_2) - \log(t_1)$ is negative.

Again the monomials are graded by their total degrees and then ties are decided by the negation of the lexicographic ordering of the variables in reverse order. For two variables, the last two orderings give the same results, but for three or more variables the orderings are different and one order cannot be obtained from the other by simply interchanging variables.

Example: Let $p(x, y, z) = (x+y+z)^3 + y^4$ be a polynomial in $k[x, y, z]$ with $k = GF(2)$ and $x >_{lex} y >_{lex} z$ implied. The polynomials are always written in decreasing order of their monomials.

- Under lexicographic order: $p(x, y, z) = x^3 + x^2y + x^2z + xy^2 + xz^2 + y^4 + y^3 + y^2z + yz^2 + z^3$.
- Under graded lexicographic order: $p(x, y, z) = y^4 + x^3 + x^2y + x^2z + xy^2 + xz^2 + y^3 + y^2z + yz^2 + z^3$.
- Under graded reverse lexicographic order: $p(x, y, z) = y^4 + x^3 + x^2y + xy^2 + y^3 + x^2z + y^2z + xz^2 + yz^2 + z^3$.

Any polynomial in $k[x]$ contains a term whose monomial is maximal with respect to a given ordering σ. We will adopt the following notation.

Definition 7.2.6.

1.) Given $f \in k[x] \setminus \{0\}$ and an admissible ordering σ for T^n, there exists a unique representation $f = \sum_{i=1}^{m} c_i t_i$ for f, with $m \in \mathbb{N}^+$, $c_i \in k \setminus \{0\}$, $t_i \in T^n$, and $t_1 >_\sigma \cdots >_\sigma t_m$.

2.) The **leading monomial** *of f is t_1. It is the maximal monomial in Supp(f) with respect to the order σ. It is denoted by $LM_\sigma(f)$ or simply by $LM(f)$ if the term ordering σ is understood.*

3.) The **leading coefficient** *is c_1, and is denoted by $LC_\sigma(f)$ or simply $LC(f)$.*

4.) The **leading term** *is $c_1 t_1$, and is the product of the leading coefficient and the leading monomial:*

$$LT(f) = LC(f) \cdot LM(f).$$

5.) The convention $LC(0) = 0$ and $LT(0) = 1$ is standard for the null-polynomial.

Reduction of Multivariate Polynomials

The structure imposed by the ordering σ on $k[x]$ allows certain types of simplification when a set F of polynomials is given. This reduction from one polynomial to another modulo F is the most computationally intensive part of the algorithm. It can be viewed as one step in a more generalized division process for multivariate polynomials. A polynomial g reduces to another polynomial h modulo the subset $F \subset k[x]$ if and only if the leading term $LT(g)$ can be deleted by subtracting an appropriate multiple of a polynomial in F. This reduction is denoted by $g \mapsto_F h$ and is defined as follows.

Definition 7.2.7. *The reduction $g \mapsto_F h$ holds if and only there exist $f \in F$ and $p \in k[x]$ such that*

$$h = g - pf$$

with $LM(g) > LM(h)$. If such a reduction is not possible, then g is called irreducible modulo F.

In other words g is irreducible modulo F if no leading term of an element in F divides the leading term of g. If g is reducible then the reduction process involves the subtraction of a multiple of a polynomial in F so that the resulting answer is in some sense smaller. We note that this answer may actually consist of more terms afterwards. Of particular significance is that the new polynomial h is equivalent to g with respect to the ideal generated by F.

Example: Consider the set $F = \{f_1, f_2\} \subset GF(2)[x, y]$ with graded lexicographic ordering, where

$$f_1 = x^2y + x^2 + y^2, \qquad f_2 = xy^2 + y^3 + 1.$$

Then for the polynomial $g = x^3y + x^2y^2 + xy + x$, the following reduction is possible:

$$g \mapsto_{f_1} h = g - xf_1 = x^3 + x^2y^2 + xy^2 + xy + x. \tag{7.2}$$

Obviously additional reductions are now possible modulo F and this leads to the following definition.

Definition 7.2.8. *A polynomial h is the normal form of g if $g \mapsto_F h$ and h is irreducible modulo F.*

Continuing with the example above:

$$h \mapsto_{f_1} h - yf_1 = x^3 + x^2y + xy^2 + y^3 + xy + x. \tag{7.3}$$

On the other hand, if f_2 is used instead on the result in (7.2) then two reductions are possible:

$$\begin{aligned} h &\mapsto_{f_2} h - xf_2 = xy^3 + x^3 + xy^2 + xy \\ &\mapsto_{f_2} y^4 + x^3 + xy^2 + xy + y \end{aligned} \tag{7.4}$$

This example shows that a normal form is not unique and that it depends on which order the elements of F are used for the reduction. Although the leading terms in (7.3) or (7.4) can no longer be reduced, the example shows that lower terms can still be eliminated by subtracting appropriate multiples of f_1 or f_2. From (7.3) we can subtract first f_1 and then f_2 to obtain:

$$\begin{aligned} &\mapsto_{f_1} x^3 + xy^2 + y^3 + x^2 + xy + y^2 + x \\ &\mapsto_{f_2} x^3 + x^2 + xy + y^2 + x + 1, \end{aligned} \tag{7.5}$$

whereas from (7.4) we can subtract f_2 to arrive at:

$$\mapsto_{f_2} y^4 + x^3 + y^3 + xy + y + 1. \tag{7.6}$$

These examples lead to the following definition.

Definition 7.2.9. *A polynomial g is* **completely reduced** *with respect to F if no term in g is divisible by any $LM(f_i)$ for all $f_i \in F$.*

The example also leads to a generalization of the division algorithm for polynomials in a single variable x, which says the following: Given $g, h \in k[x]$ with $h \neq 0$, there exist unique polynomials $q, r \in k[x]$ such that $g = hq + r$, and either $r = 0$ or $\deg(r) < \deg(h)$. For the multivariate case we need to divide the polynomial $g \in k[x_1, \ldots, x_n]$ by polynomials from a set $F = \{f_1, \ldots, f_s\}$, where each $f_i \in k[x]$ comes from the same ring of polynomials. The generalized division algorithm for the multivariate case $x = (x_1, \ldots, x_n)$ is given by in the following theorem.

Theorem 7.2.1. *Let $F = \{f_1, \ldots, f_s\}$ be an ordered set of polynomials in $k[x]$. Then for any $g \in k[x]$ there exists polynomials $p_1, \ldots, p_s \in k[x]$ such that*

$$g = p_1 f_1 + \cdots + p_s f_s + r,$$

where either $r = 0$ or r is a completely reduced polynomial.

Remark 7.2.1. *The examples in (7.2) to (7.6) show that r (the "remainder") is not unique in the multivariate case, as on the one hand*

$$r = g - (x + y + 1)f_1 - f_2 = x^3 + x^2 + xy + y^2 + x + 1,$$

and on the other

$$r = g - xf_1 - (x + y + 1)f_2 = y^4 + x^3 + y^3 + xy + y + 1.$$

Uniqueness exists when a complete reduction gives 0, as then all other reductions must lead to 0 as well.

Corollary 7.2.1. *Assume that one complete reduction of the polynomial g as given in the previous theorem is $g \mapsto_F 0$. Then all other reductions lead to 0 as well.*

Proof. Assume that

$$g - p_1 f_1 - \cdots - p_s f_s = 0 \quad \text{but that} \quad g - q_1 f_1 - \cdots - q_s f_s = r$$

with $r \neq 0$. Then r can be reduced further to 0 within F since from the two equations it follows that

$$r + (q_1 - p_1) f_1 + \cdots + (q_s - p_s) f_s = 0.$$

□

A more troublesome aspect of the reduction modulo F is that even if $g \mapsto_F 0$ and $h \mapsto_F 0$ for two polynomials $g, h \in k[x]$, this does not guarantee that either $g+h$ or $g-h$ reduce to 0 modulo F. The following simple example with $g, h \in GF(2)[x, y, z]$ and $F = \{f_1, f_2, f_3\}$ illustrates this phenomenon. Let

$$f_1 = x^2yz + z^2, \qquad f_2 = xy^2z + xyz, \qquad f_3 = x^2y^2 + z^2,$$

and

$$g = x^2y^2z + z^3 \quad \text{and} \quad h = x^2y^2z + x^2yz.$$

Since $g = zf_3$ and $h = xf_2$ for both polynomials, it is clear that $g, h \mapsto_F 0$ holds. However, $g + h = x^2yz + z^3$ is already completely reduced modulo F. The example shows that additional polynomials have to be added to F in order to obtain a basis for the ideal generated by F. This will be at the heart of the algorithm of Buchberger and will be discussed in the next section.

The reduction process occurs at the inner-most level of Buchberger's algorithm and thus deserves some comments on how to organize it efficiently. First of all, the polynomials in F have to be mutually irreducible. This is accomplished by adding one polynomial at a time to the set. Assume that $F = \{f_1, \ldots, f_n\}$ already consists of mutually irreducible polynomials and that g is a possible candidate for inclusion in F. After seeing that $g \mapsto_F h$ with $h \neq 0$, then h can be added to F. At this point it is necessary to check that none of the polynomials f_i, $i = 1, \ldots, n$ can be reduced further with the help of h, or even removed from the set F in case $f_i \mapsto_h 0$. A minor speed-up is achieved when all leading coefficients for the polynomials in F are 1. This is easily accomplished by dividing h by its leading coefficient $LC(h)$ before adding h to the set F.

It is not clear if complete reduction is always a good strategy, as the reduction of the leading monomial suffices and the additional computational cost may not be worth it. It is clear even for complete reduction that monomials have to be eliminated in decreasing order since each individual reduction step will affect the lower order terms.

The order in which the polynomials of F are used to reduce g can give different reduced forms of g. Since a division algorithm will use the polynomials $f_1, f_2, \ldots$ in the order in which they are given, it will have an influence on the result. It is desirable to reduce a new polynomial g as quickly as possible, therefore a good choice is to store the polynomials of F by increasing order of their leading terms; that is, $f_1 <_\sigma f_2 <_\sigma \cdots$.

A more efficient approach has been advocated by Faugère, which works well for homogeneous polynomials. Given a set of polynomials

$$F = (f_1, \ldots, f_s)^T,$$

viewed as a column vector, arrange the monomials in Supp(F) in the order as given by σ, that is, $t_1 >_\sigma \cdots >_\sigma t_m$. Then write the monomials also as a column vector:

$$\mathsf{X} = (t_1, \ldots, t_m)^T.$$

In this way the polynomials can be written in matrix form by:

$$\mathsf{F} = \mathsf{AX},$$

with some coefficient matrix A of size $s \times m$. The reduction of F is then the same as bringing the matrix A into the row echelon form $\tilde{\mathsf{A}}$ with the help of the Gaussian elimination. The unique row echelon form is:

$$\tilde{\mathsf{A}} = \begin{bmatrix} 1 & * & \cdots & * & 0 & * & \cdots & * & 0 & * & \cdots & * & 0 & * & \cdots & * \\ 0 & 0 & \cdots & 0 & 1 & * & \cdots & * & 0 & * & \cdots & * & 0 & * & \cdots & * \\ 0 & 0 & \cdots & 0 & 0 & 0 & \cdots & 0 & 1 & * & \cdots & * & 0 & * & \cdots & * \\ \vdots & \vdots & \ddots & \vdots & \vdots & \vdots & \ddots & \vdots & \vdots & \vdots & \ddots & \vdots & \vdots & \vdots & \ddots & \vdots \\ 0 & 0 & \cdots & 0 & 0 & 0 & \cdots & 0 & 0 & 0 & \cdots & 0 & 1 & * & \cdots & * \end{bmatrix}.$$

Any complete row of zeroes has been omitted from $\tilde{\mathsf{A}}$ and it is easy to see that $rank(\tilde{\mathsf{A}}) \leq \min(m, s)$. The completely reduced system of polynomials is then given by $\tilde{\mathsf{A}}\mathsf{X}$.

Example: In order to keep things simple, again consider polynomials in $GF(2)[x, y, z]$ with lexicographical ordering. The set F consists of the four polynomials:

$$\begin{aligned} f_1 &= x^3 + x^2y + xyz + yz^2 \\ f_2 &= x^3 + x^2y + xy^2 + y^3 \\ f_3 &= x^3 + x^2y + xyz + y^2z + z^3 \\ f_4 &= x^3 + x^2y + xy^2 + y^3 + y^2z + yz^2 + z^3 \end{aligned}$$

so that the coefficient matrix is given by

$$\mathsf{A} = \begin{bmatrix} 1 & 1 & 0 & 1 & 0 & 0 & 1 & 0 \\ 1 & 1 & 1 & 0 & 1 & 0 & 0 & 0 \\ 1 & 1 & 0 & 1 & 0 & 1 & 0 & 1 \\ 1 & 1 & 1 & 0 & 1 & 1 & 1 & 1 \end{bmatrix}$$

with the vector for the support $\mathsf{X} = (x^3, x^2y, xy^2, xyz, y^3, y^2z, yz^2, z^3)^T$. The unique row echelon form is found to be

$$\tilde{\mathsf{A}} = \begin{bmatrix} 1 & 1 & 0 & 1 & 0 & 0 & 1 & 0 \\ 0 & 0 & 1 & 1 & 1 & 0 & 1 & 0 \\ 0 & 0 & 0 & 0 & 0 & 1 & 1 & 1 \end{bmatrix}$$

so that the reduced polynomial set is

$$\tilde{F} = \{x^3+x^2y+xyz+yz^2,\ xy^2+xyz+y^3+yz^2,\ y^2z+yz^2+z^3\}. \quad (7.7)$$

A homogeneous polynomial of degree d with n variables can have

$$\binom{n+d-1}{d}$$

different terms. This number quickly becomes large for increasing degree, and for that reason only a sparse matrix implementation is feasible. Faugère reports that when trying to find a Gröbner basis, huge matrices can be encountered, and even the sparse matrix implementation for reducing a set of polynomials may run out of storage space.

S-Polynomials

Definition 7.2.10. *Let I be an ideal. A basis G for I is called a Gröbner basis if*

$$p \in I \iff p \mapsto_G 0.$$

Equivalent definitions exist. For example, G is a Gröbner basis when the only irreducible polynomial in G is $p = 0$. An arbitrary basis F in general does not form a Gröbner basis, since some combination of polynomials in F may lead to a nonzero irreducible polynomial, which then has to be added to F in order to complete the basis. Buchberger showed that it suffices consider the so-called *S-polynomials*, which are formed form pairs of polynomials in the given basis in accordance with the following definition.

Definition 7.2.11. *The S-polynomial of two polynomials f and g in F is defined by:*

$$spoly(f,g) = \frac{J}{LT(f)} \cdot f - \frac{J}{LT(g)} \cdot g,$$

where $J = lcm(LT(f), LT(g))$.

J is the smallest term that both f and g can reduce. Thus $J/LT(f)$ and $J/LT(g)$ are terms. Therefore $spoly(f,g)$ is a linear combination of f and g formed in such a way that the leading terms in both components cancel each other.

Example: Let $F = \{f_1, f_2\}$ with $f_1 = xyz + z^3$ and $f_2 = z^3 + z^2$ be polynomials with integer coefficients, using the lexicographical ordering. Then $J = xyz^3$ and

$$spoly(f_1, f_2) = \frac{xyz^3}{xyz} f_1 - \frac{xyz^3}{z^3} f_2 = z^2 f_1 - xyz f_2 = z^5 - xyz^2.$$

Example: Let $F = \{f_1, f_2\}$ with $f_1 = 3x^2yz - 2z^2$ and $f_2 = 6xz^3 + 2yz$ be polynomials, again with lexicographical ordering. This time $J = 6x^2yz^3$ and

$$spoly(f_1, f_2) = \frac{6x^2yz^3}{3x^2yz} f_1 - \frac{6x^2yz^3}{6xz^3} f_2 = 2z^2 f_1 - xy f_2 = -2xy^2z - 4z^4.$$

Theorem 7.2.2. *(Buchberger) G is a Gröbner basis if and only if*

$$spoly(p, q) \mapsto_G 0$$

for all $p, q \in G$.

Proof. Since $spoly(p, q)$ is in the ideal generated by G, the only if part is obvious.

Let $\{g_1, \ldots, g_s\}$ be a basis for G. Without loss of generality we can assume that the g_i are monic, since each g_i can be divided by its leading coefficient and it does not affect the statement or conclusion of the theorem. Assume that $f \in \langle G \rangle$ but that it cannot be reduced to zero. This means that

$$f = \sum_{i=1}^{s} h_i g_i. \tag{7.8}$$

With the given term order σ we then can find the monomial t that is maximal among the head terms $LT(h_i g_i)$. It is easily possible that $t >_\sigma LM(f)$, since when forming the sums in (7.8) high order terms can cancel each other. The monomial t is also not necessarily unique, but of all the possible candidates there will be one that is minimal with respect to the order σ. This is the one we consider.

Without loss of generality we can also say that t occurs in the first m terms of (7.8), for $i = 1, \ldots, m$. We will use induction on m to show that f can be reduced further.

When $m = 1$,

$$f = h_1 g_1 + \sum_{i=2}^{s} h_i g_i \qquad \text{and} \qquad f \mapsto_{g_1} \sum_{i=2}^{s} h_i g_i,$$

so that t was not maximal among the $h_i g_i$.

Now assume that f can be reduced further when t is in $m = k$ terms of (7.8), but that f cannot be reduced further if t is in $m = k+1$ leading terms of (7.8). Write

$$f = h_1 g_1 + h_2 g_2 + \sum_{i=3}^{k+1} h_i g_i + \sum_{i=k+2}^{s} h_i g_i \tag{7.9}$$

$$= LT(h_1)g_1 - \alpha LT(h_2)g_2 + (h_2 + \alpha LT(h_2))g_2 + \sum_{i=3}^{k+1} h_i g_i + \tilde{f},$$

where $\alpha = LC(h_1)/LC(h_2)$ since $LC(g_1) = LC(g_2) = 1$. Also, $\tilde{f}$ consists of terms smaller than t; that is,

$$\tilde{f} = (h_1 - LT(h_1))g_1 + \sum_{i=k+2}^{s} h_i g_i.$$

Consider the first two terms in the expanded form of (7.9). Since $LM(h_1 g_1) = LM(h_2 g_2) = t$ it follows that

$$LT(h_1)g_1 - \alpha\, LT(h_2)g_2 = \gcd\,(LT(h_1), LT(h_2))\; spoly(g_1, g_2).$$

By the assumption of the theorem it can be reduced to zero. Thus (7.9) has been rewritten so that it only has k terms of the form $h_i g_i$ whose leading terms contain t. By our induction hypothesis they also can be eliminated. □

The theorem can be translated immediately into an algorithm for computing a Gröbner basis.

```
procedure Groebner(F)
// given a set of polynomials F = {f_1, ..., f_k}
// return a Gröbner basis G
  G = F
  k = length(G)
  B = {(g_i, g_j) : 1 ≤ i < j ≤ k}
  while (B ≠ ∅) do
        pick (p, q) ∈ B
        B = B − {(p, q)}
        h = NormalForm(spoly(p, q), G)
        if h ≠ 0 then
              k = k + 1
              B = B ∪ {(g_i, h) : 1 ≤ i < k}
              G = G ∪ {h}
        end if
  end while
  return G
end procedure
```

One obvious question is whether or not the algorithm terminates. The answer comes from Hilbert's basis theorem, which guarantees that only finitely many different h must be added.

The correctness of the algorithm does not depend on the order the polynomials in F are given or on the order the pairs (g_i, g_j) are selected from the set B, but different choices will lead to different outcomes. In other words, the algorithm produces one of the many possible Gröbner bases for the ideal generated by F.

It should also be noted that most reductions of an S-polynomial will lead to zero. This causes a lot of unnecessary work. Buchberger himself has suggested several criteria that can tell in advance if the S-polynomial of a pair of polynomials will reduce to zero, and therefore the reduction does not need to be carried out. These improvements to the basic algorithm will be considered in the next section.

Improved Buchberger's Algorithm

Fortunately the non-uniqueness of the basic algorithm can easily be corrected.

Definition 7.2.12. *A Gröbner basis $G = \{g_1, \ldots, g_s\}$ is said to be reduced if all g_i are monic and $LM(g_i)$ does not divide $LM(g_j)$ for all $i \neq j$, $1 \leq i, j \leq s$.*

Buchberger showed already in [Buchberger, 1965] that if G and H are reduced, monic Gröbner bases generating the same ideal then $G = H$. The modified algorithm of the previous section then reads as follows.

procedure Groebner(F)
// given a set of polynomials $F = \{f_1, \ldots, f_n\}$
// return a reduced Gröbner basis G
  G = Reduce(F)
  k = length(G)
  $B = \{(g_i, g_j) \,:\, 1 \leq i < j \leq k\}$
  while ($B \neq \emptyset$) do
    pick $(p, q) \in B$
    $B = B - \{(p, q)\}$
    h = NormalForm($spoly(p, q), G$)
    if $h \neq 0$ then
      $B = B \cup \{(g_i, h) \,:\, 1 \leq i \leq k\}$
      G = Reduce($G \cup \{h\}$)
      k = length(G)
      Update(B)
    end if
  end while
  return G
end procedure

The initial step insures that we start with a reduced set of polynomials. Later, each time a new polynomial h is added to the basis G, it must be checked whether or not any polynomials in G can be reduced further or perhaps reduced to zero with the help of h. When this happens the set of pairs of functions in B must be updated.

A problem that arises is the very large number of S-polynomials that must be computed. The typical behavior of the algorithm is that at the beginning, a reduction of an S-polynomial may lead to an element that must be added to the basis. At that time more S-polynomials are added to the list for checking later. Since the algorithm will terminate, the portion of S-polynomials that reduce to zero increases as we get further into the computation. After some time, a large amount of work is carried out with little gain. Long before the algorithm terminates, the desired Gröbner basis has likely already been found. From this point on, all remaining S-polynomials will reduce to zero and the computations could be considered useless. Unfortunately, we do not know when the basis is complete and we must carry out the computations in order to verify that we do indeed have a Gröbner basis.

For this reason it is desirable to have rules for minimizing the number of S-polynomials, or to have criteria that allow us to decide *a priori* if an S-polynomial reduces to zero. Buchberger presented such rules and criteria and proved their correctness in [Buchberger, 1979] and [Buchberger, 1986]. When used, they can drastically reduce the required computation time.

In order to simplify the notation for the algorithm we use the following definition.

Definition 7.2.13. *When referring to an element g_i in the basis G, we will just use the index i. For example, a critical pair $(g_i, g_j) \in B$ is referenced from now on by (i, j), or by (j, i) since the order is irrelevant. When a new basis element g_k is added to the basis G, with this notation the basis G is a subset of the integers 1 to k. It will only be a subset since some elements can be deleted when the basis is reduced and we do not renumber the basis elements.*

All polynomials in the basis are kept in monic form. We also use the following notation:

$$T(i) = LT(g_i)$$
$$T(i,j) = lcm(LT(g_i), LT(g_j)).$$

In this notation $T(i) \cdot T(j) = T(i,j)$ means that g_i and g_j are disjoint; that is, $\gcd(LT(g_i), LT(g_j)) = 1$, or equivalently

$$\text{lcm}(LT(g_i), LT(g_j)) = LT(g_i) \cdot LT(g_j).$$

With this notation, we write

$$spoly(i,j) = \frac{T(i,j)}{T(i)} \cdot g_i - \frac{T(i,j)}{T(j)} \cdot g_j.$$

When adding an element to a basis, the following rule is helpful. If $T(i)|T(j)$ (i.e., $LT(g_i)$ divides $LT(g_j)$), then g_j can be deleted from the basis. With this criterion we can detect possible reductions without having to carry out the computation. If this criterion is applied consistently whenever a new polynomial is added to the basis, then the resulting basis will be reduced.

Two methods for deciding if an S-polynomial reduces to zero (without carrying out the computations) were given by Buchberger. They are referred to in the literature as the two criteria of Buchberger. Based on his experience Buchberger also recommended a way of selecting a pair from B so that the algorithm will run more efficiently in most cases.

- **Criterion 1:** If $T(i) \cdot T(j) = T(i,j)$ (i.e., $LT(g_i)$ and $LT(g_j)$ are relatively prime), then $spoly(g_i, g_j)$ reduces to zero and can be ignored.
- **Criterion 2:** If there exists an element g_k in the basis such that $LT(g_k)$ divides $\text{lcm}(LT(g_i, g_j))$, and if $spoly(g_i, g_k)$ and $spoly(g_j, g_k)$ have already been considered, then $spoly(g_i, g_j)$ reduces to zero and can be ignored.
- **Selection Strategy:** When selecting a pair (i, j) from the set B, choose one such that $T(i, j)$ is minimal with respect to the term order σ.

We rewrite these criteria and the selection strategy as procedures for use in the algorithms Groebner1 and Groebner2.

procedure Criterion1(i, j, B)
  if $T(i) \cdot T(j) \neq T(i,j)$ then $B = B \cup \{(i,j)\}$
end procedure

procedure Criterion2(i, j, B, G)
  if $\exists k \in G$ and $T(k)|T(i,j)$ and $(i,k) \notin B$ and $(j,k) \notin B$
    return false
  else
    return true
  end if
end procedure

```
procedure Select(B)
    find (i, j) ∈ B such that T(i, j) ≤_σ T(r, s), ∀ (r, s) ∈ B
    B = B − {(i, j)}
    return (i, j)
end procedure
```

Practical details on how to implement these procedures efficiently are only touched upon briefly here. For example, the set of pairs $(i, j) \in B$ could be an ordered list in accordance with the selection strategy so that the first pair in this list is the one to be selected. In order to discover quickly which S-polynomials have been computed already, we could use a two-dimensional boolean array b whose entry at $b(i, j)$ when set to true indicates that $spoly(g_i, g_j)$ has been computed already. Whatever is done, it must be remembered that the functions Update and Reduce will affect this array and the ordered list.

How efficiently the programming details are dealt with will have an influence on how well a program for computing a Gröbner basis will perform on large problems. In particular, the function Reduce can be very time consuming, but without it the so-called "intermediate expression swell" might be so bad that space limitations make it impossible to complete the computations. The program below uses the two criteria of Buchberger with the standard selection strategy. After the two polynomials (g_i, g_j) have been selected, the algorithm checks if it can find a polynomial $g \in G$ such that $LT(g)|\text{lcm}(LT(g_i), LT(g_j))$, and if the two pairs (g, g_i) and (g, g_j) have already been treated. If this is the case, then the pair (g_i, g_j) can be ignored; otherwise the pair is treated as in the algorithm Groebner.

```
procedure Groebner1(F)
// given a set of polynomials F = {f_1, ..., f_s}
// return a reduced Gröbner basis G
    G = Reduce(F)
    B = ∅
    for (i ∈ G and j ∈ G and i < j) B = B ∪ {(i, j)}
    while(B ≠ ∅) do
        (i, j) = Select(B)
        if Criterion2(i, j, B, G) then
            h = NormalForm(spoly(g_i, g_j), G)
            if h ≠ 0 then
                g_k = h
                G = Reduce(G ∪ {k})
                for (all i ∈ G) Criterion1(i, k, B)
            end if
```

```
            end if
        end while
        return G
end procedure
```

Although the selection strategy has no influence on the final outcome of the computation of Groebner1, it has an influence on which critical pairs can be ignored by the second criterion of Buchberger. It appears that the selection strategy, also known as the normal strategy, must be used in the algorithm Groebner1 in order not to miss some critical pairs that could be deleted by Criterion2.

It turns out that the second criterion of Buchberger can be very effective in reducing the number of S-polynomials that must be considered. Whereas in the algorithm Groebner1 the criterion is used only when a new pair is selected from the set B, Gebauer and Möller suggested in [Gebauer and Möller, 1988] to update the list B of critical pairs in accordance with Criterion2 as soon as a new basis element $h \neq 0$ is found. If g_1 and g_2 are two elements in the basis when h is added to the basis, then the critical pair (g_1, g_2) can be ignored provided that the critical pairs (g_1, h) and (g_2, h) will be computed and the following condition holds:

$$LT(h) | \mathrm{lcm}(LT(g_1), LT(g_2)).$$

The last condition is equivalent to

$$\mathrm{lcm}(LT(g_1), LT(h)) | \mathrm{lcm}(LT(g_1), LT(g_2))$$

and

$$\mathrm{lcm}(LT(g_2), LT(h)) | \mathrm{lcm}(LT(g_1), LT(g_2)).$$

In the special case that

$$\mathrm{lcm}(LT(g_1), LT(h)) = \mathrm{lcm}(LT(g_1), LT(g_2)), \tag{7.10}$$

two critical pairs could be deleted by mistake. The pair (g_1, g_2) could be deleted on behalf of h and the pair (g_1, h) could be deleted on behalf of g_2. In order to avoid this trap, careful programming is required when the ideas of Gebauer and Möller are implemented; see [Becker and Weispfenning, 1991]. The algorithm Groebner1 avoids this pitfall by ignoring the pair (g_1, g_2) by checking that the critical pairs (g_1, h) and (g_2, h) have been considered already. In the procedure Groebner2 the difficulties are handled in the procedure Update2 in a form found in [Becker and Weispfenning, 1991].

The basic algorithm Groebner2 is similar to Groebner1 except that now whenever a new function h (namely g_k) is added to the basis, the procedure Update2 is called. It checks the list of critical pairs to see which can be deleted. Furthermore, it removes elements in G when their leading term is divisible by $LT(h)$. The same method is used when the initial list G is created from the given set F of polynomials. Thus, the procedure Update2 is also used to initialize the two sets G and B.

```
procedure Groebner2(F)
// given a set of polynomials F = {f_1, ..., f_k}
// return a reduced Gröbner basis G
   G = {1}
   B = ∅
   for (i = 2 to k) Update2(G, B, i)
   while (B ≠ ∅) do
        (i, j) = Select(B)
        h = NormalForm(spoly(g_i, g_j), G)
        if h ≠ 0 then
                 k = k + 1
                 g_k = h
                 Update2(G, B, k)
        end if
   end while
   return G
end procedure
```

The procedure Update2 is similar to the one given in [Becker and Weispfenning, 1991]. The first while loop looks for each pair (g_i, h) for another pair (g_j, h), still in the list $C \cup D$, such that

$$\operatorname{lcm}(LT(g_j), LT(h)) | \operatorname{lcm}(LT(g_i), LT(h)).$$

If the pair is found, then (g_i, h) is not considered. The trap is avoided since the pair (g_i, g_j) is only considered when the pairs from B_0 are looked at later. Although pairs (g_i, h) whose leading terms are relatively prime to each other could be eliminated here by Buchberger's first criterion, Becker and Weispfenning recommend keeping them for now so that they can be used to delete additional pairs with the help of Buchberger's second criterion.

The next loop will then eliminate all pairs in the list D with relatively prime leading terms, and will put all others on the new critical pairs list B.

The following loop then goes through the old list of critical pairs. If it finds a pair (i, j) where $LT(h)|\text{lcm}(LT(g_i), LT(g_j))$, then it is removed from the critical list. The pair is also removed when $\text{lcm}(LT(g_i), LT(h))$ or $\text{lcm}(LT(g_j, LT(h))$ properly divides $\text{lcm}(LT(g_i), LT(g_j))$ and thus avoids the trap.

The last loop uses the first criterion of Buchberger to delete all functions in the original list G_0, whose leading term is a multiple of the leading term of h. Finally, the new function h is added to the new list G before it is returned to the calling procedure.

```
procedure Update2(G, B, k)
// use the two criterion of Buchberger when adding a new basis element
// return updated lists G and B
    G0 = G
    B0 = B
    C = ∅
    for (all i ∈ G0)  C = C ∪ {(i, k)}
    D = ∅
    for all (i, j) ∈ C  do
        C = C − {(i, j)}
        if (T(i) · T(k) = T(i, k) or
          T(j, k) ∤ T(i, k) for all (j, k) ∈ C ∪ D) then D = D ∪ {(i, k)}
    end for
    B = ∅
    for all (i, k) ∈ D  do
        if T(i) · T(k) ≠ T(i, k) then B = B ∪ {(i, k)}
    end for
    for all (i, j) ∈ B0 do
        if (T(k) ∤ T(i, j) or T(i, k) = T(i, j) or T(j, k) = T(i, j)) then
          B = B ∪ {(i, j)}
        end if
    G = {k}
    for all i ∈ G0 do
        if T(k) ∤ T(i) then G = G ∪ {i}
    end for
end procedure
```

7.3 Faugère's Algorithms F_4 and F_5

The main difference from Buchberger's algorithm is that instead of selecting a single critical pair $(f_i, f_j) \in B$, Faugère proposes to select a subset $C \subset B$ of critical elements. A strategy proposed by Faugère is to select all pairs in B simultaneously whose degree of $T(i, j)$ is minimal.

Let

$$d = \min\{\deg(\text{lcm}(LM(f_i), LM(f_j))) \ : \ \forall\, (f_i, f_j) \in B\}.$$

Then the selected subset is

$$\text{Select}(B) = \{(f_i, f_j) \in B \ : \ \deg(\text{lcm}(LM(f_i), LM(f_j)) = d\}. \quad (7.11)$$

Other selection strategies are possible. The reduction process is now applied to a set of polynomials. The idea was described informally in the previous section and it can be made more precise with the help of the following definitions.

Definition 7.3.1. *Let $F = \{f_1, \ldots, f_s\}$ be a set of polynomials in $k[x_1, \ldots, x_n]$. Denote by $X = \{t_1, \ldots, t_m\}$ the monomials in* Supp(F) *listed in decreasing order in accordance with the ordering σ. The matrix representation of the polynomials is:*

$$\mathsf{F} = \mathsf{AX},$$

where F and X are now to be treated as column vectors and the $s \times m$ matrix A has entries from k.

Let $\tilde{\mathsf{A}}$ be the unique row echelon form of A, with rows of zeros removed; i.e., an $\bar{s} \times m$ matrix, where $\bar{s}$ is the rank of A. Then $\tilde{F}$ is the corresponding set of polynomials coming from

$$\tilde{\mathsf{F}} = \tilde{\mathsf{A}}\mathsf{X}.$$

Define

$$\tilde{F}^+ = \{f \in \tilde{F} \mid LM(f) \notin LM(F)\},$$

so that $\tilde{F}^+$ consists of those polynomials in $\tilde{F}$ whose leading monomials are not leading monomials in F.

As an illustration, consider the example (7.7) given at the end of the previous section, which gives

$$\tilde{F}^+ = \{xy^2 + xyz + y^3 + yz^2,\ y^2z + yz^2 + z^3\}.$$

Definition 7.3.2. *Let $F = \{f_1, \ldots, f_s\} \subset k[x_1, \ldots, x_n]$ and let $p = (f_i, f_j)$ be a pair of polynomials with $i \neq j$. Let t_i and t_j be the two monomials such that*

$$t_i \cdot LM(f_i) = t_j \cdot LM(f_j) = lcm(LM(f_i), LM(f_j)).$$

Define Left(p) $= (t_i, f_i)$ and Right(p) $= (t_j, f_j)$; that is, as a pair consisting of a monomial and a polynomial.

If $C \subset B$ is a set of pairs, then extend the definition as follows:

$$Left(C) = \cup_{p \in C} Left(p) \quad and \quad Right(C) = \cup_{p \in C} Right(p).$$

We now give Faugère's F_4 algorithm. Besides the input $F \subset k[x]$, a selection function Select(B) must also be specified. In this case Select(B) returns a subset of pairs from the list B of pairs, and at the same time removes them from the set B. A possible selection function could return all pairs with smallest degree as indicated above, or it could return just one pair as in Buchberger's algorithm.

```
procedure F4(F)
// given a set of polynomials F = {f_1, ..., f_s}
// return a reduced Gröbner basis G
    G = F̃
    B = {(f_i, f_j) | f_i ≠ f_j ∈ G}
    d = 0
    while (B ≠ ∅) do
        d = d + 1
        C_d = Select(B)
        L_d = Left(C_d) ∪ Right(C_d)
        F̃+ = Reduction(L_d, G)
        for h ∈ F̃+ do
            B = B ∪ {(h, g)} | g ∈ G}
            G = G ∪ {h}
        end for
    end while
    return G
end procedure
```

The reduction procedure must take into account that we are reducing a subset of $k[x]$ by G. It expects two parameters: $L \subset T^n \times G$ and G, and then returns a reduced set of polynomials whose leading monomials have not yet appeared as leading monomials in G. In the procedure, $t * f$ is simply the multiplication of a monomial with a polynomial.

```
procedure Reduction(L_d, G)
    F_d = {t * f | (t, f) ∈ L_d}
    D = LM(F_d)
    while (D ≠ Supp(F_d)) do
        choose m ∈ Supp(F_d) \ D
        D = D ∪ {m}
        if there exists g ∈ G such that m = m' * LM(g) then
```

$F_d = F_d \cup \{m' * g\}$
end while
use Gaussian elimination to compute $\tilde{F}_d$
$\tilde{F}_d^+ = \{f \in F_d \mid LM(f) \notin LM(F_d)\}$
return $\tilde{F}_d^+$
end procedure

Faugère calls the part before the Gaussian elimination "SymbolicPreprocessing," as it can be done quickly in a strictly symbolic manner, and also in a time that is linear in the size of the input. The Gaussian elimination is the time consuming aspect since $\mathrm{Supp}(F_d)$ can grow exponentially in n as the degree of the terms increases. Unfortunately for the Gröbner bases algorithm, intermediate polynomials can have a very high degree even when the degree of the final basis elements are moderate.

In order to illustrate how the procedure works, we will use the "Cyclic-4 problem" that Faugère used as an example. Due to several misprints in [Faugère, 1999], it is difficult to follow his presentation. We hope that by giving more details it will be easier to see how the algorithm works. With the grevlex ordering for $x_1 < x_2 < x_3 < x_4$ and $F = \{f_1, f_2, f_3, f_4\}$, the polynomials are as follows:

$$\begin{aligned} f_1 &= x_1 + x_2 + x_3 + x_4, \\ f_2 &= x_1x_2 + x_2x_3 + x_1x_4 + x_3x_4, \\ f_3 &= x_1x_2x_3 + x_1x_2x_4 + x_1x_3x_4 + x_2x_3x_4, \\ f_4 &= x_1x_2x_3x_4 - 1. \end{aligned}$$

Since all terms in the given F are distinct and $\tilde{F} = F$, we start with $G = \{f_1, f_2, f_3, f_4\}$. For Select we use the pairs of minimal degree as given in (7.11). From the six pairs in B, Select finds one pair with degree three and returns $C_1 = (f_1, f_2)$, which gives $L_1 = \{(x_2, f_1), (1, f_2)\}$. With it, the function Reduction is now entered. Here:

$$\begin{aligned} &F_1 = \{x_1x_2 + x_2^2 + x_2x_3 + x_2x_4, x_1x_2 + x_2x_3 + x_1x_4 + x_3x_4\}, \\ &D = \{x_1x_2\}, \\ &\mathrm{Supp}(F_1) = \{x_1x_2, x_2^2, x_2x_3, x_1x_4, x_2x_4, x_3x_4\}. \end{aligned}$$

We now go through the elements in $\mathrm{Supp}(F_1) \setminus D$. The monomials x_2^2 and x_2x_3 are not reducible by a polynomial in G, so they are simply added to the set D. However $x_1x_4 = x_4 \cdot LM(f_1)$, and we must add x_4f_1 to the set F_1. This also introduces the monomial x_4^2 into $\mathrm{Supp}(F_1)$, but it and the other monomials in $\mathrm{Supp}(F_1)$ are not reducible by G, so after the symbolic preprocessing we have

$$\begin{aligned} F_1 = \{&x_1x_2 + x_2^2 + x_2x_3 + x_2x_4, \\ &x_1x_2 + x_2x_3 + x_1x_4 + x_3x_4, \\ &x_1x_4 + x_2x_4 + x_3x_4 + x_4^4\} \end{aligned}$$

and

$$\mathrm{Supp}(F_1) = \{x_1x_2, x_2^2, x_2x_3, x_1x_4, x_2x_4, x_3x_4, x_4^2\}.$$

In matrix form, the polynomials in F_1 correspond to the rows of the matrix

$$\mathsf{A} = \begin{bmatrix} 1 & 1 & 1 & 0 & 1 & 0 & 0 \\ 1 & 0 & 1 & 1 & 0 & 1 & 0 \\ 0 & 0 & 0 & 1 & 1 & 1 & 1 \end{bmatrix}.$$

The reduced echelon form of this matrix is

$$\tilde{\mathsf{A}} = \begin{bmatrix} 1 & 0 & 1 & 0 & -1 & 0 & -1 \\ 0 & 1 & 0 & 0 & 2 & 0 & 1 \\ 0 & 0 & 0 & 1 & 1 & 1 & 1 \end{bmatrix},$$

which corresponds to the set of polynomials

$$\tilde{F_1} = \{x_1x_2 + x_2x_3 - x_2x_4 - x_4^2, x_2^2 + 2x_2x_4 + x_4^2, x_1x_4 + x_2x_4 + x_3x_4 + x_4^4\}.$$

The first and last polynomials have leading monomials that have already appeared in F_1 so that $\tilde{F}_1^+ = \{x_2^2 + 2x_2x_4 + x_4^2\}$. This is returned to the main program, where it is added as $f_5 = x_2^2 + 2x_2x_4 + x_4^2$ to the set G.

The new set of critical pairs becomes

$$\begin{aligned} B = \{&(f_1, f_3), (f_1, f_4), (f_2, f_3), (f_2, f_4), (f_3, f_4), \\ &(f_1, f_5), (f_2, f_5), (f_3, f_5), (f_4, f_5)\} \end{aligned}$$

and Select will choose $C_2 = \{(f_1, f_3), (f_2, f_3), (f_1, f_5), (f_2, f_5)\}$ in accordance with (7.11) so that

$$L_2 = \{(x_2x_3, f_1), (x_2^2, f_1), (x_2, f_2), (x_3, f_2), (1, f_3), (x_1, f_5)\}.$$

The leading monomials for F_2 in Reduction are $D = \{x_1x_2^2, x_1x_2x_3\}$ and

$$\begin{aligned} \mathrm{Supp}(F_2) = \{&x_1x_2^2, x_2^3, x_1x_2x_3, x_2^2x_3, x_2x_3^2, x_1x_2x_4, \\ &x_2^2x_4, x_1x_3x_4, x_2x_3x_4, x_3^2x_4, x_1x_4^2\}. \end{aligned}$$

Obviously, the presentation of the Cyclic-4 example in this form is no longer feasible, as the matrices become too big to be displayed here.

For that reason Faugère had chosen a Select function in his paper that returns only one critical pair. The interested reader is advised to consult [Faugère, 1999].

The presentation as given here shows that the basic algorithm of Faugère does not take the criteria of Buchberger into account. For example, the pair (f_1, f_5) should not have been added to the critical list. The improved F_4 algorithm will do so in the form given by Becker and Weispfenning [Becker and Weispfenning, 1991] and which was presented at the end of the previous section.

In the F_5 algorithm, Faugère formalized the selection process as indicated in (7.11) by representing all critical pairs (f_i, f_j) as a quintuple of the form

$$\text{CriticalPair}(f_i, f_j) = (\text{lcm}(LM(f_i), LM(f_j)), u_i, f_i, u_j, f_j),$$

where the monomials u_i and u_j are such that

$$\text{lcm}(LM(f_i), LM(f_j)) = u_i LM(f_i) = u_j LM(f_j).$$

The list of these critical pairs are then sorted by the first entry in accordance with the given term order σ.

Given the set of polynomials $(f_1, f_2, \ldots, f_s)$, the F_5 algorithm will compute a Gröbner basis in increments; that is, for $(f_1), (f_1, f_2), \ldots,$ $(f_1, f_2 \ldots, f_s)$. Each leading monomial is assigned a "signature" in order to discover unnecessary reductions. Furthermore, F_5 uses the fact that if the given set is a regular sequence, then $f_i \neq 0 \bmod \langle f_1, f_2, \ldots, f_{i-1} \rangle$.

Cryptanalysis of HFE using Gröbner bases

Faugère used a test suite of problems to compare his implementation of the Gröbner basis algorithm to those of others. His main accomplishment is that he could solve practical problems where others implementations had failed. Part of this success comes from the design of his algorithm, but part of it is also careful and efficient programming in a lower-level language.

The HFE cryptosystem of Patarin [Patarin, 1996b] had been considered for a long time as a very promising system, even with a signature as short as 80 bits. Before Faugère's F_4 or F_5 algorithm, the best implementations of Buchberger's algorithm could only solve the system of polynomial equations for toy examples with a signature of about 20 bits. Faugère [Faugère and Joux, 2003] was able to break the first HFE challenge posted by Patarin at `http://www.minrank.org` consisting of 80 quadratic equations in 80 variables with coefficients in $GF(2)$. The quadratic equations came from a polynomial of degree 96 in the extension field K. Faugère was able to solve these equations with his F_5

algorithm, optimized for the field GF(2), in roughly 52 hours. Allan Steel implemented the F_4 algorithm for Magma and used it to solve the given challenge again in 2004 with a computing time of about 22 hours. Since the computer used by Steel was slower than the one used by Faugére, the improvement comes from a better implementation of the basic algorithm, including better memory management needed to deal with the huge systems of linear equations encountered during the computations.

The set of field equations $x_i^2 + x_i = 0$ for $i = 1, \ldots, n$ must be added to the $n = 80$ quadratic equations before computing the Gröbner basis. These additional equations show that the polynomials in the Gröbner basis are square free. For that reason, the maximal degree D of polynomials occurring during the construction of the Gröbner basis is bounded by n. Computer experiments have shown that $D = 12$ for random systems of equations over $GF(2)$ and n near 80.

For quadratic equations coming from the HFE system (4.1), even this estimate is too pessimistic. Let HFE(n, d) stand for a cryptosystem where n is the number of variables and d is the degree of the polynomial in $K[X]$ which was used in the construction (4.1). Via a large number of experiments, Faugère saw that the degree D of polynomials encountered when computing a Gröbner basis is bounded as given in the following table:

$$\begin{aligned} 4 < d \leq 16 \quad &: \quad D = 3 \\ 17 < d \leq 128 \quad &: \quad D = 4 \\ 129 < d \leq 512 \quad &: \quad D = 5 \end{aligned}$$

and that these values are independent of n, at least for $n < 160$.

The complexity of solving an HFE(n, d)-type problem with F_5 depends on the size N_D of the linear systems encountered during the computations. The complexity of solving such systems is of order $O(N_D^\omega)$, where $2 \leq \omega \leq 3$ depending on which method is used for solving the linear systems of equations. For example, $\omega = 3$ when using the standard Gaussian elimination, whereas $\omega = 2$ for a sparse matrix implementation and Wiedemann's method.

With $N_D = O(n^{D-1})$ and $\omega = 2$, Faugère gives $O(n^{2D})$ as the overall complexity of attacking an HFE(n, d) system with his F_5 algorithm. This shows that when d and D are too small, then an HFE cryptosystem with coefficients in $GF(2)$ cannot resist a direct attack.

The second challenge posted by Patarin is an HFE$^-$ system with $n = 36$, $d = 144$, and four of the 36 quadratic equations removed. As of the writing of this book this challenge has not yet been broken.

7.4 The XL method

The XL algorithm is designed to solve a set of equations of the form:

$$f_1(x_1, \dots, x_n) = 0,$$
$$\vdots$$
$$f_m(x_1, \dots, x_n) = 0,$$

using the basic concepts of an ideal.

We will assume again that the equations are quadratic, and that this set of equations has a unique solution $x_1 = a_1, \dots, x_n = a_n$. We know from Theorem 7.1.3 that there exists a set of functions h_{ij}, $i = 1, \dots, n$ and $j = 1, \dots, m$ such that

$$x_i - a_i = \sum h_{ij} f_j.$$

If we can find these h_{ij}, then we find the solution. However, we do not know how to find the h_{ij} directly, so instead we generate the ideal gradually until we have the solution. The XL algorithm actually tries to find single variable equations one-by-one and then solves such a set of equations just as in the case of Gröbner bases.

In [Courtois et al., 2000], Courtois, Klimov, Patarin and Shamir proposed a new computational method called "eXtended Linearization," or XL algorithm, to solve such systems of polynomial equations. They also presented certain heuristics of the efficiency of this method. These heuristics have subsequently been criticized by Moh [Moh, 2001]. Later, Diem [Diem, 2004] presented an argument based on what is called the maximum rank conjecture in algebraic geometry from which he derived some very different estimates of the complexity of the XL algorithm. In particular, his analysis showed that the original estimate of the complexity for the algorithm was rather optimistic.

Let M_D be the set of monomials in $x_1, \dots, x_n$ with degree less than or equal to D, and let U_D be k-vector space generated by the polynomials in the form $h \cdot f_j$, where $h \in M_{D-2}$. The XL algorithm can be generally described as follows.

1.) Fix an integer $D > 2$.

2.) Generate all polynomials $h \cdot f_j$, for all $h \in M_{D-2}$ and $j = 1, \dots, m$.

3.) Fix one variable, say x_1, and an order σ on the set of monomials. Perform Gaussian elimination on the set of all polynomials in the previous step to derive a polynomial containing only the variable x_1.

4.) Assume that the step above produces at least one univariate polynomial. Solve the corresponding polynomial equation over the finite field k, for example with Berlekamp's algorithm or a polynomial root finding method, in order to find the value of x_1.

5.) Simplify the equations by substituting in the value of the variable x_1 found in the previous step.

6.) Repeat the process to find the values of the other variables one-by-one until all are found.

Although this algorithm can be used for any field (not necessarily a finite field), it works in the space of functions (not polynomials) if the field is finite. This means that over a finite field we need to use the field equations $x_i^q = x_i$ for $i = 1, \ldots, n$ at each step of the computation. This will make the algorithm run much faster in particular when q is small.

The XL algorithm depends on the parameter D. Note that XL may not work if D is too small, since then the third step may not yield anything. Thus we need to increase D if a particular D does not work. We can do this by incrementing D by one and by using the basis of U_D to generate the basis for U_{D+1}.

For any $D > 2$, let $k[x_1, \ldots, x_n]_{\leq D}$ be the k-vector space of polynomials in the variables $x_1, \ldots, x_n$ whose total degree is less than or equal to D, and let

$$C(D) := \dim(k[x_1, \ldots, x_n]_{\leq D}) - \dim(U_D).$$

Diem's argument in [Diem, 2004] uses the key observation that we can obtain a non-trivial univariate polynomial in the third step of the XL algorithm if $C(D) \leq D$. The key idea of his work points out that an estimation of D can be derived if the maximum rank conjecture in algebraic geometry is correct.

Variants of the XL Algorithm

Several modifications to the XL algorithm have been proposed.

1.) The XL′ Variant.

The computational procedure of XL′ [Courtois and Patarin, 2003] is similar to that of XL. The main difference is that in the third step of the procedure it tries to use elimination in order to find r equations that involve only monomials in a set of r variables, say $x_1, \ldots, x_r$. In the normal XL algorithm $r = 1$. It then solves this system of r equations by brute-force for these r variables. Finally, it solves the remaining equations by substituting in the values of these variables.

2.) The FXL and XFL Variant.

The 'F' here stands for 'fix' [Courtois et al., 2000]; that is, the values of a small number of variables are guessed at random. After guessing values for each of these variables, XL is run and tested for a valid solution. In [Courtois, 2004], a new suggestion was proposed. In this suggestion, after the second step of the XL procedure as given above, the elimination procedure should be run as far as it can go before guessing another variable. This was first called "improved FXL," but in [Yang and Chen, 2004b] the name XFL was suggested.

3.) The XLF Variant.

In [Courtois, 2004], another variation was proposed. This variation tries to utilize the Frobenius relation

$$x^q = x,$$

where $q = 2^l$ is the size of a finite field for some $l \geq 1$ (the field is of characteristic two). By treating the terms

$$x_i^{2^1} = x_{i_1},\ x_i^{2^2} = x_{i_2},\ \ldots,\ x_i^{2^{l-1}} = x_{i_{l-1}}$$

as independent new variables, additional equations are derived by repeatedly squaring the original equations and by using the equivalence of identical monomials as extra equations, for example

$$x_i^2 = x_{i_1}.$$

This variant is called XLF, where here 'F' stands for 'field' or 'Frobenius equations'.

4.) The XSL variant.

XSL stands for "eXtended Sparse Linearization." This variation [Courtois and Pieprzyk, 2002] is a linearization-based method designed to solve over-defined systems of sparse quadratic equations, for example the corresponding algebraic equations that characterize the AES block ciphers. It was suggested that it may be possible to break AES using XSL, but it seems that this is a very controversial claim and few researchers appear to believe it.

5.) The XL2 variant.

The XL2 algorithm was first proposed in [Courtois and Patarin, 2003] and works over the field $GF(2)$. The basic idea is that since we work in $GF(2)$, we should then automatically add the field equations $x_i^2 = x_i$, which essentially is to say that we should work in the function ring and not the polynomial ring. This idea was reformulated in [Yang et al., 2004b], and this method then can be viewed as a way to more efficiently manage the elimination process.

Overall, we see that the complexity of the algorithm of the XL family is primarily determined by the parameter D.

7.5 Connections Between XL and Gröbner Bases

So far we have seen that essentially all existing algorithms to solve a set of multivariate equations belong to the Gröbner basis family, including the new F_4 and F_5 algorithms, or to the XL family, including its several variants. They are currently used as powerful tools to attack public key cryptosystems, block ciphers, and stream ciphers.

In [Courtois et al., 2000], it was speculated that because the XL algorithm does not calculate an entire Gröbner basis, it might thus be more efficient. However, experiments seem to indicate that it is the other way around. For example, Faugère [Faugère and Joux, 2003] used the algorithms F_4 and F_5 to successfully attack the 80-bit HFE cryptosystem, while the XL algorithm has not yet been shown to be able to do so. In a paper presented at Asiacrypt 2004, Ars, Faugère, Imai, Kawazoe, and Sugita [Ars et al., 2004] gave an argument using some of the ideas of semi-regular sequences of [Bardet et al., 2005] in order to clarify the relationship between the XL algorithm and the Gröbner basis algorithm. In particular, they showed that:

1.) In essence the XL algorithm is a Gröbner basis algorithm, and that it can even be viewed as a redundant variant of the Gröbner basis algorithm F_4.

2.) If the XL algorithm terminates, the new Gröbner basis algorithm will also terminate with a lexicographic ordering.

3.) On semi-regular sequences with finite field $GF(2)$, the degree D of the parameter needed for the XL algorithm is almost the same as the degree of the polynomials in the matrix constructed by the F_5 algorithm. The complexities of these two algorithms are determined by the size of their corresponding matrices, and F_5 is claimed to have a smaller matrix.

4.) On semi-regular sequences with finite field $GF(q)$ with $q \ll n$, it is shown that the XL algorithm terminates for a degree higher than the Gröbner basis algorithms with grevlex order, and the XL matrices are therefore much bigger compared to the matrices used in F_5.

5.) For a system of algebraic equations whose solution is unique in a given finite field, solving for this solution amounts to nothing more than calculating the reduced Gröbner basis for the ideal associated with that system.

6.) To attack HFE, a Gröbner basis algorithm manages to find structure in the multivariate system, and never exceeds a low degree. On the other hand, in the XL algorithm the degree seems to increase as the number of variables n increases.

Their conclusions indicate that in general the improved Gröbner basis algorithm seems to work better than the XL algorithms. Additionally, they also pointed out the close connection of the XL algorithm with some earlier work of Lazard [Lazard, 1983].

Complexity Estimates

Recently a lot of work has been done in estimating this parameter D, and therefore finding a complexity estimate for the XL family, the F_4 and F_5 family of algorithms; or to find a simplified form for the estimates [Yang et al., 2004b; Yang and Chen, 2004a; Yang and Chen, 2004b; Bardet et al., 2005; Ars et al., 2004]. The theoretical foundation of these estimates is based on the maximum rank conjecture, which was first used in [Diem, 2004]. In [Bardet et al., 2005], a class of semi-regular systems of polynomial equations was defined and used to study the complexity of these algorithms. This concept is an extension of the regular system of equations [Macaulay, 1916].

For an overdefined semi-regular system of equations with n variables and $n+m$ equations over a finite field k with q elements, the estimate for the computational complexity is given by

$$\min\{q^f \binom{n+d-f}{d}^2 (c_0 + c_1 \lg \binom{n+d-f}{d}) : d := \min\{D : [t^D](\tfrac{(1-t^2)^m(1-t^q)^{n-f}}{(1-t)^{n+1-f}(1-t^{2q})^m}) < 0\}\},$$

where f is the number of variables the values of which are guessed one-by-one. For the F_5 family the complexity is estimated to be:

$$\min\{q^f \binom{n-1+d-f}{d}^\omega : d := \min\{D : [t^D](\tfrac{(1-t^2)^m(1-t^q)^{n-1-f}}{(1-t)^{n-f}(1-t^{2q})^m}) < 0\}\},$$

where $2 \leq \omega \leq 3$ is a constant determined by which method is used for performing the Gaussian elimination, and f is the number of variables the values of which are guessed one-by-one.

In the next section, we will present a new idea to solve a set of multivariate polynomial equations that uses the extension field idea to transform the problem into a single equation in a single variable.

7.6 The Zhuang-Zi Algorithm

The Zhuang-Zi algorithm requires that the polynomial equations (7.1) are given with coefficients in a finite field k of size q. The algorithm tries to find all $(a_1, \ldots, a_n) \in k^n$ such that:

$$\begin{aligned} f_1(a_1, \ldots, a_n) &= 0, \\ &\vdots \\ f_m(a_1, \ldots, a_n) &= 0. \end{aligned} \tag{7.12}$$

The main idea is to lift the polynomial map

$$f : k^n \longrightarrow k^m \tag{7.13}$$

with components $f_1, \ldots, f_m$ to a map over an extension field K. In order to accomplish this we will assume that $m = n$. When we have more variables than equations ($n > m$) then we can augment the given equations with $n - m$ trivial equations of $0 = 0$. Similarly, when $n < m$ we can simply introduce additional variables $x_{n+1}, \ldots, x_m$ to make up for the shortfall. In other words, we use the larger of m and n in (7.13) when lifting the map to K. With these observations we can assume that $m = n$ holds in all cases.

Choose an irreducible polynomial $g(y)$ of degree n so that with it $K = k[y]/g(y)$ is a degree n field extension of k. Let $\phi : K \longrightarrow k^n$ be the standard k-linear map that allows us to lift $f = (f_1, \ldots, f_n)$ up to the extension field by

$$F = \phi^{-1} \circ f \circ \phi.$$

With $X = x_1 + x_2 y + \cdots + x_n y^{n-1}$ as the new intermediate, F is a polynomial in $K[X]$ and has a unique representation in $K[X]/(X^{q^n} - X)$. Using the Frobenius maps $G_i(X) = X^{q^i}$, $i = 0, \ldots, n-1$, the transformation ϕ^{-1} can be be given in matrix form. Let $\mathsf{x} = (x_1, x_2 \ldots, x_n)^T$ and $\mathsf{X} = (X, X^q, \ldots, X^{q^{n-1}})^T$. Then

$$\mathsf{X} = \mathsf{B}\mathsf{x},$$

where the entries of B are in K and can be found directly by computing $X, X^q, \ldots, X^{q^{n-1}}$. The matrix B is invertible so that ϕ^{-1} is given by

$$\mathsf{x} = \mathsf{B}^{-1}\mathsf{X}.$$

The roots of $F(X) = 0$ (those $A = a_1 + a_2 + \cdots + a_n y^{n-1} \in K$ such that $F(A) = 0$) correspond to the solutions of the polynomial equations (7.12). In most cases, the degree of $F(X)$ will be too high in order to use one of the standard methods for factoring $F(X)$ or for finding the polynomial roots of $F(X)$. The Zhuang-Zi algorithm as described below provides a mechanism for reducing the degree of the polynomial.

The Zhuang-Zi Algorithm: Given $f_1, \ldots, f_n \in k[x_1, \ldots, x_n]$, select an irreducible $g(y) \in k[y]$ of degree n so that $f = (f_1, \ldots, f_n)$ can be lifted to the unique polynomial function $F(X) \in K[X] \bmod (X^{q^n} - X)$, where $K = K[y]/g(y)$.

Step 1: Choose an integer $D > 0$ so that finding the roots of any polynomial $F(X) \in K[X]$ with $\deg(F(X)) < D$ is computationally feasible. If $\deg(F(X)) < D$ go to Step 5; otherwise go to next Step.

Step 2: Set $F_0(X) = F(X)$ and use the Frobenius maps $G_i(X) = X^{q^i}$ for $i = 1, \ldots, n-1$ to compute:

$$F_i(X) = G_i \circ F(X) = F(X)^{q^i} \bmod (X^{q^n} - X).$$

Step 3: Let N be the number of monomials that appear in any $F_i(X)$. The coefficients of $F_i(X)$ are entered in decreasing order in row $i+1$ of an $n \times N$ matrix P. Use Gaussian elimination to bring P into row echelon form $\tilde{\mathsf{P}}$. This procedure will eliminate the terms with highest degrees first and produce a new set of basis polynomials $\{S_0(X), S_1(X), \ldots, S_{t-1}(X)\}$, with $t \leq n$ and S_{t-1} being the polynomial of lowest degree coming from the last non-zero row of $\tilde{\mathsf{P}}$.

Step 4: If $\deg(S_{t-1}) < D$, then set $F(X) = S_{t-1}(X)$ and go to Step 5. Otherwise, for each $i = 0, 1, \ldots, t-1$ and each $j = 0, 1, \ldots, n-1$ compute:

$$F_{i+jt} = X^{q^j} S_i(X) \bmod (X^{q^n} - X).$$

Set n to $n\,t$ and go back to Step 3.

Step 5: Use a suitable algorithm like distinct degree factorization, Berlekamp, or any other method to find the roots of $F(X) = 0$.

Remark 7.6.1. *If no k-linear combination of the polynomials $f_1, \ldots, f_n$ is zero in $k[x_1, \ldots, x_n]$ modulo $(x_1^q - x_1, \ldots, x_n^q - x_n)$, then no K-linear combination of the polynomials $F_1, \ldots, F_n$ in Step 2 is zero in $K[X]$* mod $(X^{q^n} - X)$, *since such a linear combination of the F_i will have degree at most $q^n - 1$.*

Remark 7.6.2. *Steps 3 and 4 can introduce spurious solutions when powers of X are reduced modulo $q^n - 1$. Because of this, solutions of $F(X) = 0$ have to be checked in Step 5 in order to ensure that they satisfy the original system of equations. The spurious solutions disappear when the loop in Steps 3 and 4 are repeated until the whole system of equations becomes invariant. The overhead of these additional iterations is usually not worth the computational effort.*

Remark 7.6.3. *If f consists only of linear equations, then Step 2 with a single application of Step 3 suffices to bring $F(X)$ into its final form. If the system of linear equations is inconsistent, then $F(X)$ will be a constant; otherwise, the degree of $F(X)$ will be q^r with $r \geq 0$ indicating the reduction in rank of the corresponding $n \times n$ matrix.*

Remark 7.6.4. *If f consists of quadratic polynomials, then the only powers of X which may appear in Step 1 and Step 2 are either of the form X^{q^i} or $X^{q^i+q^j}$ (i and j can be equal). The first application of Step 4 introduces new powers of the form $X^{q^i+q^j+q^l}$ (where some of the i, j, l may be equal), corresponding to cubic terms in $k[x_1, \ldots, x_n]$. Each additional iteration of Step 4 adds another q^r to the possible exponents of X. Since there are qn different exponents possible, after at most qn iterations all possible exponents have been generated and the basis polynomials remain unchanged from one iteration to the next.*

Toy Example for the Zhuang-Zi Algorithm

In order to illustrate the algorithm we present a toy example with coefficients again from the finite field $k = GF(2^2)$, whose field operations are given in Table 2.1. The toy example is given by the polynomial map $f : k^2 \longrightarrow k^2$ with the components:

$$\begin{aligned} f_1 &= x_1^2 + x_2^2 + 1, \\ f_2 &= x_2^2 + x_1 x_2 + \alpha. \end{aligned} \tag{7.14}$$

An irreducible polynomial of degree two in $k[y]$ is:

$$g(y) = y^2 + y + \alpha^2.$$

With $X = x_0 + x_1 y$ the transformation to the big field $\phi^{-1} : k^2 \longrightarrow K$ is:

$$\begin{bmatrix} x_0 \\ x_1 \end{bmatrix} = \begin{bmatrix} 1+y & y \\ 1 & 1 \end{bmatrix} \begin{bmatrix} X \\ X^4 \end{bmatrix}$$

and the polynomial map $F = \phi^{-1} \circ f \circ \phi$ with f from (7.14) is given over K by:

$$F(X) = (y+1)X^8 + yX^5 + \alpha y + 1.$$

Since this is a trivial example, we can factor $F(X)$ immediately and obtain:

$$F(X) = (X + \alpha y + \alpha^2)(X^7 + (\alpha y + \alpha^2)X^6 + \alpha^2 y X^5 + (y + \alpha)X^4 + \alpha y X^3 + (\alpha y + \alpha)X^2 + (y + \alpha^2)X + \alpha^2),$$

from which we find that $X = \alpha y + \alpha^2$ is the only solution to $F(X) = 0$. With this we find $(x_1, x_2) = (\alpha^2, \alpha)$ is the solution to $f_1 = f_2 = 0$ from (7.14).

In general the degree of $F(X)$ is much larger. We use the Frobenius maps in order to find the polynomials of Step 2:

$$F_0(X) = F(X) = (y + 1)X^8 + yX^5 + \alpha y + 1,$$
$$F_1(X) = F^4(X) = (y + 1)X^5 + yX^2 + \alpha y + \alpha^2.$$

The echelon form gives equivalent basis polynomials:

$$F_0(X) = X^8 + (\alpha^2 y + \alpha^2)X^2 + \alpha^2 y + \alpha,$$
$$F_1(X) = X^5 + (\alpha y + 1)X^2 + \alpha y + \alpha.$$

The factorization of $F_1(X)$ would then give:

$$F_1 = (X + \alpha y + \alpha^2)(X^4 + (\alpha y + \alpha^2)X^3 + \alpha^2 y X^2 + (y + \alpha)X + \alpha y),$$

and we could again stop here. However, in order to show how the Zhuang-Zi algorithm works, we continue with Steps 3 and 4. The equations F_0 and F_1 are augmented with XF_0, X^4F_0, XF_1, X^4F_1 modulo $X^{16} - X$. The associated matrix is then put into echelon form, producing the following set of polynomials:

$$F_0 = X^{12} + \alpha^2 y X^3 + \alpha X,$$
$$F_1 = X^9 + (\alpha^2 y + \alpha^2)X^3 + (\alpha^2 y + \alpha)X,$$
$$F_2 = X^8 + (\alpha^2 y + \alpha^2)X^2 + \alpha^2 y + \alpha,$$
$$F_3 = X^6 + (\alpha y + 1)X^3 + (\alpha y + \alpha)X,$$
$$F_4 = X^5 + (\alpha y + 1)X^2 + \alpha y + \alpha,$$
$$F_5 = X^4 + (\alpha^2 y + \alpha^2)X.$$

The factorization of the polynomial of smallest degree gives:

$$F_5 = X(X + y + \alpha)(X + \alpha y + \alpha^2)(X + \alpha^2 y + 1),$$

and this demonstrates that not all solutions of F_5 are solutions of the original problem. In order to reduce the degree further, the next iteration of Steps 3 and 4 gives:

$$F_0 = X^{13} + \alpha^2 y + \alpha^2,$$

$$
\begin{aligned}
F_1 &= X^{12} + \alpha^2 y X^3 + \alpha^2 y + 1, \\
F_2 &= X^{10} + 1, \\
F_3 &= X^9 + (\alpha^2 y + \alpha^2) X^3 + \alpha, \\
F_4 &= X^8 + \alpha^2 y + \alpha^2, \\
F_5 &= X^7 + \alpha^2 y, \\
F_6 &= X^6 + (\alpha y + 1) X^3 + y + \alpha^2, \\
F_7 &= X^5 + 1, \\
F_8 &= X^4 + \alpha y + 1, \\
F_9 &= X^2 + \alpha^2 y, \\
F_{10} &= X + \alpha y + \alpha^2.
\end{aligned}
$$

We stop here as the degree of the lowest polynomial is one, and thus we have the desired solution immediately. After two more iterations the system of polynomials becomes invariant, and the final form is:

$$
\begin{aligned}
F_0 &= X^{15} + 1, \\
F_1 &= X^{14} + \alpha y + 1, \\
F_2 &= X^{13} + \alpha^2 y + \alpha^2, \\
F_3 &= X^{12} + \alpha^2 y, \\
F_4 &= X^{11} + \alpha y + \alpha^2, \\
F_5 &= X^{10} + 1, \\
F_6 &= X^9 + \alpha y + 1, \\
F_7 &= X^8 + \alpha^2 y + \alpha^2, \\
F_8 &= X^7 + \alpha^2 y, \\
F_9 &= X^6 + \alpha y + \alpha^2, \\
F_{10} &= X^5 + 1, \\
F_{11} &= X^4 + \alpha y + 1, \\
F_{12} &= X^3 + \alpha^2 y + \alpha^2, \\
F_{13} &= X^2 + \alpha^2 y, \\
F_{14} &= X + \alpha y + \alpha^2.
\end{aligned}
$$

Of course, this trivial example could have been solved more easily by finding the Gröbner basis for

$$\left\{x_1^2 + x_2^2 + 1, x_2^2 + x_1x_2 + \alpha, x_1^4 - x_1, x_2^4 - x_2\right\},$$

which is

$$\left\{x_1 + \alpha x_2, x_2^2 + \alpha^2\right\}.$$

Non-trivial examples can also be constructed where the Zhuang-Zi algorithm can find the solution easily, whereas a Gröbner basis algorithm will fail due to the requirements for memory. The idea is to select a function $F(X) : K \longrightarrow K$ of low enough degree that can be factored easily, while the corresponding map $f : k^n \longrightarrow k^n$ is complicated. The degree of terms in $f_1, \ldots, f_n$ depends on which powers of X have been selected in F. Terms of the form X^{q^i} will lead to linear terms, $X^{q^i+q^j}$ to quadratic terms, $X^{q^i+q^j+q^l}$ to cubic terms, and so on. By keeping the exponents $i, j, l, \ldots$ small enough, we can find a polynomial $F(X)$ that can be factored easily, while the components of f are quadratic, cubic, or even higher degree in $k[x_1, \ldots, x_n]$. At the same time, few terms in $F(X)$ will produce a lot of terms in f. The idea is reminiscent of what has been suggested for the HFE public key encryption in [Patarin, 1996b].

For a specific example, let $k = GF(2^3)$ and let $K = k[y]/g(y)$ be a degree n extension of k by some irreducible polynomial $g(y) \in k[y]$. A computer algebra system like Magma [Computational Algebra Group, 2005] can easily factor a polynomial of the form:

$$F(X) = X^{72} + a_1X^{65} + a_2X^{64} + a_3X^{16} + a_4X^9 + a_5X^8 + a_6X^2 + a_7X + a_8, \tag{7.15}$$

where the coefficients a_j, $j = 1, \ldots, 8$ are chosen at random from the finite field $k = GF(2^3)$. We treat k as a subfield of K via the standard embedding. With $q = 8$, all powers of X in (7.15) can be written in the form X^{8^i} or $X^{8^i+8^j}$ so that that (7.15) represents a quadratic polynomial map $f : k^n \longrightarrow k^n$. Depending on n and the selection of $a_1, \ldots, a_8$, we obtain zero, one, or more linear factors in $F(X)$. Each factor $X + \alpha$ gives rise to a solution of the corresponding polynomial equations $f_i(x_1, \ldots, x_n) = 0$, for $i = 1, \ldots, n$. These solutions can also be found directly with a Gröbner bases program like Faugère's F_4. As expected, the computing time for the Gröbner bases method increases exponentially with n. On the other hand, the time complexity for factoring (7.15) by Berlekamp's algorithm is $O(nd^3)$, where d is the degree of the polynomial $F(X)$.

The quadratic polynomials $f_i(x_1, \ldots, x_n)$ for the above example have too many terms in order to be displayed here. It could be argued that the comparison is unfair since we obfuscate the problem and force Faugère's algorithm to work on a system with a huge number of terms. However, the Gröbner algorithm in Magma is very efficient and even changing some terms in (7.15) made it difficult to find examples where F_4 failed even with a large n.

This non-trivial example shows that the Zhuang-Zi algorithm sometimes has an advantage over the best Gröbner bases algorithm, even if only the first step is used. Other non-trivial examples can be constructed by using Steps 3 and 4 in reverse. That is, start with a set of polynomials $F_0(X), F_1(X), \ldots, F_t(X)$ with $F_t(X)$ of low enough degree, and construct the previous set of polynomials, which leads to the given set of polynomials via Steps 3 and 4. This can be done so that none of the polynomials in the preceding set is of low enough degree.

In summary, the Zhuang-Zi algorithm shows that some polynomial equations in $k[x_1, \ldots, x_n]$ can be solved easily if these equations are lifted up into K. Of course, the method cannot solve all polynomial equations easily, and it remains a challenge to characterize in advance the polynomials in $k[x_1, \ldots, x_n]$ where the Zhuang-Zi algorithm has an advantage over the Gröbner bases algorithm. It is clear that the Zhuang-Zi algorithm is just at the infant stage of its development and much more work needs to be done in order to deal with different situations. On the other hand, the basic idea behind it provides a way to think about related problems in MPKCs, in particular the problem of security of MPKCs and the complexity of solving polynomial equations in a finite field.

Chapter 8

FUTURE RESEARCH

So far we have presented the main developments of MPKCs during the last ten years. These developments have produced many interesting new ideas, tools and constructions from the point of view of theory and of practical applications. In this chapter we present some thoughts on the future of research in multivariate public key cryptography. In particular, we will present some of the critical questions that must be resolved to move the research of MPKCs to the next level.

8.1 Construction of MPKCs

Though the idea of MPKCs was initiated in the mid-1980s by several people using essentially the same idea of multivariate triangular map constructions, none of these constructions really worked. The real breakthrough should be attributed to the work by Matsumoto and Imai in 1988 [Matsumoto and Imai, 1988], though the potential of this new mathematical idea was not fully realized until the work of Patarin in 1995 [Patarin, 1995], and later work by him and others.

The new idea of Matsumoto and Imai could be called the "Big Field" construction. In this construction we build something in a degree n extension field (Big Field) K over a small finite field k. We then move it down to a vector space over the small finite field, where the extension field is isomorphic to a vector space over the small finite field. The key is the identification map $\phi : K \longrightarrow k^n$, the standard k-linear isomorphism between K and k^n, which we have used throughout this book. The identification map can be used to build the connection of maps in these two different spaces as depicted in Figure 8.1.

It is well-known in mathematics that any finite field k has a degree n extension that can be identified as k^n, an n dimensional vector space

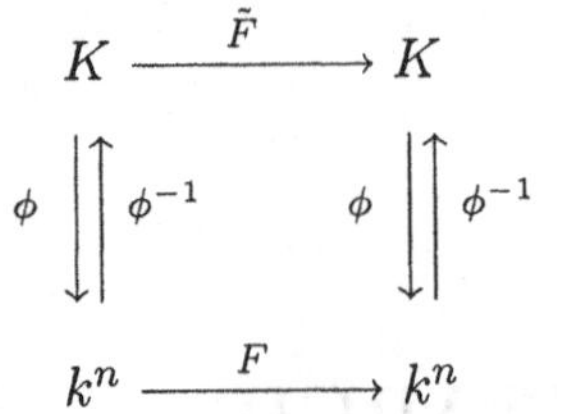

$$\begin{array}{ccc} K & \xrightarrow{\tilde{F}} & K \\ \phi \downarrow \uparrow \phi^{-1} & & \phi \downarrow \uparrow \phi^{-1} \\ k^n & \xrightarrow{F} & k^n \end{array}$$

Figure 8.1. Identifying maps on a k-vector space with those on extension fields K/k.

over k. The Matsumoto-Imai construction was the first time that this idea was utilized and it turned out to be a very powerful idea.

Through this identification map, this idea allows us to move freely between the setting of a vector space over a small finite field and the setting of an extension field, the Big Field. Therefore, we may consider problems from the view of both these two settings and deal with it in whichever setting is easier as dictated by the underlying structures. This is also the idea behind the new Zhuang-Zi algorithm [Ding et al., 2006a], where we lift the problem of solving a set of multivariate polynomial equations over a small finite field to solving a set of single variable equations over an extension field.

Great efforts are still being devoted to develope MPKCs using new mathematical ideas and structures. For example, in early 2006 a new family of cryptosystems called the Medium Field Equations (MFE) was presented in CT-RSA 2006 [Wang et al., 2006], which combines the extension field idea (though it is called a "Medium Field") with some matrix equalities involving the determinants of the related matrices. However, MFE was defeated by a method of high order linearization equations [Ding et al., 2006b]. This attack is a generalization of the linearization equation attack of Patarin, where the multivariate equations used contain high order terms in the ciphertext variables but only linear terms in the plaintext variables. This allows us to defeat the system using the same procedure as in [Patarin, 1995].

Recently Patarin [Patarin, 2006] proposed a new idea of probabilistic MPKCs. In the case of a signature scheme, for example in the verification process, a verifier does not require that all the public equations be satisfied by the document (or its hash) and its signature, but rather a signature is accepted as legitimate as long as a sufficient number (or the majority) of the public equations are satisfied by the pair. In the concrete example proposed, the construction of the example is related to the idea of internal perturbation [Ding, 2004a], where a certain special internal perturbation is added that will probabilistically affect only a small portion of public key polynomials each time in the verification process.

Recently an interesting paper was written to group different mathematical ideas in the construction of MPKCs [Wolf and Preneel, 2005c]. It gives quite a reasonable (though debatable) classification of the mathematical ideas behind various constructions.

From what we have seen, it is evident that what could really drive the development of the design in MPKCs are indeed new mathematical ideas that bring new mathematical structures and insights that can be used in the construction of MPKCs. Therefore, we believe we should study and search for further mathematical ideas and structures that could be used to construct MPKCs.

One particularly interesting problem would be to make the TTM cryptosystems work. The TTM construction by Moh [Moh, 1999a] uses a very different but very elegant mathematical idea, though it is also based on triangular maps (or tame transformations, as they are called in TTM). However, the realization of this idea requires some highly non-trivial algebraic identities in terms of quadratic functions. There are several very nice constructions by Moh and his collaborators, but all of these constructions, including the latest [Moh et al., 2004], recently defeated in [Nie et al., 2006], have failed in the sense that they are all suffer from the weakness hidden in the special polynomials used in the constructions. We believe that one of the main reasons behind this is that all the current constructions of TTM are in fact certain ad hoc constructions, where the principles behind the nice and elegant formulas used are not explicitly described, and therefore cannot be easily modified or fixed. The key problem in the TTM construction can be formulated as a problem of building two invertible "non-trivial" nonlinear triangular maps (one upper triangular and one lower triangular), G_1 and G_2, such that at least one, say G_1, is of high degree, while the composition of these two maps $G = G_1 \circ G_2$ gives a "generic" quadratic map. Though we still do not know whether or not this idea will indeed eventually work, we believe that to make it work a systematic approach to solving the related problems must be developed. This will likely require some deep insights and possibly the usage of some intrinsic combinatorial structures from algebraic geometry.

From the point of view of practical applications, there are two critical problems that deserve more attention in designing new MPKCs. The first one is the problem of the public key size. For a MPKC with m polynomials and n variables, the public key size normally has

$m(n+2)(n+1)/2$ terms, where m is at least 25 and n is at least 30. Compared with all other public key cryptosystems, for example RSA, one disadvantage is that in general a MPKC has a relatively large public key (tens of Kbytes). This is not a problem from the point view of modern computers such as the PCs we use, but it could be a problem if we want to use it for small devices with limited memory resources. This would also be a problem if a device with limited communication abilities needs to send the public key for each transaction, for example in the case of authentication.

One idea is to do something like in [Tsujii et al., 1989], where a cryptosystem is built with a very small number of variables (five) but with a higher degree (four) over a much bigger base field (32 bits). In other words, we can try high degree constructions with fewer variables but over a much bigger field.

In [Wolf and Preneel, 2005a; Wolf and Preneel, 2005b], the concept of equivalent keys is proposed, but it is used to study the key size problem for the private keys and not the public key. Therefore it does not help in this aspect. Similar work is also done for triangular-type or STS-type cryptosystems [Hu et al., 2005]. In general, any new idea for how to reduce the public key size or in how to manage it in practical applications would be really helpful.

A second idea is that of using sparse polynomials constructions. In the Oil-Vinegar construction, while choosing a set of specific Oil-Vinegar polynomials, we are given a large amount of freedom in making random selections among a large set of possible such polynomials in terms of the coefficients of these polynomials. In particular, for an Oil-Vinegar polynomial

$$\sum_{i=1}^{o}\sum_{j=1}^{v} a_{ij}x_iy_j + \sum_{i=1}^{v}\sum_{j=1}^{v} b_{ij}y_iy_j + \sum_{i=1}^{o} c_ix_i + \sum_{j=1}^{v} d_jy_j + e,$$

we should randomly choose $a_{ij}, b_{ij}, c_i, d_j, e$ from the finite field k. A natural question we may ask is if it is possible to choose these in a way so as to improve the efficiency of the resulting system without affecting the security of the system in any substantial way. This idea was used in the TTM constructions [Moh, 1999a], which in some way may be viewed as an accident rather than intentional. It is also one of the main reasons why some of these cryptosystems can be defeated.

It is very reasonable to assume that many people probably thought about the idea of using sparse polynomials at one time or the other, but the first explicit usage of such constructions should be attributed to the works of Yang and Chen [Yang and Chen, 2003], where they used it in the construction of the generation of the TTS signature schemes.

The idea of sparse polynomials, in the case of Oil-Vinegar polynomial, is to choose a very small number of nonzero coefficients among all the possible choices for the set $a_{ij}, b_{ij}, c_i, d_j, e$. This will make the process of inverting the Oil-Vinegar map much faster. Some of the early TTS constructions were broken exactly because of the usage of sparse polynomials [Ding and Yin, 2004], which brought unexpected weakness to the system. However, we believe that the idea of using sparse polynomials is an excellent idea, especially from the point view of practical applications. For example, they make the TTS signing process incredibly fast compared with all other schemes [Yang et al., 2006]. The new constructions of TTS proposed in [Yang and Chen, 2005a] again use sparse polynomials, and have not yet been broken.

From the theoretical point of view, one critical question that needs to be addressed carefully is that of whether or not the use of specific sparse polynomials has any substantial impact on the security of the given cryptosystem. We believe this is a very interesting and challenging question. The answer to this problem will help us to establish the principles for how we should choose sparse polynomials that do not affect the security of the given cryptosystem. At this moment we still do not know what these principles should exactly be. An unexpected consequence of answering this problem is that it might also shed some light on the problem mentioned above about reducing the size of the public key.

From a very general mathematical point view, a very interesting and important question is the classification problem. One critical construction used for the cipher of a MPKC is based on the formula

$$\bar{F} = L_1 \circ F \circ L_2,$$

where L_1, L_2 are invertible affine maps and F is a quadratic map. This means that given a specific F, we can build a large family of MPKCs by choosing different L_1 and L_2. We can then define an equivalence relation among all such maps by defining each equivalence class as the set of all quadratic maps that are derived from the same F. The classification problem is to classify all the equivalence classes. Once we can find all these classes we can then ask, among all such classes, which are the good ones we can use for the constructions of the quadratic maps for the MPKCs.

This problem is actually closely related to the IP problem, and in [Faugère and Perret, 2006] a good mathematical frame work is already proposed using the language of group actions. This classification problem can be viewed as the mathematical problem of finding the orbit space of the action of the direct product of two affine groups, one from

L_1 and one for L_2, on the space of quadratic maps, where the first one, L_1, acts linearly, and the other one, L_2, acts nonlinearly. This is actually a very hard mathematical problem, which as far as we know does not yet have any good answers even for small n and q.

8.2 Attack on MPKCs and Provable Security

Several major methods have been developed to attack the MPKCs. They can be roughly grouped into the following two categories.

- **Structure-based** – These attacks rely solely on the specific structures of the corresponding MPKC. Here we may use several methods, for example, the rank attack, the invariant subspace attack, the differential attack, the extension field structure attack, the low degree inverse, and others.

- **General Attack** – This attack uses the general method of solving a set of multivariate polynomial equations, for example using the Gröbner basis method, including the Buchberger algorithm, its improvements (such as F_4 and F_5), the XL algorithm, and the new Zhuang-Zi algorithm.

Of course, we may also combine both methods to attack a specific MPKC.

It is clear that for a given multivariate cryptosystem, we should first try the general attack. If this does not work, then we may then look for methods that use the weaknesses of the underlying structure.

Though a lot of work has been done in analyzing the efficiency of different attacks, we still do not fully understand the full potential or the limitations of some of the attack algorithms, such as the MinRank algorithm, Gröbner basis algorithms, the XL algorithm, and the new Zhuang-Zi algorithm. For example, we still know very little about how these general attacks will work on the internal perturbation type systems such as PMI+ [Ding et al., 2005; Ding and Gower, 2006], though we do have some experimental data to give us some ideas about how things work. Another interesting question is to find out exactly why and how the improved Gröbner basis algorithms like F_4 and F_5 work on HFE and its simple variants with low parameter D . The question is why the hidden structure of HFE can be discovered by these algorithms.

Much work is still needed to understand both the theory and practice of how efficiently general attack algorithms work and how to implement them efficiently. From the theoretical point of view, to answer these problems, the foundation again lies in modern algebraic geometry as in [Diem, 2004]. One critical step would be to prove the maximum rank conjecture pointed in [Diem, 2004], which is currently the theoretical basis used to estimate the complexity of the XL algorithm and the F_4 and F_5 algorithms for example. Another interesting problem is a conjecture on a semi-regular system of polynomial equations presented in [Bardet et al., 2005].

One more important problem we would like to emphasize is the efficient implementation of general algorithms. Even for the same algorithm, the efficiency of various implementations can be substantially different. For example, one critical problem in implementing F_4 or F_5, or the XL type algorithms, is that the programs tend to use a large amount of memory for any nontrivial problem. Often the computation fails not because of time constraints but because the program runs out of memory. Therefore, efficient implementations of these algorithms with good memory management should be studied and tested carefully.

Recently, Chen, Yang, and Chen [Chen et al., 2006] developed a new XL implementation with a Wiedemann solver that is as close to optimal as might be possible. They showed that in a few cases the simple FXL algorithm can even outperform the more sophisticated F_4 and F_5 algorithms. In general, any new idea or technique in implementing these algorithms efficiently could have very serious practical implications.

In order to convince industry to actually use MPKCs in practical applications, the first and the most important problem is the concern of security. Industry must be convinced that MPKCs are indeed secure. A good answer to this problem is to prove that a given MPKC is indeed secure with some reasonable theoretical assumptions; that is, we need to solve the problem of provable security of MPKCs. From this point of view, the different approaches taken in attacking MPKCs present a very serious problem in terms of provable security. Many people have spent a considerable amount of time thinking about this problem, but there are still no substantial results in this area. However, recently there have been some works related to this topic [Dubois et al., 2006; Faugère and Perret, 2006]. One possible approach should be from the point view of algebraic geometry; that is, we need to study further all the different attacks and somehow put them into one theoretical framework using some (maybe new) abstract notion. This would allow us to formulate some reasonable theoretical assumptions, which is the foundation of any type of provable security. This is likely a very hard problem.

Currently algebraic attacks are a very popular research topic in cryptography, and in particular in attacking symmetric block ciphers like AES [Courtois and Pieprzyk, 2002] and stream ciphers [Armknecht and Krause, 2003]. We would like to point out that the origin of such an idea is actually from MPKCs, and in particular Patarin's linearization equation attack method. From recent developments we see that there is a trend that the research of MPKCs will interact very closely with that of symmetric ciphers and stream ciphers. We believe some of the new ideas we have seen in MPKCs will have much more broad applications in the area of algebraic attacks.

One more critical question that should be brought up concerns the fundamental idea behind all constructions: that of factorization of compositions of maps. We have seen that essentially all the constructions of public keys are made in the following form:

$$\bar{F} = L_1 \circ F \circ L_2,$$

where the key construction is the map F which is easy to "invert" for anyone, and the two (sometimes only one) linear maps are used to "hide" the map F to make the system secure. If we can decompose the map $\bar{F}$ as above, then we can defeat the system completely. Therefore, we see linear maps being used to protect nonlinear maps (not the other way around). We should ask how much security is provided to a cryptosystem by this hiding via linear maps.

There are certain related discussions in the work of [Faugère and Perret, 2006] to appear in Eurocrypt 2006 where this problem is viewed from the point of view of representations of groups. Specifically, they view L_1 and L_2 as elements of two separate groups which act independently on the map. For now we know very little about this question mainly because we know very little about decomposition of maps. One big problem is the Jacobian conjecture. We also know very little about finding the orbit space as was mentioned above. The real problem behind all of this is that we do not know which maps are invertible. This is closely related to the Jacobian conjecture and it would decide the equivalence classes of the maps.

8.3 Practical Applications

Currently, a very popular notion in the computing world is the phrase "ubiquitous computing." This phrase describes a world where computing in some form is virtually everywhere, usually in the form of some small computing device such as RFID, wireless sensors, PDA, and others. Some of these devices often have very limited computing power, batteries, memory capacity, and communication capacity. Still, because of its ever growing importance in our daily lives, the security of such a system will become an increasingly important concern. It is clear that public key cryptosystems like RSA cannot be used in these settings due to the complexity of the computations.

In some way MPKCs may provide an alternative in this area. In particular, there are many alternative multivariate signature schemes such as Sflash, Rainbow, TTS and TRMC. Recently in [Yang et al., 2006], it is shown that systems like TTS have great potential for application in small computing devices, where it can work on chips with only one tenth of the gates necessary for RSA. Due to its high efficiency, a very important direction in application of MPKCs is to seek new applications where the classical public key cryptosystems like RSA cannot work satisfactorily. This will also likely be the area where MPKCs will find a real impact in practical applications. Nevertheless, it still requires a substantial amount of work to efficiently implement these schemes on small chips.

8.4 Underlying Mathematics

Though the main motivation of MPKCs is its potential in practical applications, from the theoretical point view the study actually is nothing but a study of functions over a vector space of a finite field. Such functions can be presented as multivariate polynomials, which means that in some sense we are just studying algebraic geometry over a finite field. However, the subtle point here is that we are not actually studying the polynomial ring $k[x_1, \ldots, x_n]$ over a finite field k, but rather we are studying the function ring

$$k[x_1, \ldots, x_n]/(x_1^q - x_1, \ldots, x_n^q - x_n),$$

where the highest degree of each variable can only be $q-1$, and the highest total degree can only be $n(q-1)$.

For any set S in the space k^n, let $I(S)$ be the ideal of functions in $k[x_1, \ldots, x_n]/(x_1^q - x_1, \ldots, x_n^q - x_n)$ that vanish on S. For any ideal R in $k[x_1, \ldots, x_n]/(x_1^q - x_1, \ldots, x_n^q - x_n)$, let $V(R)$ be the set of points in k^n on which all functions in R vanish. Then we have

$$V(I(S)) = S.$$

From the point of view of the function ring (as opposed to the polynomial ring), this means that the corresponding algebraic geometry is actually totally trivial in the sense that any set is an algebraic variety. At first glance this statement seems to imply that there is not much we can do here.

However, the truth is quite to the contrary, as the subtlety of the research in this area comes exactly from the fact that we are dealing with functions, not formal polynomials. One example is the concept of degree. In the case of polynomial ring, the order and the properties derived from the degree of polynomials form very powerful mathematical tools to solve problems. However, in the case of the function ring, the degree becomes something much more subtle because we do not have the additive property of degree in polynomial multiplication. This is also exactly the reason why building a degree two MPKC is a very delicate problem, and also why it is possible to build "truly" quadratic invertible multivariate maps. What we deal with here is actually the computational and combinatorial structure of these polynomial functions, which are highly nontrivial. The most powerful tools here are often still abstract mathematical tools, old and new. This is certainly an area where we do not yet have very many well developed mathematical tools and ideas. What is happening now is that researchers are trying to adapt tools used to work in polynomial rings for use in working with function rings. Though these are certainly good tools, completely new ideas and methods are certainly needed.

From what we have seen, it is evident that the research in MPKCs has already presented new mathematical challenges that demand new mathematical tools and ideas. In the future, we expect to see a mutually beneficial interaction between MPKCs and algebraic geometry to grow rapidly. We further believe that MPKCs will provide excellent motivation and critical problems in the development of the theory of functions over finite fields. There is no doubt that the area of MPKC will welcome the new mathematical tools and insights that will be critical for its future development.

The area of functions over finite fields is certainly also a central topic in coding theory. However, it seems that in coding theory the main concerns concentrate on the related combinatorial structures, and not on the algebraic structure from the point of view of polynomials and their structure. As we pointed out above, the underlying structures behind certain stream ciphers and symmetric block ciphers, such as AES and others, are certainly also structures of the functions on the space of a finite field. Though the main concerns used to be linear and differential related attacks, currently algebraic attacks that use the polynomial structure of the ciphers are a central topic in cryptanalysis. We foresee that the theory of functions on a space over a finite field will play an increasingly important role in the unification of the research in all these related areas.

Appendix A
Basic Finite Field Theory

We summarize here the basic results from finite field theory that are used throughout this book. A more detailed discussion, including proofs, can be found in any standard textbook about finite fields, such as [Lidl and Niederreiter, 1997].

Definition A.0.1. *A non-empty set k with two operations plus $+$ and multiplication $*$ is called a field if the following conditions are satisfied.*

1.) ***Commutativity:*** *For all $\alpha_1, \alpha_2 \in k$*

$$\alpha_1 + \alpha_2 = \alpha_2 + \alpha_1,$$
$$\alpha_1 * \alpha_2 = \alpha_2 * \alpha_1.$$

2.) ***Associativity:*** *For all $\alpha_1, \alpha_2, \alpha_3 \in k$*

$$\alpha_1 + (\alpha_2 + \alpha_3) = (\alpha_1 + \alpha_2) + \alpha_3,$$
$$\alpha_1 * (\alpha_2 * \alpha_3) = (\alpha_1 * \alpha_2) * \alpha_3.$$

3.) ***Distributive Property:*** *For all $\alpha_1, \alpha_2, \alpha_3 \in k$*

$$\alpha_1 * (\alpha_2 + \alpha_3) = (\alpha_1 * \alpha_2) + (\alpha_1 * \alpha_3).$$

4.) ***Additive & Multiplicative Identities:*** *There exists an additive identity $0 \in k$ and a multiplicative identity $1 \in k$ such that for all $\alpha \in k$*

$$0 + \alpha = \alpha,$$
$$1 * \alpha = \alpha.$$

5.) ***Additive & Multiplicative Inverses:*** *For each $\alpha \in k$ there exists an element in k, denoted by $-\alpha$, such that*

$$\alpha + (-\alpha) = 0.$$

For each $\alpha \in k - \{0\}$ there exists an element in $k - \{0\}$, denoted by α^{-1} or $1/\alpha$, such that

$$\alpha * \alpha^{-1} = 1.$$

Definition A.0.2. *Let k be a field of cardinality q. If $q < \infty$ then we say that k is a finite field.*

Theorem A.0.1. *Let k be a finite field of cardinality q. The nonzero elements of k form a multiplicative cyclic group of order $q - 1$.*

Let p be a prime number and define $k = \{0, 1, 2, \ldots, p-1\}$. Define $+$ and $*$ on k with the usual definition of addition and multiplication modulo p. This finite field is most often denoted by $GF(p)$, $\mathbb{Z}/p\mathbb{Z}$, or $\mathbb{F}_p$. If p is not a prime then this k is not a finite field. This can be seen easily since if $p = ab$ with $a, b > 1$, then neither a nor b have multiplicative inverses. There are finite fields with non-prime number of elements, but their cardinality must be a prime power.

Theorem A.0.2. *Let k be a finite field with q elements. Then $q = p^m$ for some prime p and positive integer m.*

Let k be a finite field with $q = p^m$ elements. Then p is called the characteristic of k. The characteristic of a finite field k can be also be defined as the least positive integer n such that $\sum_{i=1}^{n} 1 = 0$. With this definition it is not difficult to prove that the characteristic of a finite field must be prime.

Definition A.0.3. *Let k and k' be fields and let $\iota : k \longrightarrow k'$ be a bijective map such that for all $\alpha_1, \alpha_2 \in k$*

$$\iota(\alpha_1 + \alpha_2) = \iota(\alpha_1) + \iota(\alpha_2),$$
$$\iota(\alpha_1 * \alpha_2) = \iota(\alpha_1) * \iota(\alpha_2).$$

The map ι is called a field isomorphism between k and k'. We say that k and k' are isomorphic fields and we write $k \cong k'$. If $k = k'$, then ι is called a field automorphism.

Two isomorphic fields are treated as essentially the same field, since they share common additive and multiplicative structures.

Theorem A.0.3. *Let k and k' be finite fields with $|k| = q$ and $|k'| = q'$. If $q = q'$, then $k \cong k'$.*

We shall now discuss how to construct extension fields from a given finite field k. The ring of polynomials in the variable x with coefficients in k is denoted by $k[x]$.

Definition A.0.4. *A non-constant polynomial $f(x) \in k[x]$ is irreducible if $f(x) = g(x)h(x)$ implies either $g(x)$ or $h(x)$ is constant.*

Theorem A.0.4. *Let k be an finite field. There exists irreducible polynomials of degree n in $k[x]$ for every positive integer n.*

For any irreducible polynomial $g(x) \in k[x]$, we can construct the quotient ring $k[x]/g(x)$ as the set of equivalence classes in $k[x]$ modulo the ideal $\langle g(x) \rangle$. The additive and multiplicative structure of this ring is inherited from $k[x]$.

Theorem A.0.5. *Let k be a finite field with cardinality q, and let $g(x)$ be an irreducible polynomial in $k[x]$ of degree n. The ring $k[x]/g(x)$ is a finite field with q^n elements.*

The field $k[x]/g(x)$ is called a degree n finite extension of k. This result combined with Theorems A.0.2 and A.0.3 implies that any finite field can be seen as a finite extension of $GF(p)$, where p is a prime number.

Let k be a finite field with q elements and let $K = k[x]/g(x)$ be a degree n extension of k, where $g(x) \in k[x]$ is irreducible of degree n. We say that k is a subfield of K, and k can be identified with the image of the constant polynomials of $k[x]$ in $k[x]/g(x)$.

Each element in the set of equivalence classes in the field $k[x]/g(x)$, namely the extension field K, can be uniquely represented by a polynomial $a(x)$ of degree less than the degree of $g(x)$. When the degree of $g(x)$ is n,

$$a(x) = a_0 + a_1 x + a_2 x^2 + \cdots + a_{n-1} x^{n-1}.$$

In this case, the addition of any such two elements in the extension field is just the usual addition of two polynomials. The multiplication of two such elements $a(x)$ and $b(x)$ in the extension field in given as:

$$a(x) * b(x) = a(x)b(x) \bmod g(x);$$

or if

$$a(x)b(x) = p(x)g(x) + r(x),$$

where the degree of $r(x)$ is less than n, then

$$a(x) * b(x) = r(x).$$

Theorem A.0.6. *Let k be a finite field of cardinality q and let $f : k \longrightarrow k$ be any function from k to k. There exists a unique polynomial $\bar{f}(x) \in k[x]$ of degree $m < q$ such that $f(\alpha) = \bar{f}(\alpha)$ for all $\alpha \in k$.*

Theorem A.0.6 can be proved directly with the Lagrange interpolation polynomial.

Theorem A.0.7. *Let k be a finite field of cardinality q, and let K be a finite extension of k of degree n.*

1.) For any $\alpha \in k$

$$\alpha^q = \alpha.$$

For any $\alpha \in k - \{0\}$

$$\alpha^{q-1} = 1.$$

2.) For any $X, Y \in K$

$$(X+Y)^q = X^q + Y^q.$$

3.) Let ι be an automorphism of K such that $\iota(\alpha) = \alpha$ for all $\alpha \in k$. Then

$$\iota(X) = X^{q^i},$$

for some $0 \leq i < n$.

Definition A.0.5. *Let K be a degree n extension of k. Let $Gal(K/k)$ be the set of all field automorphism of K such that they keep k invariant. This set forms a group with the multiplication defined as the composition of maps.*

This group is called the Galois group of K over k. The maps $\sigma_i(X) = X^{q^i}$ for $i = 0, 1, \ldots, n-1$ mentioned in Theorem A.0.7 are called the Frobenius automorphisms of K. The following is a well known theorem in mathematics.

Theorem A.0.8. *The set $\{\sigma_i \,|\, i = 0, 1, \ldots, n-1\}$ forms a group under function composition and it is exactly the Galois group of K over k, $Gal(K/k)$.*

Let $g(x) \in k[x]$ be irreducible of degree n and let $K = k[x]/g(x)$ be the corresponding degree n extension of k. Let $\phi : K \longrightarrow k^n$ be the map defined by

$$\phi(\alpha_0 + \alpha_1 x^1 + \cdots + \alpha_{n-1}x^{n-1}) = (\alpha_0, \alpha_1, \ldots, \alpha_{n-1}).$$

It is easy to check that ϕ is a k-vector space isomorphism between K and k^n.

Lemma A.0.1. *[Lidl and Niederreiter, 1997] Let $L : k^n \longrightarrow k^n$ be a polynomial map*

$$L(x_1, \ldots, x_n) = (L_1, \ldots, L_n),$$

where each $L_i = L_i(x_1, \ldots, x_n)$ is a polynomial in $k[x_1, \ldots, x_n]$ of total degree at most one. Then the map $\bar{L} : K \longrightarrow K$ defined by

$$\bar{L} = \phi^{-1} \circ L \circ \phi$$

is of the form

$$\bar{L}(X) = \sum_{i=0}^{n-1} \alpha_i X^{q^i},$$

for some $\alpha_i \in K$. Furthermore, the map $L \longrightarrow \bar{L}$ is a bijection.

Let $r < n$ and define the projection map $\pi : k^n \longrightarrow k^r$ by

$$\pi(\alpha_1, \ldots, \alpha_n) = (\alpha_1, \ldots, \alpha_r).$$

Corollary A.0.1. *Let $r < n$ and let $L : k^r \longrightarrow k^n$ be a polynomial map*

$$L(x_1, \ldots, x_r) = (L_1, \ldots, L_n),$$

where each $L_i = L_i(x_1, \ldots, x_r)$ is a polynomial in $k[x_1, \ldots, x_r]$ of total degree at most one. Then the map $\bar{L} : K \longrightarrow K$ defined by

$$\bar{L} = \phi^{-1} \circ L \circ \pi \circ \phi$$

is of the form

$$\bar{L}(X) = \sum_{i=0}^{n-1} \alpha_i X^{q^i},$$

for some $\alpha_i \in K$.

Lemma A.0.2. *[Kipnis and Shamir, 1999] Let $Q : k^n \longrightarrow k^n$ be a polynomial map*

$$Q(x_1, \ldots, x_n) = (Q_1, \ldots, Q_n),$$

where each $Q_i = Q_i(x_1, \ldots, x_n)$ is a polynomial in $k[x_1, \ldots, x_n]$ of total degree at most two. Then the map $\bar{Q} : K \longrightarrow K$ defined by

$$\bar{Q} = \phi^{-1} \circ Q \circ \phi$$

is of the form

$$\bar{Q}(X) = \sum_{i=0}^{n-1} \sum_{j=0}^{i} \alpha_{ij} X^{q^i + q^j} + \sum_{i=0}^{n-1} \beta_i X^{q^i} + \delta,$$

for some $\alpha_{ij}, \beta_i, \delta \in K$, *if* $q > 2$;

$$\bar{Q}(X) = \sum_{i=0}^{n-1}\sum_{j=0}^{i-1} \alpha_{ij} X^{q^i+q^j} + \sum_{i=0}^{n-1} \beta_i X^{q^i} + \delta,$$

for some $\alpha_{ij}, \beta_i, \delta \in K$, *if* $q = 2$. *Furthermore, the map* $Q \longrightarrow \bar{Q}$ *is a bijection.*

Lemma A.0.3. *[Ding and Schmidt, 2005a] Let* $R : k^n \times k^n \longrightarrow k^n$ *be a polynomial map*

$$R(x_1, \ldots, x_n, y_1, \ldots, y_n) = (R_1, \ldots, R_n),$$

where each $R_i = R_i(x_1, \ldots, x_n, y_1, \ldots, y_n)$ *is a polynomial in* $k[x_1, \ldots, x_n, y_1, \ldots, y_n]$ *of total degree at most two. Then the map* $\bar{R} : K \times K \longrightarrow K$ *defined by*

$$\bar{R} = \phi^{-1} \circ R \circ (\phi \times \phi)$$

is of the form:

$$\bar{R}(X, Y) = \sum_{i=0}^{n-1}\sum_{j=0}^{i} \alpha_{ij} X^{q^i+q^j} + \sum_{i=0}^{n-1}\sum_{j=0}^{n-1} \beta_{ij} X^{q^i} Y^{q^j} + \sum_{i=0}^{n-1}\sum_{j=0}^{i} \delta_{ij} Y^{q^i+q^j} + \sum_{i=0}^{n-1} \gamma_i X^{q^i} + \sum_{i=0}^{n-1} \epsilon_i Y^{q^i} + \nu,$$

for some $\alpha_{ij}, \beta_{ij}, \delta_{ij}, \gamma_i, \epsilon_i, \nu \in K$ *if* $q > 2$;

$$\bar{R}(X, Y) = \sum_{i=0}^{n-1}\sum_{j=0}^{i-1} \alpha_{ij} X^{q^i+q^j} + \sum_{i=0}^{n-1}\sum_{j=0}^{n-1} \beta_{ij} X^{q^i} Y^{q^j} + \sum_{i=0}^{n-1}\sum_{j=0}^{i-1} \delta_{ij} Y^{q^i+q^j} + \sum_{i=0}^{n-1} \gamma_i X^{q^i} + \sum_{i=0}^{n-1} \epsilon_i Y^{q^i} + \nu,$$

for some $\alpha_{ij}, \beta_{ij}, \delta_{ij}, \gamma_i, \epsilon_i, \nu \in K$ *if* $q = 2$. *Furthermore, the map* $R \longrightarrow \bar{R}$ *is a bijection.*

References

Adams, William W. and Loustaunau, Phillippe (1994). *An Introduction to Gröbner Bases*, volume 3 of *Graduate Studies in Mathematics*. AMS.

Akkar, Mehdi-Laurent, Courtois, Nicolas, Duteuil, Romain, and Goubin, Louis (2003). A fast and secure implementation of Sflash. In Desmedt, Y.G., editor, *Public Key Cryptography - PKC 2003: 6th International Workshop on Practice and Theory in Public Key Cryptography, Miami, FL, USA, January 6-8, 2003*, volume 2567 of *LNCS*, pages 267–278. Springer.

Armknecht, Frederik and Krause, Matthias (2003). Algrebraic attacks on combiners with memory. In *Crypto 2003, August 17-21, Santa Barbara, CA, USA*, volume 2729 of *LNCS*, pages 162–176. Springer.

Ars, Gwénolé, Faugère, Jean-Charles, Imai, Hideki, Kawazoe, Mitsuru, and Sugita, Makoto (2004). Comparison between XL and Gröbner basis algorithms. In Lee, Pil Joong, editor, *Advances in Cryptology, ASIACRIPT–2004, 10th International Conference on the Theory and Application of Cryptology and Information Security, Jeju Island, Korea, December 5-9, 2004*, volume 3329 of *LNCS*, pages 338–353. Springer.

Bardet, M., Faugère, J.-C., Salvy, B., and Yang, B.-Y. (2005). Asymptotic expansion of the degree of regularity for semi-regular systems of equations. Proc. 8th Conférence des Méthodes Effectives en Géométrie Algebrique (MEGA '05, May 27- June 1, Porto Conte, Sardinia, Italy).

Becker, Thomas and Weispfenning, Volker (1991). *Gröbner Bases, a computational approach to commutative algebra*, volume 141 of *Graduate Texts in Mathematics*. Springer.

Berlekamp, Elwyn R., McEliece, R.J., and Tilborg, H.C.A. Van (1978). On the inherent intractability of certain coding problems. *IEEE Transactions on Information Theory*, IT-24(3):384–386.

Bollobás, Béla (2001). *Random Graphs*, volume 73 of *Cambridge Studies in Advanced Mathematics*. Cambridge University Press, 2nd edition.

Braeken, An, Wolf, Christopher, and Preneel, Bart (2005). A study of the security of unbalanced oil and vinegar signature schemes. In Menezes, Alfred J., editor, *The Cryptographer's Track at RSA Conference 2005*, volume 3376 of *LNCS*, pages 29–43. Springer. http://eprint.iacr.org/2004/222/.

Buchberger, Bruno (1965). *Ein Algorithmus zum Auffinden der Basiselemente des Restklassenrings nach einem nulldimensionalen Polynomideal.* Universität Innsbruck.

Buchberger, Bruno (1979). A criterion for detecting unnecessary reductions in the construction of Gröbner bases. In *Proc. EUROSAM 79*, volume 72 of *LNCS*, pages 3–21. Springer.

Buchberger, Bruno (1986). Gröbner bases: An algorithmic method in polynomial ideal theory. In Bose, N. K., editor, *Recent Trends in Multidimensional Systems Theory*, pages 184–232. D. Reidel Publishing Company.

Chen, Chia-Hsin Owen, Yang, Bo-Yin, and Chen, Jiun-Ming (2006). Testing the limits of faug'ere-lazard solvers: The story of an xl implementation. Preprint, Tamkang University, Taiwan.

Chen, Jiun-Ming and Moh, Tzuong-Tsieng (2001). On the Goubin-Courtois attack on TTM. Cryptology ePrint Archive. http://eprint.iacr.org/2001/072.

Chen, Jiun-Ming, Yang, Bo-Yin, and Peng, Bor-Yuan (2002). Tame transformation signatures with topsy-yurvy hashes. In *IWAP'02*, pages 1–8. http://dsns.csie.nctu.edu.tw/iwap/proceedings/proceedings/sessionD/7.pdf.

Chen, Kefei (1996). A new identification algorithm. In Dawson, Ed and Govic, Jovan, editors, *Cryptography: policy and algorithms : international conference, Brisbane, Queensland, Australia, July 3-5, 1995*, volume 1029 of *LNCS*, pages 244–249. Springer-Verlag.

Computational Algebra Group, University of Sydney (2005). The MAGMA computational algebra system for algebra, number theory and geometry. http://magma.maths.usyd.edu.au/magma/.

Coppersmith, Don, Stern, Jacques, and Vaudenay, Serge (1997). The security of the birational permutation signature schemes. *J. Cryptology*, 10(3):207–221.

Courtois, Nicolas (2001). The security of hidden field equations (HFE). In Naccache, C., editor, *Progress in cryptology, CT-RSA*, volume 2020 of *LNCS*, pages 266–281. Springer.

Courtois, Nicolas (2004). Algebraic attacks over $GF(2^k)$, application to HFE challenge 2 and sflash^{v2}. In Bao, Feng, Deng, Robert, and Zhou, Jianying, editors, *Public Key Cryptography - PKC 2004: 7th International Workshop on Theory and Practice in Public Key Cryptography, Singapore, March 1-4, 2004*, volume 2947 of *LNCS*, pages 201 - 217. Springer.

Courtois, Nicolas, Daum, Magnus, and Felke, Patrick (2003a). On the security of HFE, HFEv- and Quartz. In *PKC 2003*, volume 2567 of *LNCS*, pages 337–350. Springer.

Courtois, Nicolas, Goubin, Louis, Meier, Willi, and Tacier, Jean-Daniel (2002). Solving underdefined systems of multivariate quadratic equations. In Naccache, David and Paillier, Pascal, editors, *Public Key Cryptography — PKC 2002*, volume 2274 of *LNCS*, pages 211–227. Springer.

Courtois, Nicolas, Goubin, Louis, and Patarin, Jacques (2003b). Sflash^{v3}, a fast asymmetric signature scheme. http://eprint.iacr.org/2003/211.

Courtois, Nicolas, Klimov, Alexander, Patarin, Jacques, and Shamir, Adi (2000). Efficient algorithms for solving overdefined systems of multivariate polynomial equations. In Preenel, B., editor, *Advances in cryptology, Eurocrypt 2000*, volume 1807 of *LNCS*, pages 392–407. Springer.

Courtois, Nicolas and Patarin, Jacques (2003). About the XL algorithm over $GF(2)$. In Joye, M., editor, *Topics in Cryptology - CT-RSA 2003: The Cryptographers'*

Track at the RSA Conference 2003, San Francisco, CA, USA, April 13-17, 2003, volume 2612 of *LNCS*, pages 141–157. Springer.

Courtois, Nicolas and Pieprzyk, Josef (2002). Cryptanalysis of block ciphers with overdefined systems of equations. In Zheng, Y., editor, *Advances in Cryptology - ASIACRYPT 2002: 8th International Conference on the Theory and Application of Cryptology and Information Security, Queenstown, New Zealand, December 1-5, 2002*, volume 2501 of *LNCS*, pages 267–287. Springer.

Cox, David A., Little, John, and O'Shea, Don (2005). *Using algebraic geometry*, volume 185 of *Graduate Texts in Mathematics*. Springer. 2nd ed.

Daum, Magnus and Felke, Patrick (2002). Some new aspects concerning the analysis of HFE type cryptosystems. Presented at Yet Another Conference on Cryptography (YACC2002) at Porquerolles Island of France.

Delsarte, Philippe, Desmedt, Yvo, Odlyzko, Andrew M., and Piret, Phillipe (1985). Fast cryptanalysis of the Matsumoto-Imai public key scheme. In Beth, T., Cot, N., and Ingemarsson, I., editors, *Advances in Cryptology: Proceedings of EUROCRYPT 84 - A Workshop on the Theory and Application of Cryptographic Techniques, Paris, France, April 1984*, volume 209 of *LNCS*, pages 142–155. Springer.

Dickson, Leonard Eugene (1909). Definite forms in a finite field. *Trans. Amer. Math. Soc.*, 10:109–122.

Diem, Claus (2004). The XL-algorithm and a conjecture from commutative algebra. In Lee, Pil Joong, editor, *Advances in cryptology, ASIACRYPT 2004 : 10th International Conference on the Theory and Application of Cryptology and Information Security, Jeju Island, Korea, December 5-9, 2004*, volume 3329 of *LNCS*, pages 323–338. Springer.

Diene, Adam, Ding, Jintai, Gower, Jason E., Hodges, Timothy J., and Yin, Zhijun (2006). Dimension of the linearization equations of the matsumoto-imai cryptosystems. In Ytrehus, Ø., editor, *The International Workshop on Coding and Cryptography (WCC 2005), Bergen, Norway*, volume 3969 of *LNCS*, pages 242–251. Springer.

Diffie, Whitfield and Hellman, Martin (1976). New directions in cryptography. *IEEE Transactions on Information Theory*, 22(6):644–654.

Ding, Jintai (2004a). Cryptanalysis of HFEV. preprint, Department of Mathematics, University of Cincinnati.

Ding, Jintai (2004b). A new variant of the Matsumoto-Imai cryptosystem through perturbation. In Bao, F., Deng, R., and Zhou, J., editors, *Public Key Cryptosystems, PKC 2004*, volume 2947 of *LNCS*, pages 305–318. Springer.

Ding, Jintai and Gower, Jason E. (2006). Inoculating multivariate schemes against differential attacks. In et al., M. Yung, editor, *PKC 2006*, volume 3958 of *LNCS*, pages 290–301. Springer. http://eprint.iacr.org/2005/255.

Ding, Jintai, Gower, Jason E., and Schmidt, Dieter (2006a). Zhuang-Zi: A new algorithm for solving multivariate polynomial equations over a finite field. Preprint, University of Cincinnati.

Ding, Jintai, Gower, Jason E., Schmidt, Dieter, Wolf, Christopher, and Yin, Zhijun (2005). Complexity estimates for the F_4 attack on the perturbed matsumoto-imai cryptosystem. In Smart, N.P., editor, *Cryptography and Coding 2005*, volume 3796 of *LNCS*, pages 262–277. Springer. http://math.uc.edu/~aac/pub/pmi-groebner.pdf.

Ding, Jintai and Hodges, Timothy (2004). Cryptanalysis of an implementation scheme of TTM. *J. Algebra Appl.*, 3:273–282. http://eprint.iacr.org/2003/084.

Ding, Jintai, Hu, Lei, Nie, Xuyun, Li, Jianyu, and Wagner, John (2006b). High order linearization equation (hole) attack on multivariate public key cryptosystems. preprint, University of Cincinnati.

Ding, Jintai and Schmidt, Dieter (2004). The new TTM implementation is not secure. In Feng, Kegin, Niederreiter, Harald, and Xing, Chaoping, editors, *Workshop on Coding Cryptography and Combinatorics, CCC2003 Huangshan (China)*, volume 23 of *Progress in Computer Science and Applied Logic*, pages 113–128. Birkhauser Verlag.

Ding, Jintai and Schmidt, Dieter (2005a). Cryptanalysis of HFEV and the internal perturbation of HFE. In Vaudenay, Serge, editor, *Public key cryptography: PKC 2005: 8th International Workshop on Theory and Practice in Public Key Cryptography, Les Diablerets, Switzerland, January 23-26, 2005*, volume 3386 of *LNCS*, pages 288–301. Springer.

Ding, Jintai and Schmidt, Dieter (2005b). Rainbow, a new multivariable polynomial signature scheme. In Ioannidis, John, Keromytis, Angelos D., and Yung, Moti, editors, *Third International Conference Applied Cryptography and Network Security (ACNS 2005)*, volume 3531 of *LNCS*. Springer.

Ding, Jintai and Yin, Zhijun (2004). Cryptanalysis of TTS and Tame–like signature schemes. In *Third International Workshop on Applied Public Key Infrastructures.*

Dobbertin, Hans (2002). Analysis of HFE schemes based on power functions. Invited talk at YACC 2002, Conference on Cryptography, Porquerolles Island, France, 3.-7.06.2002.

Dubois, Vivien, Granboulan, Louis, and Stern, Jacques (2006). An efficient provable distinguisher for hfe. Preprint, Ecole normale supérieure.

Faugère, Jean-Charles (1999). A new efficient algorithm for computing Gröbner bases (F_4). *Journal of Pure and Applied Algebra*, 139:61–88.

Faugère, Jean-Charles (2002). A new efficient algorithm for computing Gröbner bases without reduction to zero (F_5). In *International Symposium on Symbolic and Algebraic Computation — ISSAC 2002*, pages 75–83. ACM Press.

Faugère, Jean-Charles (2003). Algebraic cryptanalysis of (HFE) using Gröbner bases. Technical report, Institut National de Recherche en Informatique et en Automatique. http://www.inria.fr/rrrt/rr-4738.html, 19 pages.

Faugère, Jean-Charles and Joux, Antoine (2003). Algebraic cryptanalysis of hidden field equation (HFE) cryptosystems using Gröbner bases. In Boneh, Dan, editor, *Advances in cryptology - CRYPTO 2003*, volume 2729 of *LNCS*, pages 44–60. Springer.

Faugère, Jean-Charles and Perret, Ludovic (2006). Polynomial equivalence problems: Algorithmic and theoretical aspects. In *Eurocrypt 2006*, LNCS. Springer. To appear.

Felke, Patrick (2005). On certain families of HFE-type cryptosystems. In *International Workshop on Coding and Cryptography, WCC 2005.*

Fell, Harriet and Diffie, Whitfield (1986). Analysis of a public key approach based on polynomial substitution. In *Advances in cryptology—CRYPTO '85 (Santa Barbara, Calif.)*, volume 218 of *LNCS*, pages 340–349. Springer.

Feller, William (1968). *An Introduction to Probability Theory and Its Applications (Third Edition)*, volume I. Wiley & Sons.

Fouque, Pierre-Alain, Granboulan, Louis, and Stern, Jacques (2005). Differential cryptanalysis for multivariate schemes. In Cramer, Ronald, editor, *Advances in Cryptology - EUROCRYPT 2005*, volume 3494 of *LNCS*, pages 341–353. Springer.

Gabidulin, Ernst M. (1985). Theory of codes with maximum rank distance. *Problems of Information Transmission*, 21:1–12.

Garey, Michael R. and Johnson, David S. (1979). *Computers and intractability, A Guide to the theory of NP-completeness.* W.H. Freeman.

Gebauer, Rüdiger and Möller, H. Michael (1988). On an installation of Buchberger's algorithm. *Journal of Symbolic Computation*, 6:257–286.

Geddes, Keith O., Czapor, Stepen R., and Labahn, George (1992). *Algorithms for Computer Algebra.* Amsterdam, Netherlands: Kluwer.

Geiselmann, Willi, Meier, W., and Steinwandt, Rainer (2003). An attack on the isomorphisms of polynomials problem with one secret. *Int. Journal of Information Security*, 2(1):59–64.

Geiselmann, Willi, Steinwandt, Rainer, and Beth, Thomas (2001). Attacking the affine parts of SFlash. In Honary, B., editor, *Cryptography and Coding - 8th IMA International Conference*, volume 2260 of *LNCS*, pages 355–359. Springer.

Gilbert, Henri and Minier, Marine (2002). Cryptanalysis of SFLASH. In Knudsen, L., editor, *Advances in Cryptology - EUROCRYPT 2002*, volume 2332 of *LNCS*, pages 288–298. Springer.

Goubin, Louis and Courtois, Nicolas (2000). Cryptanalysis of the TTM cryptosystem. In Okamoto, Tatsuaki, editor, *Advances in Cryptology – ASIACRYPT 2000, International Conference on the Theory and Application of Cryptology and Information Security, Singapore, December 3-7, 2000*, volume 1976 of *LNCS*, pages 44–57. Springer.

Hasegawa, S. and Kaneko, T. (1987). An attacking method for a public key cryptosystem based on the difficulty of solving a system of non-linear equations. In *Proc. 10th Symposium on Information Theory and Its applications*, pages JA5–3.

Hu, Yuh-Hua, Wang, Lih-Chung, Chou, Chun-Yen, and Lai, Feipei (2005). Similar keys of multivariate quadratic public key cryptosystems. In Desmedt, Yvo G., Wang, Huaxiong, Mu, Yi, and Li, Yongqing, editors, *Cryptology and Network Security: 4th International Conference, CANS 2005, Xiamen, China, December 14-16, 2005.*, volume 3810 of *LNCS*, pages 211 – 222. Springer.

Imai, Hideki and Matsumoto, Tsutomu (1985). Algebraic methods for constructing asymmetric cryptosystems. In Calmet, Jacques, editor, *Algebraic Algorithms and Error-Correcting Codes – 3rd International Conference(AAECC-3)*, volume 229 of *LNCS*, pages 108–119. Springer.

Kasahara, Masao and Sakai, Ryuichi (2004a). A construction of public-key cryptosystem based on singular simultaneous equations. In *Symposium on Cryptography and Information Security — SCIS 2004*, pages 74–80. The Institute of Electronics, Information and Communication Engineers.

Kasahara, Masao and Sakai, Ryuichi (2004b). A construction of public key cryptosystem for realizing ciphtertext of size 100 bit and digital signature scheme. *IEICE Trans. Fundamentals*, E87-A(1):102–109.

Kemeny, John G. and Snell, J. Laurie (1960). *Finite Markov Chains.* D. Van Nostrand Company, Inc.

Kipnis, Aviad, Patarin, Jacques, and Goubin, Louis (1999). Unbalanced oil and vinegar signature schemes. In Stern, Jacques, editor, *EUROCRYPT '99 : International Conference on the Theory and Application of Cryptographic Techniques, Prague, Czech Republic, May 2-6, 1999*, volume 1592 of *LNCS*, pages 206–222. Springer.

Kipnis, Aviad, Patarin, Jacques, and Goubin, Louis (2003). Unbalanced oil and vinegar signature schemes — extended version. 17 pages, citeseer/231623.html, 2003-06-11.

Kipnis, Aviad and Shamir, Adi (1998). Cryptanalysis of the oil & vinegar signature scheme. In Krawczyk, H., editor, *Advances in Cryptology - CRYPTO'98: 18th Annual International Cryptology Conference, Santa Barbara, California, USA, August 1998*, volume 1462 of *LNCS*, pages 257–267. Springer.

Kipnis, Aviad and Shamir, Adi (1999). Cryptanalysis of the HFE public key cryptosystem by relinearization. In Wiener, M., editor, *Advances in cryptology – Crypto '99*, volume 1666 of *LNCS*, pages 19–30. Springer.

Knuth, Donald (1981). *The Art of Computer Programming*. Adison Wesley, 2nd edition.

Kreuzer, Martin and Robbiano, Lorenzo (2000). *Computational Commutative Algebra*. Springer.

Lazard, Daniel (1983). Gröbner bases, gaussian elimination and resolution of systems of algebraic equations. In Hulzen, J. A., editor, *Computer algebra (London, 1983)*, volume 162 of *LNCS*, pages 146–156. Springer.

Levy-dit-Vehel, Francoise and Perret, Ludovic (2003). Polynomial equivalence problems and applications to multivariate cryptosystems. In Johansson, Thomas and Maitra, Subhamoy, editors, *Progress in Cryptology - INDOCRYPT 2003*, volume 2904 of *LNCS*, pages 235–251. Springer.

Lidl, Rudolf and Niederreiter, Harald (1997). *Finite Fields*. Cambridge University Press.

Macaulay, F.S. (1916). *The algebraic theory of modular systems*, volume xxxi of *Cambridge Mathematical Library*. Cambridge University Press.

Matsumoto, Tsutomu and Imai, Hideki (1983). A class of asymmetric crypto-systems based on polynomials over finite rings. In *IEEE Intern. Symp. Inform. Theory, St. Jovite, Quebec, Canada, Sept. 26-30, 1983*, pages 13, 1–132. IEEE.

Matsumoto, Tsutomu and Imai, Hideki (1988). Public quadratic polynomial-tuples for efficient signature verification and message encryption. In Guenther, C. G., editor, *Advances in cryptology – EUROCRYPT '88*, volume 330 of *LNCS*, pages 419–453. Springer.

Matsumoto, Tsutomu, Imai, Hideki, Harashima, H., and Miyagawa, H. (1985). High speed signature scheme using compact public key. National Conference of system and information of the Electronic Communication Association of year Sowa 60, S9-5.

Michon, Jean Francis, Valarcher, Pierre, and Yune, Jean Baptiste (2004). Hfe and bdds: a practical attempt at cryptanalysis. In Feng, Kegin, Niederreiter, Harald, and Xing, Chaoping, editors, *Workshop on Coding Cryptography and Combinatorics, CCC2003 Huangshan (China)*, volume 23 of *Progress in Computer Science and Applied Logic*, pages 237–246. Birkhauser Verlag.

Moh, Tzuong-Tsieng (1999a). A fast public key system with signature and master key functions. *Comm. in Algebra*, 27:2207–2222. http://www.usdsi.com/ttm.html.

Moh, Tzuong-Tsieng (1999b). A fast public key system with signature and master key functions. *Proceedings of CrypTEC'99, InternationalWorkshop on Cryptographic Techniques and E-commerce*, pages 63–69.

Moh, Tzuong-Tsieng (2001). On the method of "XL" and its inefficiency to TTM. Cryptology ePrint Archive, Report 2001/047. http://eprint.iacr.org/2001/047.

Moh, Tzuong-Tsieng, Chen, Jiun-Ming, and Yang, Bo-Yin (2004). Building Instances of TTM Immune to the Goubin-Courtois Attack and the Ding-Schmidt Attack. Cryptology ePrint Archive. http://eprint.iacr.org/2004/168.

Nagata, Masayoshi (1972). *On Automorphism Group of* $K[x, y]$, volume 5 of *Lectures on Mathematics*. Kyoto University, Kinokuniya, Tokyo.

NESSIE (1999). European project IST-1999-12324 on New European Schemes for Signature, Integrity and Encryption. http://www.cryptonessie.org.

Nie, Xuyun, Hu, Lei, Li, Jianyu, Updegrove, Crystal, and Ding, Jintai (2006). Breaking a new instance of TTM cryptosystems. In *4th International Conference on Applied Cryptography and Network Security (ACNS'06), 6-9 June, 2006, Singapor*, LNCS. Springer. To appear.

Okamoto and Nakamura (1986). Evaluation of public key cryptosystems proposed recently. In *Proc. 1986's Symposium of cryptography and information security*, volume D1.

Okeya, Katsuyuki, Takagi, Tsuyoshi, and Vuillaume, Camille (2005). On the importance of protecting in sflash against side channel attacks. *IEICE Transactions*, 88:123–131.

Ong, H., Schnorr, Claus P., and Shamir, Adi (1984). Signatures through approximate representations by quadratic forms. In *Advances in cryptology, Crypto '83*, pages 117–131. Plenum Publ.

Ong, H., Schnorr, Claus P., and Shamir, Adi (1985). Efficient signature schemes based on polynomial equations. In Blakley, G. R. and Chaum, D., editors, *Advances in cryptology , Crypto '84*, volume 196 of *LNCS*, pages 37–46. Springer.

Patarin, Jacques (1995). Cryptanalysis of the Matsumoto and Imai public key scheme of Eurocrypt'88. In Coppersmith, D., editor, *Advances in Cryptology – Crypto '95*, volume 963 of *LNCS*, pages 248–261.

Patarin, Jacques (1996a). Asymmetric cryptography with a hidden monomial. In Koblitz, N., editor, *Advances in cryptology, CRYPTO '96*, volume 1109 of *LNCS*, pages 45–60. Springer.

Patarin, Jacques (1996b). Hidden Field Equations (HFE) and Isomorphism of Polynomials (IP): Two new families of asymmetric algorithms. In Maurer, U., editor, *Eurocrypt'96*, volume 1070 of *LNCS*, pages 33–48. Springer. Extended Version: http://www.minrank.org/hfe.pdf.

Patarin, Jacques (1997). The oil and vinegar signature scheme. *Dagstuhl Workshop on Cryptography, September 1997.*

Patarin, Jacques (2000). Cryptanalysis of the Matsumoto and Imai public key scheme of Eurocrypt'88. *Designs, Codes and Cryptography*, 20:175–209.

Patarin, Jacques (2006). Probabilistic multivariate cryptography. preprint.

Patarin, Jacques, Courtois, Nicolas, and Goubin, Louis (2001). Flash, a fast multivariate signature algorithm. In Naccache, C., editor, *Progress in cryptology, CT-RSA*, volume 2020 of *LNCS*, pages 298–307. Springer.

Patarin, Jacques and Goubin, Louis (1997). Asymmetric cryptography with S-boxes. In Y., Han, T., Okamoto, and S., Qing, editors, *Proceedings of ICICS'97*, volume 1334 of *LNCS*, pages 369–380. Springer.

Patarin, Jacques, Goubin, Louis, and Courtois, Nicolas (1998). C^*_{-+} and HM: variations around two schemes of T. Matsumoto and H. Imai. In Ohta, K. and Pei, D., editors, *Advances in Cryptology - ASIACRYPT'98: International Conference on the Theory and Application of Cryptology and Information Security, Beijing, China, October 1998*, volume 1514 of *LNCS*, pages 35–50. Springer.

Perret, Ludovic (2005). A fast cryptanalysis of the isomorphism of polynomials with one secret problem. In Cramer, Ronald, editor, *Advances in Cryptology – EUROCRYPT 2005: 24th Annual International Conference on the Theory and Applications of Cryptographic Techniques, Aarhus, Denmark, May 22-26, 2005*, volume 3494 of *LNCS*, pages 354–370. Spinger.

Pollard, John M. and Schnorr, Claus P. (1987). An efficient solution of the congruence $x^2 + ky^2 = m \pmod{n}$. *IEEE Trans. Inform. Theory*, 33(5):702–709.

Rivest, Ronald, Shamir, Adi, and Adleman, Leonard M. (1982). A method for obtaining digital signatures and public key cryptosystems. secure communications and asymmetric cryptosystems. In Simmons, G, editor, *AAAS Sel. Sympos. Ser.*, volume 69, pages 217–239. Westview Press.

Rivest, Ronald L., Shamir, Adi, and Adleman, Leonard M. (1978). A method for obtaining digital signatures and public-key cryptosystems. *Communications of the ACM*, 21(2):120–126.

Shallit, J.O., Frandsen, G.S., and Buss, J.F. (1996). The computational complexity of some problems of linear algebra. BRICS series report, Aarhus, Denmark, RS-96-33. Available at http://www.brics.dk/RS/96/33.

Shamir, Adi (1993). Efficient signature schemes based on birational permutations. In Stinson, Douglas R., editor, *Advances in cryptology - CRYPTO '93 (Santa Barbara, CA, 1993)*, volume 1462 of *LNCS*, pages 257–266. Springer.

Shestakov, Ivan P. and Umirbaev, Ualbai U. (2003). The Nagata automorphism is wild. *Proc. Natl. Acad. Sci. USA*, 100:12561–12563.

Shor, Peter (1999). Polynomial-time algorithms for prime factorization and discrete logarithms on a quantum computer. *SIAM Rev.*, 41(2):303–332.

Smale, Steve (1998). Mathematical problems for the next century. *Math. Intelligencer*, 20(2):7–15.

Stern, J. and Chabaud, F. (1996). The cryptographic security of the syndrome decoding problem for rank distance codes. In Kim, Kwangjo and Matsumoto, Tsutomu, editors, *Advances in cryptology-ASIACRYPT '96 : International Conference on the Theory and Applications of Cryptology and Information Security, Kyongju, Korea, November 3-7, 1996*, volume 1163 of *LNCS*, pages 368–381. Springer-Verlag.

Tsujii, Shigeo, Fujioka, Atsushi, and Hirayama, Yuusuke (1989). Generalization of the public key cryptosystem based on the difficulty of solving a system of non-linear equations. In *ICICE Transactions (A) J72-A*, volume 2, pages 390–397. English version is appended at http://eprint.iacr.org/2004/336.

Tsujii, Shigeo, Fujioka, Atsushi, and Itoh, T. (1987). Generalization of the public key cryptosystem based on the difficulty of solving a system of non-linear equations. In *Proc. 10th Symposium on Information Theory and Its applications*, pages JA5–3.

Tsujii, Shigeo, Kurosawa, K., Itoh, T., Fujioka, Atsushi, and Matsumoto, Tsutomu (1986). A public key cryptosystem based on the difficulty of solving a system of nonlinear equations. *ICICE Transactions (D) J69-D*, 12:1963–1970.

Tsujii, Shigeo, Tadaki, Kohtaro, and Fujita, Ryou (2004). Piece In Hand Concept for Enhancing the Security of Multivariate Type Public Key Cryptosystems: Public Key Without Containing All the Information of Secret Key. Cryptology ePrint Archive, Report 2004/366. http://eprint.iacr.org/2004/366.

Tsujii, Shigeo, Tadaki, Kohtaro, and Fujita, Ryou (2006). Proposal for piece in hand matrix ver.2: General concept for enhancing security of multivariate public key cryptosystems. Cryptology ePrint Archive. http://eprint.iacr.org/2006/051.

Vandersypen, Lieven, Steffen, Matthias, Breyta, Gregory, Yannoni, Costantino, Sherwood, Mark, and Chuang, Isaac (2001). Experimental realization of Shor's quantum factoring algorithm using nuclear magnetic resonance. *Nature*, 414:883–887.

von zur Gathen, Joachim and Gerhard, Jürgen (2003). *Modern Computer Algebra*. Cambridge University Press, 2nd edition.

Wang, Lih-Chung, Hu, Yuh-Hua, Lai, Feipei, Chou, Chun-Yen, and Yang, Bo-Yin (2005). Tractable rational map signature. In Vaudenay, Serge, editor, *Public Key*

Cryptography - PKC 2005: 8th International Workshop on Theory and Practice in Public Key Cryptography, volume 3386 of *LNCS*, pages 244–257. Springer.

Wang, Lih-Chung, Yang, Bo-Yin, Hu, Yuh-Hua, and Lai, Feipei (2006). A "medium-field" multivariate public-key encryption scheme. In Pointcheval, David, editor, *Topics in Cryptology – CT-RSA 2006: The Cryptographers' Track at the RSA Conference 2006, San Jose, CA, USA, February 13-17, 2005*, volume 3860 of *LNCS*, pages 132–149. Springer.

Wolf, Christopher, Braeken, An, and Preneel, Bart (2004). Efficient cryptanalysis of rse(2)pkc and rsse(2)pkc. In Blundo, Carlo and Cimato, Stelvio, editors, *Security in Communication Networks: 4th International Conference, SCN 2004, Amalfi, Italy, September 8-10, 2004*, volume 3352 of *LNCS*, pages 294–309. Springer.

Wolf, Christopher and Preneel, Bart (2005a). Equivalent keys in HFE, C*, and variations. In Dawson, Ed and Vaudenay, Serge, editors, *Progress in Cryptology - Mycrypt 2005: First International Conference on Cryptology in Malaysia, Kuala Lumpur, Malaysia, September 28-30, 2005*, volume 3715, pages 33–49. Springer. Extended version: http://eprint.iacr.org/2004/360/, 15 pages.

Wolf, Christopher and Preneel, Bart (2005b). Large superfluous keys in multivariate quadratic asymmetric systems. In Vaudenay, Serge, editor, *Public Key Cryptography - PKC 2005: 8th International Workshop on Theory and Practice in Public Key Cryptography, Les Diablerets, Switzerland, January 23-26, 2005*, volume 3386 of *LNCS*, pages 275–287. Springer.

Wolf, Christopher and Preneel, Bart (2005c). Taxonomy of public key schemes based on the problem of multivariate quadratic equations. Cryptology ePrint Archive, http://eprint.iacr.org/2005/077.

Wu, Zhiping, Ding, Jintai, Gower, Jason E., and Ye, Dingfeng (2005). Perturbed hidden matrix cryptosystems. In Osvaldo Gervasi, etc, editor, *Computational Science and Its Applications - ICCSA (2)*, volume 3481 of *LNCS*, pages 595–602. Springer.

Yang, Bo-Yin and Chen, Jiun-Ming (2003). A more secure and efficacious TTS signature scheme. *ICISC 2003*. http://eprint.iacr.org/2003/160.

Yang, Bo-Yin and Chen, Jiun-Ming (2004a). All in the XL family: Theory and practice. In Park, Choonsik and Chee, Seongtaek, editors, *Proc. 7th International Conference on Information Security and Cryptology (ICISC '04, Dec. 2-3, Seoul, Korea)*, volume 3506 of *LNCS*, pages 67–86. Springer.

Yang, Bo-Yin and Chen, Jiun-Ming (2004b). Fast computation theoretical analysis of XL over small fields. In Wang, Huaxiong, Pieprzyk, Josef, and Varadharajan, Vijay, editors, *Information Security and Privacy: 9th Australasian Conference, ACISP 2004, Sydney, Australia, July 13-15, 2004.*, volume 3108 of *LNCS*, pages 277 – 288. Springer.

Yang, Bo-Yin and Chen, Jiun-Ming (2005a). Building secure tame-like multivariate public-key cryptosystems– the new TTS. In Boyd, Colin and Nieto, Juan M. González, editors, *Information Security and Privacy: 10th Australasian Conference–ACISP 2005*, volume 3574 of *LNCS*, pages 518–531. Springer.

Yang, Bo-Yin and Chen, Jiun-Ming (2005b). Perturbed Matsumoto-Imai Plus on the 8051. Private communication.

Yang, Bo-Yin and Chen, Jiun-Ming (February 2004c). TTS: Rank attacks in tame-like multivariate PKCs . http://eprint.iacr.org/2004/061.

Yang, Bo-Yin, Chen, Jiun-Ming, and Chen, Yen-Huang (2004a). TTS: High-speed signatures on a low-cost smart card. In Joye and Quisquater, editors, *Cryptographic Hardware and Embedded Systems: CHES 2004*, volume 3156 of *LNCS*, pages 371–385. Springer.

Yang, Bo-Yin, Chen, Jiun-Ming, and Courtois, Nicolas (2004b). On asymptotic security estimates in XL and Gröbner bases-related algebraic cryptanalysis. In Lopez, Javier, Qing, Sihan, and Okamoto, Eiji, editors, *ICICS 2004*, volume 3269 of *LNCS*, pages 401–413. Springer.

Yang, Bo-Yin, Cheng, Chen-Mou, Chen, Bor-Rong, and Chen, Jiun-Ming (2006). Implementing minimized multivariate public-key cryptosystems on low-resource embedded systems. In *3rd International Conference Security in Pervasive Computing (SPC) YORK, UK, APRIL 19 - 20, 2006*, LNCS. Springer. To appear.

Youssef, Amr M. and Gong, Guang (2001). Cryptanalysis of Imai and Matsumoto scheme B asymmetric cryptosystem. In Rangan, C. Pandu and Ding, Cunsheng, editors, *Progress in Cryptology — INDOCRYPT 2001*, volume 2247 of *LNCS*, pages 214–222. Springer.

Index